S0-ELC-930

THE PLASTIC METHODS OF STRUCTURAL ANALYSIS

THE PLASTIC METHODS OF STRUCTURAL ANALYSIS

by

B. G. NEAL

M.A., Ph.D. (Cantab.), A.M.I.C.E.

Professor of Civil Engineering and Chairman of Engineering Department, University College of Swansea.

Formerly Fellow of Trinity Hall and Lecturer in Engineering, University of Cambridge

NEW YORK

JOHN WILEY & SONS INC.

440 FOURTH AVENUE

1956

First Published . . . 1956

Printed in Great Britain by Butler & Tanner Ltd., Frome and London

CONTENTS

PREFACE

FROM THE FIRST BEGINNINGS of structural analysis, Hooke's Law was almost invariably taken as the starting-point, presumably on account of the analytical simplicity promised by the linear relationship *ut tensio sic vis*. Apart from Euler's work on the buckling of columns, the early investigations in this field were concerned predominantly with the distribution of stress in structures in the absence of buckling effects. The classical methods of elastic analysis, based on the pioneer work of Navier, were developed by Maxwell, Mohr, Castigliano and others towards the end of the nineteenth century. These methods were primarily concerned with trusses, but when steel and reinforced concrete frames came into use at the beginning of the present century corresponding techniques were developed for their analysis. However, the classical methods proved to be very cumbersome in their application to framed structures, mainly because practical frames are usually highly redundant. Even despite the comparatively recent development of the Moment Distribution method, the elastic analysis of all but the simplest of frames remains a matter of considerable labour.

The conventional methods of design for steel frames are based on the results of elastic analyses. Owing to the complexity of these analyses, it has been necessary to introduce many simplifying assumptions in developing workable design methods. In applying these methods, the members of a frame are proportioned so that when the working loads are applied the maximum stress computed by a simplified elastic procedure does not exceed a permissible working stress whose value is well within the elastic limit. However, the extensive use which is made of mild steel as a structural material is due primarily to its marked ductility, a property associated with behaviour outside the elastic range. When tested in tension or compression, mild steel behaves elastically until the stress reaches a well-defined yield limit, and then quite large extensions or contractions occur while the stress remains constant, this behaviour being termed plastic flow. It was the ability of mild steel to undergo plastic flow and thereby to accommodate without distress the concentrations of stress which occur at rivet holes and other more or less sudden changes of section that led to its widespread adoption in place of brittle materials such as cast iron. Nevertheless, these localized effects are disregarded in conventional design calculations, which are concerned solely with behaviour in the elastic range. In fact, since the results of elastic analyses cease to be applicable when the yield stress is attained at the most highly stressed cross-section, elastic design methods are of necessity based on the assumption that a frame will become unsafe when yield first occurs, and a margin of safety is ensured by specifying a working stress which is well below the yield limit.

The load-carrying capacity of a redundant frame of mild steel, which carries load by virtue of the resistance to bending of the members, is, however, rarely exhausted when the yield stress is first attained. If the possibility is excluded of an individual member of the frame failing by buckling, collapse can only occur when plastic flow is taking place at several cross-sections simultaneously, this type of failure being termed plastic collapse. In general, there will be a range of partly plastic behaviour of a frame after it has yielded and before collapse occurs, and plastic collapse will only take place at a load which is somewhat greater than the load at which the frame first yields. Redundant mild steel frames thus possess a reserve of strength above the yield load, which can be looked upon as another manifestation of the ability of mild steel to undergo favourable readjustments of stress distribution by plastic flow. This capacity to carry loads in excess of the yield load cannot be utilized in elastic designs, so that while it is unquestionable that safe designs result from the use of the conventional elastic methods, these designs cannot be economical in the use of steel. Full use of the available strength of a mild steel frame can only be achieved by a plastic design method based on the calculation of the plastic collapse load. In such a method, a margin of safety is provided by proportioning the members so that the working loads would have to be multiplied by a specified load factor before collapse would occur, and in addition it is necessary to ensure that none of the members would fail by buckling before the attainment of the plastic collapse load.

The plastic methods of structural analysis have been developed for the purpose of calculating the plastic collapse loads for frames, with the ultimate objective of establishing a rational and economical design procedure. It might be thought that any attempt to evaluate the plastic collapse load for a given frame would require an investigation of the complete behaviour of the frame between the first yielding and its ultimate collapse. However, it is a remarkable fact that if buckling effects are absent it is possible to calculate the collapse load directly by the plastic methods without considering the intervening elastic-plastic range. Moreover, the calculations involved in applications of the plastic methods are much simpler than the corresponding calculations which must be made for an elastic analysis of the same structure. This analytical simplicity is a notable advantage of the plastic methods of analysis, for it implies that plastic design methods need not involve such extensive simplifications as are necessary in the elastic methods. Each simplifying assumption which is introduced in developing a design method must perforce be of a conservative nature, with a consequent loss of economy.

The plastic methods of analysis rest on certain basic assumptions concerning the behaviour of structural members in flexure which are closely obeyed by mild steel, and may also prove to be obeyed by other ductile materials. Quite apart from their applicability to mild steel frames, the plastic methods are of considerable intrinsic interest, since they represent one analytical extreme in which conditions of collapse are discussed, whereas the elastic methods represent the other extreme in which the entire structure obeys Hooke's law. Furthermore, the plastic methods of analysis may now be said to be as fully developed as the elastic methods. It therefore seemed that an account of the plastic methods would be of considerable value, not only because of its interest to practising structural engineers and research workers, but also because a study of the plastic methods and their contrast with the elastic methods of analysis can constitute an important addition to an undergraduate teaching curriculum.

The first four chapters of the book are concerned solely with a presentation of the plastic methods of analysis, and the material contained therein could profitably be included in the later stages of an undergraduate course in the Theory of Structures. The remaining four chapters deal with topics which are closely associated with the plastic methods, and are suitable for more advanced study, although selected material from these chapters should be included in an undergraduate course. To assist in the use of the book, examples have been provided at the end of each chapter.

Although the plastic methods of analysis are now fully developed, the conditions under which the members of frames fail by instability after they have partially yielded are not yet completely understood. A comprehensive plastic design procedure for all types of frame, containing rules which would ensure that no individual member would fail by any form of buckling before the plastic collapse load was attained, cannot therefore be devised at the present time. Nevertheless, there are many practical structures in which the members are well restrained against instability by their cladding, and the plastic methods have been used in recent years for the design of such structures. In fact, the use of the plastic methods has been permitted in Great Britain since 1948, when an appropriate clause was included for the first time in a revision of British Standard No. 449, *The Use of Structural Steel in Building*. No attempt has been made to describe the present state of the investigations into the problems of elastic-plastic buckling, although references are given to this work whenever appropriate.

Much of the book was written while the author was a member of the staff of the Engineering Department at the University of Cambridge, where he worked under Professor J. F. Baker, F.R.S., who has directed the extensive research work into the plastic theory conducted at Cambridge since 1943, and earlier at Bristol. The author takes great pleasure in acknowledging a debt of gratitude to Professor Baker, and to the many members of his research team with whom he worked for several years. He was also fortunate in spending a year at Brown University, where he worked in the Graduate Division of Applied Mathematics under Professor W. Prager and began a close collaboration with Professor P. S. Symonds, which was continued for some time at Cambridge.

The author is greatly indebted to the various members of the assistant staff of the Cambridge University Engineering Department, who traced the figures and typed the first draft of the manuscript, and to Miss Anne Jones of the Engineering Department of the University College of Swansea who prepared the final version.

Thanks are also due to the British Welding Research Association for granting permission to reproduce data concerning the plastic moduli of British Standard Beams which appear in *Appendix B*. Extracts from British Standard No. 4 (1932), which also appear in this appendix, and from British Standard No. 449 (1948) are given by permission of the British Standards Institution, 2 Park Street, London, W.1.

B. G. N.

SWANSEA.
April 1956.

CHAPTER

1

Basic Hypotheses

1.1. Introduction

THE PLASTIC METHODS of structural analysis which are described in this book find their principal application at present to the design of statically indeterminate or redundant framed structures of mild steel, which carry load by virtue of the resistance of their members to bending action. The multi-storey, multi-bay rectangular frame and the single or multi-bay pitched roof portal shed type of frame are familiar examples of this type of structure, and the definition also includes simply supported and continuous beams. For such structures Baker [1]* has pointed out that, as compared with the orthodox elastic design technique, more rational designs can undoubtedly be achieved by the use of the plastic methods, and striking economies in the total weight of steelwork can often be made. The reasons for these advantages of the plastic methods can only be appreciated by comparing their fundamental basis with that of the orthodox elastic methods.

In applications of the elastic methods the stresses due to the working loads must not exceed certain permissible working stresses. The values of the working loads and the permissible working stresses are laid down in the appropriate specification, which for normal British practice is British Standard No. 449. The working loads are supposed to represent the maximum loads which may be expected to arise in normal usage during the lifetime of the structure, and the values of the working stresses are intended to ensure an adequate margin of safety to account for unpredictable overloads, defective workmanship and materials, and so on. To take a specific example of particular interest, the permissible working stress for beams which are not too slender, so that failure by buckling will not occur, is laid down in B.S.449

* *References are given at the end of the chapter.*

(1948 edition) as 10 tons per sq. in. for steel having a yield stress guaranteed to be at least 15·25 tons per sq. in. Thus in an elastically designed steel framed structure the extreme fibre stress at the most highly stressed cross-section under the working loads will be 10 tons per sq. in., and it follows that these loads could be increased by a factor of 1·525 before yield would occur at this section. The effect of a further increase in the loads cannot be determined by elastic analysis, and so it can only be concluded from such an analysis that there is a *safety factor* of 1·525 against the occurrence of yield.

It is, however, possible to analyse the behaviour of a frame under further increases in the loads by the plastic theory, provided that certain idealisations of the properties of the steel are made. Broadly speaking, it is found that as the loads increase yielding spreads quite rapidly through the most highly stressed section. When this section has yielded fully it is transmitting a bending moment which for a British Standard beam is about 1·15 times greater than the bending moment which was developed at first yield. The plastic theory then postulates that a hinging action can occur at this section, the hinge rotation taking place while the bending moment transmitted across the hinge remains constant. While this hinging action is in operation, the longitudinal fibres of the beam are extending or contracting while the stress in them remains constant at the yield value, so that each fibre may be said to be flowing in a completely plastic manner. The hinge is termed a *plastic hinge*, and the bending moment which is developed at the plastic hinge is termed the *fully plastic moment*. A fundamental hypothesis of the plastic theory is that a plastic hinge can undergo rotation of any magnitude provided that the bending moment stays constant at the fully plastic value.

Behaviour of simply supported beam

If the loads on a framed structure are increased above the values which first cause yield, a plastic hinge will thus soon develop at the most highly stressed section. In the case of a statically determinate structure, collapse would then occur. Consider for example a simply supported beam which is subjected to a central concentrated load. The maximum bending moment, which is directly proportional to the load, occurs at the centre of the beam, and when the load reaches a certain value the fully plastic moment will be attained at this section. According to the plastic theory, the central hinge which then forms can undergo an indefinitely large amount of rotation while the bending moment, and thus the load, remains constant, so that excessive deflections could suddenly develop at this load. This behaviour is termed *plastic collapse.* The crucial point is that the formation of the assumed central plastic hinge transforms the structure into a mechanism. Thus if a frictionless hinge were inserted at the centre of the beam, the application of only an infinitesimal load would be sufficient to cause large deflections; the fact that the plastic hinge transmits a definite constant bending moment

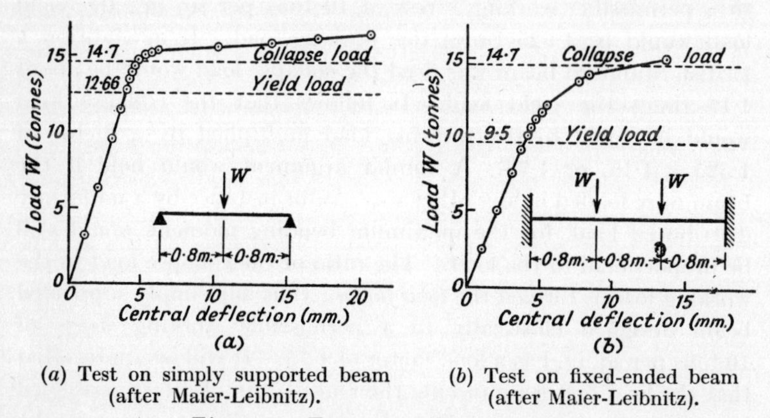

(*a*) Test on simply supported beam (after Maier-Leibnitz).

(*b*) Test on fixed-ended beam (after Maier-Leibnitz).

Fig. 1.1. *Tests on mild steel beams.*

implies that the mechanism motion only takes place when the load is such as to produce this fully plastic bending moment at the hinge.

Actual tests on mild steel beams show that the plastic hinge assumption is not strictly accurate. Thus Fig. 1.1(*a*) shows the load versus central deflection relation observed by Maier-Leibnitz [2] in testing a simply supported I-beam spanning 1·6 metres which was subjected to a central concentrated load. The calculated load at which the yield stress was first attained in the most highly stressed fibres beneath the load, termed the *yield load*, was 12·66 tonnes, and the collapse load predicted by the plastic theory was 14·7 tonnes. The indefinite growth of deflection under constant load which would have occurred at the calculated collapse load if a plastic hinge formed beneath the load did not in fact occur; instead, large deflections developed with small increases in the load. The beam actually failed by buckling at a load of 16·90 tonnes, but collapse had then already effectively occurred due to the development of unacceptably large deflections within a narrow range of loading, whose lower limit was indicated closely by the collapse load predicted by the plastic theory. This is the essential justification of the plastic theory; on its simplified assumptions a good estimate can be made of the load at which failure is about to occur due to the imminence of large deflections.

If a simply supported beam of this type were designed elastically to a permissible working stress of 10 tons per sq. in., the yield load would be 1·525 times the working load. If in addition a British Standard beam was used the collapse load would be about 1·15 times the yield load. It follows that the collapse load would be equal to the working load multiplied by a factor of 1·525 × 1·15, or 1·75. A similar argument would hold if the beam were loaded in any other way, as for instance by a uniformly distributed load, for the maximum bending moment would still be proportional to the load. The ratio of the collapse load to the working load is termed the *load factor*; thus any simply supported beam designed elastically to a permissible working stress of 10 tons per sq. in. has a load factor of 1·75. It will be appreciated that the load factor represents the real margin of safety possessed by a structure against failure by plastic collapse, for it is the factor by which the working loads would need to be increased before the structure actually failed in this way.

Behaviour of fixed-ended beam

Turning now to redundant structures, it can be said that in general the formation of the first plastic hinge will not usually cause collapse. Consider as a specific example a fixed-ended beam which carries a uniformly distributed load. In the elastic range the greatest bending moment is that developed at each end of the beam. Thus if the load is increased steadily yield first occurs at the ends of the beam, and after a further small increase of load plastic hinges form at these sections. However, the formation of these hinges does not cause collapse, for it is found that the central bending moment at this load is still less than the value which causes yield, so that the structure has not at this stage been reduced to a mechanism. Thus if the load is increased still further the plastic hinges at the ends of the beam undergo rotation while the bending moment at each hinge remains constant at the fully plastic value, and the central bending moment increases first to the yield value and soon after to the fully plastic value. During this process the rate of increase of deflection per unit increase of load becomes larger than it was during the elastic range, but there is no possibility of the growth of excessive deflections, as these are prevented by the continuity of the central portion of the beam. It is only when a plastic hinge also forms at the centre of the beam that large deflections can build up, for the structure has then been reduced to a mechanism. Plastic collapse then takes place, the load remaining constant while indefinitely large deflections build up due to the rotation at the central plastic hinge and the additional rotations at the end plastic hinges.

As will be seen in Chapter 2, the ratio of the collapse load to the yield load in this case is $1{\cdot}15 \times \frac{4}{3}$, in contrast with the ratio 1·15 found for the case of the simply supported beam. The additional factor of $\frac{4}{3}$ represents the extra load-carrying capacity due to the fact that the central bending moment is less than the yield value when the plastic hinges first form at the ends of the beam. It follows that if the beam were designed elastically to a permissible working stress of 10 tons per sq. in. the load factor against collapse would be $1{\cdot}525 \times 1{\cdot}15 \times \frac{4}{3}$, or 2·34.

Test results confirm that for continuous beams and statically indeterminate frames there is in general a greater difference between the yield and collapse loads than for simply supported beams, as predicted by the plastic theory. Thus in the investigation already referred to, Maier-Leibnitz [2] tested beams of 2·4 metres span which were effectively fixed at both ends and subjected to symmetrical two-point loading. This loading case is very similar to the case of a uniformly distributed load, and in particular the predicted ratio of the collapse load to the yield load is exactly the same. The load versus central deflection curve which was obtained is shown in Fig. 1.1(*b*). In this case the calculated yield and collapse loads were 9·5 tonnes and 14·7 tonnes, respectively. It will be seen that above the yield load the slope of the load-deflection curve decreased owing to the formation of plastic hinges at the ends of the beam, but excessive deflections did not begin to build up until the predicted plastic collapse load was approached. Thus the predicted collapse load again provided a good estimate of the load at which large deflections were imminent. The same kind of gradual decrease in the slope of load-deflection curves is observed when frames are tested, and it is nearly always found that rapid growth of deflection does not occur until the predicted collapse load is approached. A typical load-deflection curve obtained in a test on a full-scale pitched roof portal frame is given in Fig. 5.13 (p. 201).

The last point shown in Fig. 1.1(*b*) corresponds to a load of 15·0 tonnes. When a load of 15·32 tonnes was applied the beam buckled laterally; this is not surprising in view of the fact that the mid-third of the beam was in pure bending. However, this buckling again only occurred after the plastic collapse load had been surpassed.

Illogical nature of elastic design methods

The fundamental defect in the orthodox elastic design technique can now be exposed. It has been seen that a simply supported beam carrying a uniformly distributed load, designed elastically to a permissible working stress of 10 tons per sq. in., has a load factor against collapse of 1·75, whereas if a similarly loaded but fixed-ended beam were designed elastically to the same working stress its load factor against collapse would be 2·34. As was pointed out by Baker,[1] this position is indefensible, for there can be no reason for providing the fixed-ended beam with a greater margin of safety than the simply supported beam. The implication is that the section of the fixed-ended beam is stronger than necessary, and could be reduced until the load factor was the same as for the simply supported beam, with a resulting economy in the weight of steel used. Viewed from another standpoint, both the elastically designed beams have the same safety factor of 1·525, whereas their real margins of safety as represented by their load factors of 1·75 and 2·34 are considerably different. Thus the safety factor based on stresses does not give a true picture of the relative strengths of the beams. These statements can be given generality; the particular examples of a simply supported and a fixed-ended beam were only used for the sake of clarity. In fact, the load factor against the collapse of any statically determinate structure which is composed of British Standard beams and designed elastically to a permissible working stress of 10 tons per sq. in. will be 1·75, and for a corresponding statically indeterminate structure the load factor will almost invariably be in excess of this value.

The plastic design procedure

The rational procedure for the design of a steel frame is clearly to proportion the members so that the frame has some definite load factor against collapse, this load factor being the same for each frame of a given type which is designed. This is the course adopted in the plastic design methods. In the 1948 edition of B.S.449 a clause was introduced which permitted this procedure for the first time. This clause is as follows:

Clause 29(*c*). " *Fully rigid design.* This method, as compared with the methods for simple and semi-rigid design, will give the greatest rigidity and economy in the weight of steel used, when applied in appropriate cases. For the purpose of such design accurate methods of structural analysis shall be employed leading to a load factor of 2, based on the calculated or otherwise ascertained failure load of the structure or any of its parts, and due regard shall be paid to the accompanying deformations under working loads, so that deflections and other movements are not in excess of the limits implied in this British Standard."

The selection of the appropriate load factor to use in conjunction with the plastic design methods is a very complicated question, being bound up with the values which are specified for the working loads. The load factor of 2 which is specified in this clause is in fact greater than the load factor of 1·75 which any simply supported beam designed elastically to a working stress of 10 tons per sq. in. is known to possess. It might well be argued that since failures of simply supported beams designed in this way are virtually unknown, a load factor of 1·75 certainly provides an adequate safety margin and should therefore be specified. Another viewpoint is that since most redundant frames designed elastically have load factors in excess of 2, as for instance the case of the fixed-ended beam with a load factor of 2·34 which has been cited, the specification of a load factor of 2 strikes a rough average of the load factors of existing statically determinate and redundant frames. The respective merits of these viewpoints cannot be debated here; more detailed discussions are to be found in an article in *Engineering* [3] and in *The Steel Skeleton*, vol. 2.[4] It is worthy of note, however, that even if a load factor of 2 is used, economy results whenever the corresponding elastic design would have a load factor in excess of this value, as is the case in the great majority of redundant frames.

Scope of the plastic methods

The plastic methods of structural analysis which are described in this book are concerned almost entirely with the failure of framed structures by plastic collapse, the only exception being described in Chapter 8, where a somewhat similar type of failure which can occur when a frame is subjected to variable repeated loading is discussed. It is therefore implicitly assumed throughout that no part of the structure will fail by buckling before the plastic collapse load is reached. The problems of buckling of columns under the conditions actually arising in rigid frames when the members have partially yielded, and of lateral instability and other forms of buckling under similar conditions, have been the subject of an extensive investigation by J. F. Baker and his associates at Cambridge. As a result, several design rules have been formulated which will enable frames to be designed so that the possibility of failure by certain types of buckling before the attainment of the plastic collapse load is excluded.[4] A considerable amount of work on the question of column instability has also been carried out at Lehigh University.[5, 6]

The advantages of designing plastically in achieving more rational and economical designs would be of little value if the design technique were itself more complicated than the elastic technique. However, this is not the case; indeed the plastic methods have the further important advantages of remarkable simplicity and rapidity of calculation. This stems from the fact that it is possible to determine the plastic collapse load for a given structure and the corresponding collapse mechanism directly, without considering the order in which the various plastic hinges form as the loads are brought up to their collapse values.

Tests of actual steel framed structures have revealed that failure does in fact occur in very much the manner predicted by the plastic theory. Although hinges do not actually form, it can be seen that large changes of slope occur over small lengths of the members at the predicted plastic hinge positions. A slight hardening action usually occurs at the hinges, so that the growth of large deflections is accompanied by slight increases in the loads. In general, the predicted plastic collapse load indicates with a fair degree of precision the critical load at which large deflections are imminent. It thus appears that the collapse loads predicted by the plastic theory are close approximations to the actual failure loads of real structures, and since these actual failure loads are never sharply defined there is little point in demanding more exact theoretical values for what are in reality imperfectly defined quantities.

It is often thought that in this respect the plastic theory is less exact than the elastic theory, for while the plastic theory is based on approximations to the real behaviour of beams the elastic theory assumes Hooke's law, which is in fact obeyed by steel and other structural materials to a very high degree of accuracy, at least within a certain stress range. However, the apparent accuracy of the elastic methods is completely illusory, as far as anything but structures tested in the laboratory are concerned. The real structure which the elastic analysis is supposed to have described commonly differs from the hypothetical one in many ways; members do not fit exactly and are forced into place, supports settle unequally, joints assumed to be rigid are in fact flexible, and welding produces residual stresses. Moreover a rigorous analysis of stresses at joints, rivets, bearings, etc., is not even attempted, although it is well known that the stresses at such points may easily reach values much larger than the nominal maximum stresses in members on which the design is based. A further point to be borne in mind is that for all but the simplest of frames, exact methods of elastic analysis are extremely tedious and are rarely used in practice; instead, simplifying assumptions are made to reduce the calculations to an acceptable length. The stresses calculated in this way cannot be expected to bear any close relation to those existing in the actual structure. For these reasons elastic design methods are seen to be comparatively empirical in nature.

The accuracy of the plastic methods is very little affected by the presence of residual stresses, the flexibility of joints, the sinking of supports, or stress concentrations. However, this is of less fundamental importance than the fact that they deal directly and with reasonable accuracy with one type of real structural failure, and it seems axiomatic that only a method which does this can produce rational designs.

The basic physical property utilized in the plastic methods is ductility, in the sense that the material at each plastic hinge is assumed to be capable of deforming without fracture by large amounts compared with the elastic range of deformations, under constant or slowly rising loads. Of course, it is well known that mild steel does possess good ductility; indeed the popularity of mild steel is due in no small part to this property, for it enables local stress concentrations at joints, holes, etc., to be accommodated without fracture. However, in elastic designs the ductility of mild steel is only utilized passively in this way, whereas in plastic designs its use is inherent in the design technique. Many tests on various types of continuous beams and frames of mild steel have shown that mild steel does possess ample ductility for the purpose of developing those hinge rotations demanded by the plastic theory. The few tests which have been made with light alloys indicate that some of the aluminium and magnesium alloys can be expected to have the required ductility. However, the theory must not be applied to new materials, such as high strength alloys, until sufficient assurance as to their ductility is obtained by tests on structures of these materials. A. L. L. Baker[7] has investigated the possibility of designing reinforced concrete frames by a "plastic theory" which recognizes the limited ductility of the typical reinforced concrete beam. This theory, with its necessary emphasis on the calculation of deflections and slope changes for the purpose of ascertaining whether rupture would commence, is essentially different from the simple plastic theory based on the plastic hinge assumption.

Apart from the question of rupture, other types of failure may have to be considered beyond that with which the plastic methods are directly concerned. The problems of buckling of members which have partially yielded have already been referred to; it is obvious that in some circumstances consideration of these effects might dominate a design problem. There will also be cases in which there is risk of failure by fatigue under very large numbers of load cycles, or of brittle fracture under impact loading or low temperature conditions. These also are outside the scope of the simple plastic theory. Furthermore, the plastic methods do not attempt to deal accurately with deflections. Instead, they are concerned with collapse loads, at which deflections would in theory become infinite and in practice would soon grow to excessive values far beyond the elastic range of magnitude. If the design specification includes limitations on the deflections which can be tolerated under working loads, the plastic methods must be supplemented by a check on the values of these deflections.

These limitations on the plastic methods are stated for the sake of putting them in proper perspective, but they should not be emphasized unduly, for limitations of the same sort are common to any of the design methods yet available, each of which concerns itself especially with one particular type of failure. The plastic methods are suitable for use whenever static strength is of prime importance, as is the case for many framed structures. If some other consideration governs the design problem, their use would of course be inappropriate.

Historical survey

It may be of interest to give here a brief outline of the historical development of the plastic methods. Although the behaviour of beams loaded beyond the elastic range had been studied much earlier, apparently the first published papers on the possibility of directly utilizing the ductility of metals to improve the design of engineering structures were those of Kazinczy [8] in Hungary in 1914 and of Kist [9] in Holland in 1917. Kazinczy carried out some tests on fixed-ended beams, and came to the conclusion that failure only took place when yielding had occurred at three cross-sections, at which a hinging action could be said to occur. As a result of his recognition of the fundamentally important concept of the plastic hinge, Kazinczy may fairly claim to be the originator of the plastic methods. In the following decade there was a strong growth of interest in the new approach in Europe, and in 1926 a small book by Grüning [10] was published in Germany. In this book certain general results concerning the failure conditions of pin-jointed trusses were established, but the analysis was very complicated and was unaccompanied by experimental confirmation. Tests on simple and continuous beams were carried out by several investigators, notably Maier-Leibnitz [11, 2] in 1928 and 1929. These tests of Maier-Leibnitz and the accompanying theoretical interpretations may be said to have first put the plastic methods for continuous beams on a firm quantitative footing. A chapter in F. Bleich's treatise [12] on steel framed structures, published in Berlin in 1932, was devoted to a review of the plastic methods as they had then been developed for beams and simple portal frames. Girkmann [13] published a paper in 1931 which suggested an approximate method by which multi-storey, multi-bay rectangular frames could be designed. The results of the principal tests on continuous beams and simple frames which had been reported by himself and by other workers were collected together by Maier-Leibnitz [14] in 1936; this paper may be consulted for an excellent bibliography of this work.

Interest in these developments was stimulated in the U.S.A. when van den Broek [15] published a paper in 1940 in which the basic ideas of the plastic theory were restated. In this paper, which was later followed by a book on the same topic,[16] the technique of designing by the plastic methods was termed *limit design.* At about this time the analytical developments which were taking place on the continent were directed towards refining the simple calculations based on the plastic hinge concept rather than determining how this concept could be used directly to calculate the collapse loads of more complicated frames. Baker was the first to realize that the simple plastic theory might well prove to be the key to a simple and rational method for the design of complex frames. With this in mind, Baker and his associates began a comprehensive series of tests on portal frames in 1938.[17] A study of the behaviour of frame members which fail by buckling after they have partially yielded was also commenced. The present status of this work is described in *The Steel Skeleton*, vol. 2.[4] The problem of column instability has also been studied extensively at Lehigh University,[5, 6] where several other recent investigations have been carried out which have done much to clear up some of the outstanding difficulties which would otherwise prevent the application of the plastic methods to practical designs.[18, 19]

A method for calculating plastic collapse loads for more complex structures, such as two- or three-bay portal frames, was described by Baker in 1949.[20] Up to this time the basic principles on which the calculation of plastic collapse loads depended had been accepted from the earliest beginnings as intuitively obvious axioms, as indeed they were for the comparatively simple structures which were usually considered, and experimental investigations had substantiated these axioms. Formal statements and proofs of the principles have now been supplied independently by Greenberg and Prager [21] and by Horne.[22] The value of these contributions was not only that the plastic methods were placed on a firm foundation but that they stimulated the development of further general methods for the calculation of plastic collapse loads. As a result of these developments, the basic mathematical theory of the plastic methods, once the fundamental hypotheses are accepted, may be said to be as complete and rigorous as the conventional methods for the analysis of redundant structures in elastic structural theory.

The remainder of this chapter is devoted to a discussion of the fundamental postulates underlying the simple plastic theory. In Section 1.2 the basic hypotheses concerning the type of relation between bending moment and curvature which must hold for each member of a frame are stated, and it is shown how the concept of plastic hinge action arises. It is evident that some explanation for the existence of a fully plastic moment for a member, at which plastic hinge action can be assumed to occur, is required. This explanation is given in Section 1.4 after a preliminary review of the stress-strain properties of mild steel has been made in Section 1.3. Finally, the question of the applicability of the plastic methods to light alloy frames is discussed briefly in Section 1.5.

1.2. Basic Hypotheses

The principal developments which have taken place in the plastic methods have been largely confined to plane frames. In structures of this type the members all lie in one plane and are subjected to loads which also lie in this plane. The load carrying capacity of a plane frame is due primarily to the ability of the joints to transmit bending moments and the resistance to bending of the members, although these are also called upon to carry axial and shear forces. Consequently, the basic hypotheses of the plastic theory are concerned with the relation between bending moment and curvature for frame members. These hypotheses are in close accordance with the actual properties of the rolled steel members which are so often used in the construction of such frames. It is for this reason that the plastic methods have been developed almost exclusively for this type of structure. Corresponding hypotheses could, of course, be stated for the behaviour of the members of plane trusses, in which the joints are not called upon to transmit bending moments and the loads are carried primarily by axial tensile and compressive forces in the members. These hypotheses would deal with the relation between axial force and extension for the truss members. Unfortunately, as discussed in detail in Appendix A, these hypotheses are not obeyed even approximately by compression members, and so the development of plastic methods of analysis based on these assumptions would be of purely theoretical interest.

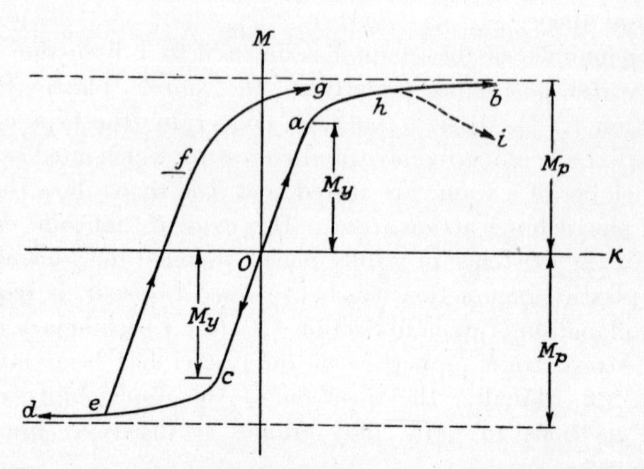

Fig. 1.2. *Assumed bending moment-curvature relation.*

The basic hypotheses underlying the plastic methods for plane frames are summarized diagrammatically in Fig. 1.2. This figure shows the type of relation between bending moment M and curvature κ which is assumed to hold at any cross-section of a typical member of a rigid frame.

In discussing this relation it will be assumed that the sign conventions adopted for bending moment and curvature are consistent. Thus if a positive bending moment is defined as causing tension in the outermost fibres on one side of the beam, a positive curvature will cause extension in the same fibres.

If a bending moment of the sense chosen as positive is applied to the previously unloaded and unstrained beam, as at O in the figure, the curvature at first increases linearly with the bending moment along *Oa*. This is the ordinary elastic range, which is terminated when a bending moment M_y is attained at *a*. This bending moment M_y is such as to cause the yield stress to be attained in the most highly stressed outer fibres of the beam, and is therefore referred to as the yield moment. When a bending moment above this value is applied, the curvature begins to increase more rapidly per unit increase of bending moment along *ab*. Physically, this corresponds to the spread of yield from the outermost fibres inwards towards the neutral axis of the beam, as will be seen in more detail in Section 1.3. As the bending moment increases above the value M_y, the curvature increases more and more rapidly with increase of bending moment, and finally tends to infinity as a limiting value of the bending moment is approached. This limiting value of the bending moment is termed the fully plastic moment, and is denoted by M_p. The attainment of the fully plastic moment may be thought of as corresponding physically to the development of the full yield stress down to the neutral axis of the beam in both tension and compression, although as will be seen in Section 1.3 this implies infinite strains in the beam and is therefore unrealistic. If the bending moment had been applied initially in the opposite sense, the behaviour would have been precisely similar in character along *Ocd*, except that negative increments of bending moment would have been accompanied by negative increments of curvature.

If instead of applying a positive bending moment to the stress-free beam, the beam had first been bent in the negative sense along *Oce* and then subjected to an increasing bending moment in the positive sense, the subsequent behaviour would be as indicated by the line *efg*. The relationship between the increments of bending moment and curvature would at first be linear along *ef*, the slope of this line being identical with the slope of the original elastic line *Oa*. When the bending moment is increased above the value at *f* more rapid increases of curvature ensue along *fg*, and eventually the bending moment again tends to the limiting value M_p. The behaviour would be of a similar character if the bending moment was reduced after loading along *Oab*. In this case the bending moment would eventually tend to the limiting value $-M_p$ as the curvature tended to infinity in the negative sense.

The fundamental concept of plastic hinge action can now be

put forward. Suppose that at some section in a frame member the bending moment attains the positive fully plastic moment M_p, while the bending moment on either side of this section decreases in magnitude, as would be the case, for example, at the cross-section under a concentrated load. As the bending moment reaches the value M_p, the curvature becomes indefinitely large, so that a finite change of slope can occur over an indefinitely small length of the member at this cross-section. Thus the behaviour at a section where M_p is attained can be described by imagining a hinge to be inserted in the member at this section, the hinge being capable of completely resisting rotation until the fully plastic moment M_p is attained, and then permitting positive rotation of any magnitude while the bending moment remains constant at the value M_p. The sense in which the hinge rotation is to be regarded as positive is, of course, the same as the sense in which the curvature is regarded as positive. The actual magnitude of the hinge rotation which would occur in any given case would of course be determined by the rigidity of the remainder of the structure under the particular load changes taking place while the hinge action was developing. If the bending moment is reduced below M_p, elastic unloading occurs and the hinge rotation remains constant. Precisely corresponding statements can be made concerning the hinge action which is supposed to occur when the negative fully plastic moment $-M_p$ is attained.

It is a fundamental concept of the plastic methods described in this book that plastic hinges are imagined to occur at those sections where the fully plastic moment is developed. This concept cannot be justified rigorously from a theoretical standpoint. A true hinge action would necessitate infinite strains in the beam, and the strain-hardening which necessarily occurs causes the fully plastic moment to be exceeded. The justification of the concept of plastic hinges is to be found not in rigorous theory but in the results of many tests on beams and framed structures. These show that something very much like hinge action occurs in structures at sections where the fully plastic moment is developed, and that the failure of rigid plane frames can in fact be considered to correspond with the development of plastic hinges at a certain critical number of positions. The indications from actual test results are in fact that at those cross-sections where the simple plastic theory assumes the formation of plastic hinges extremely large curvatures occur, so that large changes of slope occur within short lengths of the members.

It will now be evident that the type of behaviour indicated by the dotted line *ahi* in Fig. 1.2, in which an increase of curvature is accompanied by a decrease of bending moment, must be specifically excluded. If the bending moment-curvature relation was of this form it would not be possible to postulate the development of a plastic hinge as the fully plastic moment M_p was attained, for this demands infinitely large curvature.

To sum up, the basic hypotheses are that at any section of a member of a rigid plane frame the bending moment must lie between the positive and negative fully plastic moments $\pm M_p$, and that increments of bending moment always cause increments of curvature which are of the same sign. These hypotheses can be summarized as follows:

$$-M_p \leqslant M \leqslant M_p \quad . \quad . \quad . \quad 1.1$$

$$\frac{dM}{d\kappa} \geqslant 0 \quad . \quad . \quad . \quad . \quad 1.2$$

As a corollary to these hypotheses, it is assumed that as the fully plastic moment is attained at any cross-section the curvature becomes indefinitely large, and a plastic hinge forms which can undergo rotation through any arbitrary angle while the bending moment remains constant at the fully plastic value.

1.3. Stress-strain relation for mild steel

The basic hypotheses required by the plastic theory are concerned with the relation between bending moment and curvature for frame members. However, it is obvious that close relations must exist between the bending moment-curvature relation and the stress-strain relation of the material in tension and compression. These relations must be discussed, both in order to understand more fully the basic phenomena, and for the sake of the obvious practical advantages which will be gained if the bending moment-curvature relation, and more especially the value of the fully plastic moment, can be calculated for any given shape of cross-section from a knowledge of the familiar stress-strain diagram of the material. As already remarked, the plastic methods have been developed primarily for beams and frames of mild steel, since this material is widely used for such structures and has properties which are in close accordance with the basic hypotheses. It is therefore pertinent to discuss flexural behaviour in terms of the stress-strain relation for mild steel.

The stress-strain relation for a specimen of annealed mild steel in tension has the typical form shown in Fig. 1.3(*a*). The relation is almost exactly linear in the elastic range until the upper yield stress is reached at *a*. The stress then drops abruptly to the lower yield stress, and the strain then increases at constant stress up to the point *b*, this behaviour being termed purely plastic flow. Beyond the point *b* further increases of stress are required

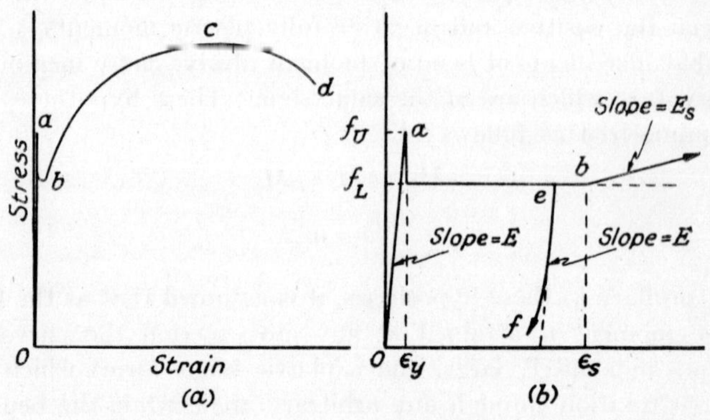

(*a*) Overall stress-strain relation for mild steel in tension.

(*b*) Initial stress-strain relation for mild steel in tension.

Fig. 1.3. *Stress-strain relation for mild steel.*

to produce further strain increases, and the material is said to be in the strain-hardening range. Eventually a maximum stress is reached at *c*, beyond which increases of strain occur with decreases of stress until rupture occurs at *d*. This latter behaviour is associated with the formation of a neck in the specimen. The maximum stress is of the order of magnitude of 20–30 tons per sq. in. and the strain at fracture is of the order of 25–50%.

Both at yield and near fracture increases of strain are accompanied by decreases of stress or load on the specimen. This type of behaviour cannot be followed on the usual form of tensile testing machine, and can only be recorded accurately on a type of machine in which the strain is imposed, as for instance by a screw mechanism, and the stress measured by means of a weighbar in series with the specimen.

The yield range *Oab* of the stress-strain relation is of the most interest from the point of view of the plastic theory. Since the strain at the onset of strain-hardening at *b* is generally of the order of 1–2%, the yield range can be examined more conveniently if the strain scale is considerably enlarged, as in Fig. 1.3(*b*). In this figure the upper and lower yield stresses are defined as f_U and f_L, respectively. The slope of the initial elastic line *Oa* is Young's Modulus, *E*, and the slope of the initial portion of the strain-hardening line beyond *b* is defined as E_s. The strains at the yield point *a* and at the onset of strain-hardening *b* are defined as ε_y and ε_s, respectively. If the stress is reduced after yield a relation such as *ef* is observed, the initial slope being Young's Modulus. The deviation from linearity in such an unloading relation is associated with the Bauschinger [23] effect.

If the stress is increased again after a reduction of this sort, yield occurs at the lower yield stress along *eb*. This indicates the effect of *cold-working* in destroying the upper yield stress, which only reappears after further heat treatment. As discussed in Chapter 6, the values of the various constants defined in Fig. 1.3(*b*) depend markedly on the composition of the steel and its heat treatment, except for the value of Young's Modulus, which shows very little variation. For a typical annealed mild steel the ratio $\frac{\varepsilon_s}{\varepsilon_y}$ is of the order of 10–20, and the ratio $\frac{E}{E_s}$ is of the order of 20–50, so that the stress-strain relation is very flat after yield. The lower yield stress f_L is usually of the order of 15–20 tons per sq. in. and the ratio $\frac{f_U}{f_L}$ of the order of 1·25.

It is difficult to determine the actual stress-strain relation of mild steel in the elastic range near the yield point, because there are always unavoidable eccentricities of loading which cause bending stresses, even when special precautions are taken to ensure that the load is applied along the axis of the specimen. However, the careful tests of Smith [24] and Morrison [25] showed that within the limits of experimental accuracy the initial departure from linearity usually observed below the yield point could always be ascribed to yielding in the most highly stressed fibres caused by the eccentricity of loading. It was therefore concluded that the yield point, proportional limit and elastic limit were all coincident. These tests also showed that the values of the upper yield stress showed no more variation from specimen to specimen of the same material than those of the lower yield stress. The unpredictable variations in the values of the upper yield stress reported by other observers were therefore concluded to be due to variations in the eccentricity of loading, and not, as had been suggested, due to the upper yield phenomenon being a form of instability akin to the elevation of the boiling point of a liquid in the absence of nuclei on which bubbles can form.

For a given steel the stress-strain relation in compression is practically identical with that for tension up to the point *b* where strain-hardening begins, as shown by Smith,[24] Morrison [25] and others.

The yield phenomenon for mild steel is known to be very complex in character. It is well known that yielding is accompanied by the formation of Lüders' lines making an angle of about 45° with the axis of the specimen. It seems probable, as suggested by Muir and Binnie,[26] that the material within the Lüders' lines has undergone a considerable amount of slip, corresponding to a jump in the strain from *a* to *b* in Fig. 1.3(*b*). This view is supported by the strain-etch figures obtained by Jevons,[27] which showed very clearly the spread of yielded zones from the ends of tensile specimens, where yield was always initiated owing to the stress concentration at the shoulders of the test pieces. Acceptance of this view implies that the longitudinal strain in a yielded fibre varies discontinuously along the fibre, and a stress-strain relation such as that shown in Fig. 1.3(*b*) can only represent average strains over a finite length and not local strains.

The stress-strain relation is often idealized by the neglect of strain-hardening and the Bauschinger effect on unloading, thus leading to a relation of the kind shown in Fig. 1.4(*a*). Although the upper yield effect is a very real one, it disappears on cold-working and is often not exhibited by the material of rolled steel sections. Moreover, it will be seen later that it has no effect on the value of the fully plastic moment. If it is disregarded, the stress-strain relation becomes that of Fig. 1.4(*b*), which is often termed the ideal plastic relation.

The neglect of strain-hardening in these idealized relations is somewhat difficult to justify in view of the fact that the strains will certainly enter the strain-hardening range in many members in actual structures. However, by neglecting the increase of

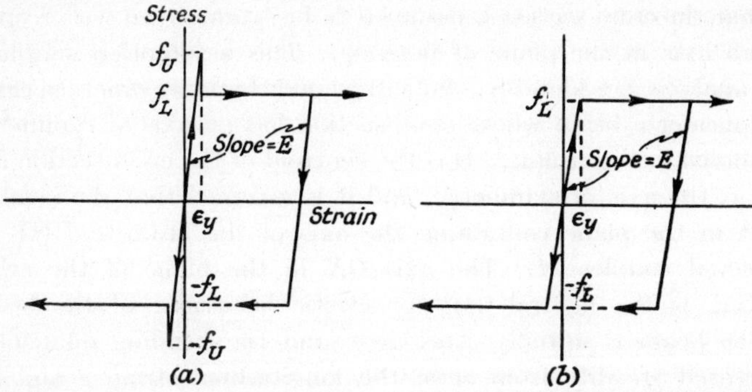

(*a*) Idealized stress-strain relation. (*b*) Ideal plastic stress-strain relation

Fig. 1.4. *Stress-strain relations neglecting strain-hardening.*

stress during strain-hardening, errors will be introduced which are on the safe side, and it will be seen in Chapter 5 that these errors are usually very small.

1.4. Evaluation of fully plastic moments for mild steel beams

If it is assumed that the stress-strain relation for the material of a given beam is one of the idealized relations shown in Fig. 1.4, the changes in the distribution of stress across the section of a beam under a steadily increasing bending moment can be followed without difficulty, and it is possible to explain the hypothesis of a fully plastic moment quite readily. For the sake of simplicity it will be assumed that the stress-strain relation is of the ideal plastic type shown in Fig. 1.4(*b*), with no upper yield stress. It is further assumed that this relation is obeyed for each individual fibre of the beam. In view of the discontinuous nature of the yielding process implied by this relation, this assumption requires experimental verification ; many investigators, notably Roderick and Phillips,[28] have provided evidence in its favour. Apart from these assumptions the usual assumptions which are made in the Bernoulli-Euler theory of bending are made. Thus it is assumed in the first place that the beam is bent by pure terminal couples, so that shear and axial forces are not present. The deformations and strains are assumed to be small, so that stresses other than the longitudinal normal stresses can be neglected. Plane sections are assumed to remain plane, so that the longitudinal strain varies linearly with distance from some neutral axis. A final assumption is that the cross-section is assumed to be symmetrical with respect to an axis in the plane of bending. This assumption simplifies the analysis considerably, and corresponds to many practical cases.

Consider a beam whose cross-section has an axis of symmetry, as shown in Fig. 1.5(a). O is the centroid of the cross-section and OY is the axis of symmetry, and it is assumed that the beam is bent in the plane containing the axis of the beam and OY by terminal couples M. The axis OX in the plane of the cross-section is the neutral axis for elastic behaviour of the beam. If the beam is initially stress-free, and the bending moment is increased steadily from zero, the longitudinal strain ε and the longitudinal normal stress σ both vary linearly across the section

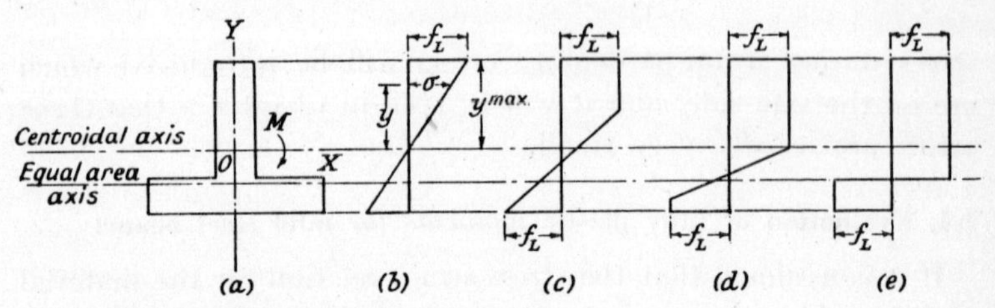

(a) Cross-section.
(b) Stress distribution at first yield.
(c) Stress distribution with yielding on one side of neutral axis.
(d) Stress distribution with yielding on both sides of neutral axis.
(e) Fully plastic stress distribution.

Fig. 1.5. *Stress distributions in elastic-plastic beam.*

with distance y from the neutral axis while the greatest stress is still less than the yield stress f_L, so that the behaviour is entirely elastic. The distribution of stress when the yield stress f_L is just attained in the most highly stressed fibres is shown in Fig. 1.5(b). Within the elastic range of behaviour the longitudinal stress σ and strain ε vary across the section according to the usual elastic relations

$$\varepsilon = y\left(\frac{1}{R} - \frac{1}{R_0}\right) \quad . \quad . \quad . \quad . \quad 1.3$$

$$\sigma = E\varepsilon = \frac{My}{I} \quad . \quad . \quad . \quad . \quad 1.4$$

where R is the radius of curvature after the application of the couples M, R_0 is the initial radius of curvature, and I is the moment of inertia of the cross-section about the axis OX.

According to equation 1.4, the greatest stress occurs in the fibre for which y has its greatest value, defined as y^{max} in Fig. 1.5(b). Thus the greatest stress σ^{max} which occurs when the behaviour is entirely elastic is given by

$$\sigma^{max} = \frac{My^{max}}{I} = \frac{M}{Z}$$

where
$$Z = \frac{I}{y^{max}}$$

The quantity Z is the familiar *elastic modulus* of the cross-section. Thus if the bending moment at which the yield stress is first attained, as in Fig. 1.5(b), is denoted by M_y, it follows that this *yield moment* is given by

$$M_y = Zf_L \quad . \quad . \quad . \quad . \quad 1.5$$

When the bending moment is increased beyond the yield moment, the strain in the outermost fibres increases beyond the yield value ε_y shown in Fig. 1.4(b), so that plastic flow ensues in a region including these outermost fibres. The distribution of stress across the section is then shown in Fig. 1.5(c), which shows the case in which the yield stress f_L is just attained on the lower surface of the beam. The neutral axis no longer passes through the centroid O, but assumes a position dictated by the fact that the resultant normal force on the cross-section must be zero. It is assumed in the simple theory that the lines of demarcation in the cross-section between the central elastic core and the outer plastic region or regions are straight lines parallel to OX; this is not strictly true, as pointed out by Hill.[29]

In the case of a cross-section with a second axis of symmetry, which must then coincide with the centroidal axis OX, both plastic zones form simultaneously and have equal areas, and the neutral axis still passes through the centroid.

A further increase of bending moment above the value corresponding to the distribution of stress shown in Fig. 1.5(c) causes yield to spread inwards from the lower surface of the beam, as well as spreading further in from the upper surface, as shown in Fig. 1.5(d). Ultimately the two zones of yield meet, the distribution of stress then being as shown in Fig. 1.5(e). This is the condition of full plasticity, and the corresponding bending moment is the fully plastic moment. When this distribution of stress is attained the curvature will have become infinitely large, for the strain is required to change from ε_y to $-\varepsilon_y$ for an infinitely small change of y at the neutral axis. Equation 1.3 still holds true if the origin of y is taken at the neutral axis, and it is seen that R is thus required to be zero.

On this basis, the plastic hinge hypothesis is at once explained, for if the curvature at any section in a beam became infinitely large a finite change of slope could take place over an infinitely short length of the beam at this section, corresponding to a hinge action. In practice, however, the condition of full plasticity shown in Fig. 1.5(e) cannot be attained, for above a certain curvature the strains in the outer fibres would become sufficiently large to cause strain-hardening. Moreover, additional radial stresses would be called into play, for a curved fibre subjected only to tensile or compressive forces at its ends applied in directions perpendicular to the cross-sections of the fibre would not be in radial equilibrium. Thus the fully plastic moment as calculated from the fully plastic stress distribution must be regarded as an approximate indication of the moment at which something very much like a hinge action will occur in practice, with large curvatures developing at the cross-section where this moment is attained. As seen in Fig. 1.1(a), a true hinge action does not in fact occur in actual mild steel beams. All that can be said is that within a narrow range of loading the slope of the load-deflection curve is reduced rather abruptly. For this beam Maier-Leibnitz [2] calculated the fully plastic moment from the fully plastic stress distribution, using a value for f_L which was the mean of the yield stresses observed from four tensile specimens cut from the flanges of the beam. The calculated collapse load of 14·7 tonnes shown in the figure corresponds to the attainment of this fully plastic moment at the centre of the beam, and this load provides a reasonable indication of the somewhat ill-defined load above which the beam developed large deflections. Experimental evidence of this kind vindicates the concept of a plastic hinge developing at the fully plastic moment; since this moment is rarely defined sharply in actual beams an approximate indication of its value is sufficient for practical purposes.

The value of the fully plastic moment can be calculated readily in terms of the yield stress f_L and the shape and dimensions of the cross-section. Since the resultant axial force is zero the neutral axes in the fully plastic condition must divide the cross-section into two equal areas, so that the resultant axial tensile and compressive forces are both equal to $\frac{1}{2}Af_L$, A being the total area of the cross-section. If the two equal areas into which the cross-section is divided have centroids G_1 and G_2 at distances $\bar{y}_1$ and $\bar{y}_2$ from the neutral axis respectively, as in Fig. 1.6(*a*), the resultant forces will act through G_1 and G_2 and the fully plastic moment will be given by

$$M_p = \tfrac{1}{2}Af_L(\bar{y}_1 + \bar{y}_2) \quad . \quad . \quad . \quad 1.6$$

Thus if the *plastic section modulus* Z_p is defined by the relation $M_p = Z_p f_L$, it follows that

$$Z_p = \tfrac{1}{2}A(\bar{y}_1 + \bar{y}_2) \quad . \quad . \quad . \quad 1.7$$

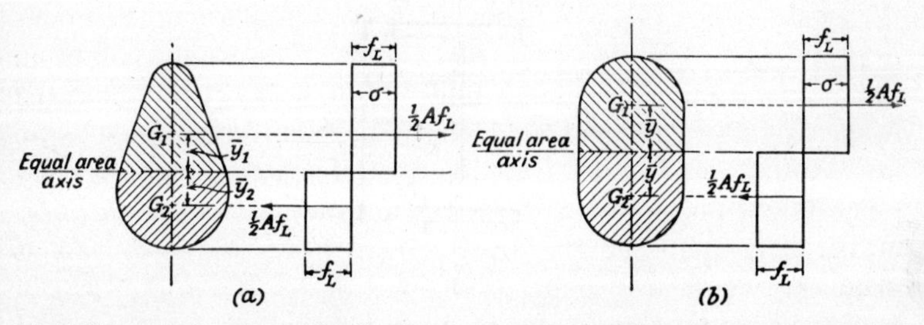

(*a*) Fully plastic stress distribution for section with single axis of symmetry.

(*b*) Fully plastic stress distribution for section with two axes of symmetry.

Fig. 1.6. *Fully plastic stress distributions.*

When the cross-section has a second axis of symmetry, as in Fig. 1.6(*b*), $\bar{y}_1$ and $\bar{y}_2$ are both equal to $\bar{y}$, say, so that

$$M_p = A\bar{y}f_L \quad . \quad . \quad . \quad . \quad 1.8$$

For a rectangular section of breadth b and depth h, bent about an axis parallel to the sides of breadth b, $A = bh$ and $\bar{y} = \frac{h}{4}$, so that the fully plastic moment is

$$M_p = \tfrac{1}{4}bh^2 f_L, \quad . \quad . \quad . \quad . \quad 1.9$$

a result first obtained by Saint-Venant.[30] To determine the yield moment M_y at which the yield stress is just attained in the outermost fibres it is noted that $I = \frac{1}{12}bh^3$ and $y^{max} = \frac{1}{2}h$, so that $Z = \frac{1}{6}bh^2$. Thus

$$M_y = \tfrac{1}{6}bh^2 f_L \quad . \quad . \quad . \quad . \quad 1.10$$

For this cross-section the ratio $\frac{M_p}{M_y}$ is thus 1·5. In general, the ratio $\frac{M_p}{M_y}$ is termed the *shape factor*, and is denoted by α, so that

$$\alpha = \frac{M_p}{M_y} = \frac{Z_p}{Z} \quad . \quad . \quad . \quad . \quad 1.11$$

This ratio depends solely on the shape of the cross-section.

A commercial I-section can be idealized by regarding the flanges as rectangles of breadth b and thickness t_2 and the web as a rectangle of depth $(h - 2t_2)$ and thickness t_1, as shown in

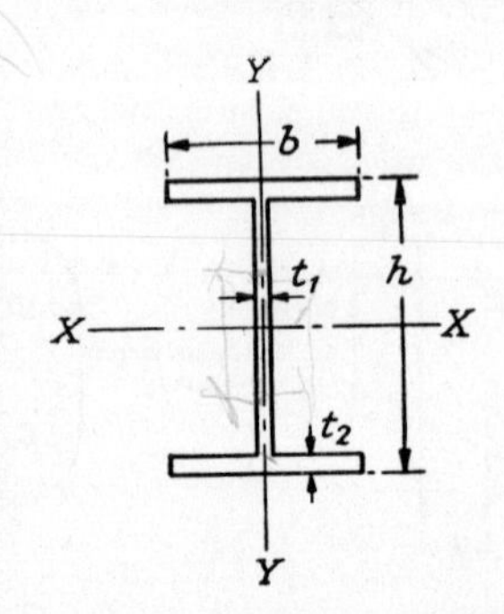

Fig. 1.7. *Idealized I-section.*

Fig. 1.7. For this idealized section it is easy to show that the elastic and plastic section moduli, Z and Z_p, are given by

$$Z = \frac{1}{h}[\tfrac{1}{3}bt_2^3 + bt_2(h - t_2)^2 + \tfrac{1}{6}t_1(h - 2t_2)^3] \quad . \quad 1.12$$

$$Z_p = bt_2(h - t_2) + \tfrac{1}{4}t_1(h - 2t_2)^2 \quad . \quad . \quad . \quad 1.13$$

Taking as an illustration a 6 in. × 3 in. × 12 lb. British Standard beam, with $h = 6$ in. and $b = 3$ in., it is found from Section Tables that the average web and flange thicknesses are $t_1 = 0{\cdot}23$ in. and $t_2 = 0{\cdot}377$ in., respectively. From equations 1.12 and 1.13 it is then found that $Z = 6{\cdot}89$ in³. and $Z_p = 7{\cdot}94$ in³., so that the shape factor α has the value $\frac{7{\cdot}94}{6{\cdot}89}$, or 1·15. The value of Z which is given in Section Tables is 7·00 in³., the difference between this value and that obtained from equation 1.12 being due to the difference between the actual and assumed shapes of the section. Horne [31] has calculated accurate values of Z_p for British Standard beams based on the actual shape of the cross-section, and his value for this section is 8·08 in³., so that the shape factor is actually $\frac{8{\cdot}08}{7{\cdot}00}$, or 1·154. These accurate values of Z_p are tabulated in Appendix B, which also gives the values of Z_p for British Standard beams bent about their minor axes. Further information will be found in these tables from which the effect of axial thrust on the

TABLE 1.1

Plastic section moduli and shape factors for structural sections

Section	Plastic section modulus Z_p	Shape factor α
Rectangular	$\frac{1}{4}bh^2$	1·5
Approximation to I-section	Axis X—X $bt_2(h - t_2) + \frac{1}{4}t_1(h - 2t_2)^2$	About 1·15 for British Standard beams
	Axis Y—Y $\frac{1}{2}b^2t_2 + \frac{1}{4}(h - 2t_2)t_1^2$	About 1·67 for British Standard beams
Solid circular	$\frac{1}{6}d^3$	$\frac{16}{3\pi} = 1{\cdot}70$
Thin-walled hollow tube	$\frac{1}{6}d^3\left[1 - \left(1 - \frac{2t}{d}\right)^3\right]$ $t \ll d$; td^2	$t = \frac{1}{10}d$, $\alpha = 1{\cdot}40$ $t \ll d$, $\alpha = 1{\cdot}27$
Approximation to channel section	Axis X—X As for I-section about X—X	About 1·17 for British Standard channels
	Axis Y—Y $\frac{1}{2}bht_1$ for case in which $ht_1 = 2(b - t_1)t_2$	$t_1 = 0{\cdot}15b$, $\alpha = 1{\cdot}80$

fully plastic moment can be calculated; the basis on which the appropriate formulae were derived is discussed in Chapter 6.

The value of the shape factor for British Standard beams bent about their major axes varies from 1·12 to 1·18, with an average of 1·15. For American wide flange sections the average shape factor is about 1·14.

Formulae for the values of Z_p and α for some of the commoner structural sections are collected together in Table 1.1. The form of the expression for Z_p for the case of a channel bent about its minor axis depends on the position of the equal area axis YY. For most channel sections the area of the web is not greatly different from the sum of the areas of the two flanges, and so the equal area axis may lie just within or just outside the web. For the special case in which this axis coincides with the inner edge of the web a simple formula for Z_p can be given, and for the sake of simplicity this is the only result given in the table for a channel section bent about its minor axis.

It is of interest to note that if the existence of an upper yield stress f_U had been taken into account, so that the idealized stress-strain diagram of Fig. 1.4(*a*) was assumed, the stress distributions of Figs. 1.5(*b*), (*c*) and (*d*) would have been modified to show a linear distribution of stress in the elastic core rising to a maximum value f_U at the elastic-plastic boundaries, with the stress then dropping abruptly to the value f_L which would be constant throughout the plastic zones. Thus the yield moment would have the value Zf_U instead of Zf_L, as given by equation 1.5. However, the stress distribution of Fig. 1.5(*e*) would be unchanged, for the width of the elastic core has dropped to zero at full plasticity. It follows that the value of the fully plastic moment, which is calculated from the fully plastic stress distribution, is unaffected by the value of the upper yield stress, and depends solely on the value of the lower yield stress f_L.

A point of considerable importance is that the fully plastic moment represents a definite limit on the value of the bending moment, regardless of the possible presence of residual stresses induced, for example, by previous bending into the partially plastic range. This follows from the fact that the stress in any fibre cannot exceed f_L; on this basis the fully plastic stress distribution clearly corresponds to the greatest possible bending moment which can be developed. Moreover, it is only when this distribution is attained that the curvature can become infinite, so that a plastic hinge can form, for with any other distribution there must be an elastic core with a correspondingly finite rate of change of stress and thus of strain with distance from the neutral axis. It follows that irrespective of any residual stress distribution across a section before loading, a plastic hinge can only form when the fully plastic moment is attained.[32]

The foregoing analyses assumed that the only stresses acting were the longitudinal normal stresses due to bending. However, in nearly all practical cases there will be shear and axial forces acting at the cross-section in addition to the bending moment. These will modify the value of the fully plastic moment to an extent which is often negligible, but in any event these effects may be calculated and allowed for where necessary. A full discussion of these and other related effects will be given in Chapter 6.

It will also be appreciated that since the value of the fully plastic moment is proportional to the lower yield stress, those factors which affect the lower yield stress will also affect the value of the fully plastic moment. Thus the value of the fully plastic moment will depend on such influences as the composition and heat treatment of the material, the rate of loading, and such effects as strain-ageing. The effect of these factors will also be discussed in Chapter 6.

1.5. Plastic hinge assumption for other structural materials

There is no reason why the plastic methods should not be applied to framed structures of ductile metals other than mild steel provided that their properties are in reasonably close accordance with the basic hypotheses which have been stated in Section 1.2. Relatively little attention has been paid to this question, but there is little doubt that certain of the light alloys possess suitable properties. For instance, Panlilio [33] carried out tensile and compressive tests on a representative selection of aluminium and magnesium alloys, together with tests of beams on two and three supports. A typical load-deflection curve for a simply supported 3 in. extruded I-beam (0·153 in. web) of 61ST aluminium alloy is shown in Fig. 1.8. It will be seen that the behaviour of the beam could be described very well by the plastic hinge hypothesis. Indeed, the contrast with the load-deflection relation for a mild steel beam shown in Fig. 1.1(*a*) is quite favourable.

Owing to the fact that the light alloys have values of Young's Modulus which are much smaller than the value for mild steel, the deflections developed prior to collapse will be considerably greater than those of similar steel frames, and there is a greater likelihood of buckling failure occurring before the attainment of the predicted plastic collapse load. Careful consideration should be given to these matters before the simple plastic theory is applied

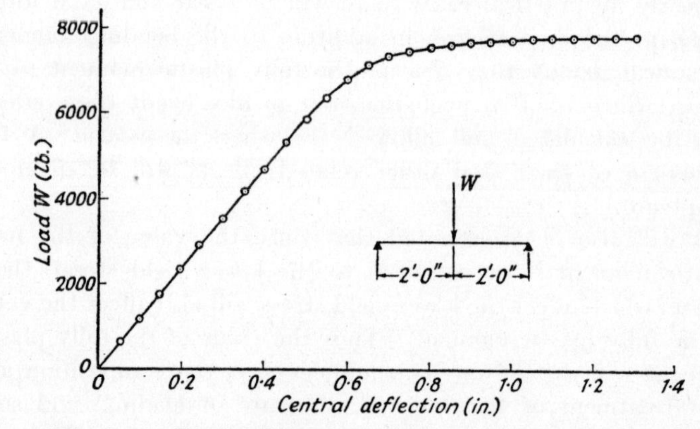

Fig. 1.8. *Load-deflection relation for light alloy beam.*

to light alloy frames. A further point is that the ductility of light alloys is often lower than that of mild steel. For instance, the elongation at fracture in a tensile test is of the order of 25–50% for many mild steels, but of the order of 10% or even less for some light alloys. In applying the simple plastic theory sufficient ductility must be available in order that the assumption of large plastic hinge rotations occurring at approximately constant bending moment without fracture should not be violated. In some of his tests Panlilio found that fractures occurred at deflections corresponding to quite small rotations of the plastic hinges.

References

1. J. F. Baker. A review of recent investigations into the behaviour of steel frames in the plastic range. *J. Instn. Civ. Engrs.*, **31**, 188 (1949).

2. H. MAIER-LEIBNITZ. Versuche mit eingespannten und einfachen Balken von I-form aus St 37. *Bautechnik*, **7,** 313 (1929).

3. ANONYMOUS. Load factors for structural steelwork. *Engineering*, **167,** 38 (1949).

4. J. F. BAKER, M. R. HORNE and J. HEYMAN. *The Steel Skeleton*, Vol. 2. Cambridge University Press. To be published.

5. R. L. KETTER, L. S. BEEDLE and B. G. JOHNSTON. Column strength under combined bending and thrust. *Weld. J.*, Easton, Pa., **31,** 607-s (1952).

6. R. L. KETTER, E. L. KAMINSKY and L. S. BEEDLE. Plastic deformation of wide-flange beam-columns. *Proc. Amer. Soc. Civ. Engrs.*, Sep. No. 330 (1953).

7. A. L. L. BAKER. Further research in reinforced concrete, and its application to ultimate load design. *Proc. Instn. Civ. Engrs.*, **2** (Part III), 269 (1953).

8. G. KAZINCZY. Kisérletek befalazott tartókkal [Experiments with clamped girders]. *Betonszemle*, **2,** 68 (1914).
[For some notes on the early work of Kazinczy see N. J. HOFF. *Weld. J.*, Easton, Pa., **33,** 14-s (1954).]

9. N. C. KIST. Leidt een Sterkteberekening, die Uitgaat van de Evenredigheid van Kracht en Vormverandering, tot een goede Constructie van Ijzeren Bruggen en gebouwen? *Inaugural Dissertation*, Polytechnic Institute, Delft (1917).

10. M. GRÜNING. *Die Tragfähigkeit statisch unbestimmten Tragwerke aus Stahl bei beliebig häufig wiederholter Belastung.* Julius Springer, Berlin (1926).

11. H. MAIER-LEIBNITZ. Beitrag zur Frage der tatsächlichen Tragfähigkeit einfacher und durchlaufender Balkenträger aus Baustahl St. 37 und aus Holz. *Bautechnik*, **6,** 11 (1928).

12. F. BLEICH. *Stahlhochbauten, ihre Theorie, Berechnung, und Bauliche Gestaltung.* Vol. I, Chap. XI, Julius Springer, Berlin (1932).

13. K. GIRKMANN. Bemessung von Rahmentragwerken unter Zugrundelegung eines ideal-plastischen Stahles. *S.B. Akad. Wiss. Wien.* (*Abt. IIa*), **140,** 679 (1931).

14. H. MAIER-LEIBNITZ. Versuche, Ausdeutung und Anwendung der Ergebnisse. *Prelim. Pubn. 2nd Congr. Intern. Assn. Bridge and Struct. Engng.*, 97, Berlin (1936).

15. J. A. VAN DEN BROEK. Theory of limit design. *Trans. Amer. Soc. Civ. Engrs.*, **105,** 638 (1940).

16. J. A. VAN DEN BROEK. *Theory of Limit Design.* John Wiley (N.Y.), Chapman & Hall (London), (1948).

17. J. F. BAKER and J. W. RODERICK. An experimental investigation of the strength of seven portal frames. *Trans. Inst. Weld.*, **1,** 206 (1938).

18. B. G JOHNSTON, C. H. YANG and L. S. BEEDLE. An evaluation of plastic analysis as applied to structural design. *Weld J.*, Easton, Pa., **32,** 224-s (1953)

19. L. S. Beedle. Plastic strength of steel frames. *Proc. Amer. Soc. Civ. Engrs.*, **81,** Paper No. 764 (1955).
20. J. F. Baker. The design of steel frames. *Struct. Engr.*, **27,** 397 (1949).
21. H. J. Greenberg and W. Prager. On limit design of beams and frames. *Trans. Amer. Soc. Civ. Engrs.*, **117,** 447 (1952). [First published as Tech. Rep. A18-1, Brown Univ. (1949).]
22. M. R. Horne. Fundamental propositions in the plastic theory of structures. *J. Instn. Civ. Engrs.*, **34,** 174 (1950).
23. J. Bauschinger. Die Veränderungen der Elastizitätsgrenze. *Mitt. mech.-tech. Lab. tech. Hochschule*, München (1886).
24. C. A. M. Smith. Compound stress experiments. *Proc. Instn. Mech. Engrs.*, Parts 3 and 4, 1237 (1909).
25. J. L. M. Morrison. The yield point of mild steel with particular reference to the effect of size of specimen. *Proc. Instn. Mech. Engrs.*, **142,** 193 (1939).
26. J. Muir and D. Binnie. The overstraining of steel by bending. *Engineering*, **122,** 743 (1926).
27. J. D. Jevons. Stress distribution in mild steel as indicated by special etching. *Engineering*, **123,** 155 (1927).
28. J. W. Roderick and I. H. Phillips. The carrying capacity of simply supported mild steel beams. *Research (Engng. Struct. Suppl.) Colston Papers*, **2,** 9 (1949).
29. R. Hill. *Plasticity*, Chap. IV. Oxford University Press (1950).
30. L. Navier. *Resumé des Leçons*, 3rd Ed., 175. Dunod (Paris), (1864). [See article by Saint-Venant.]
31. M. R. Horne. The plastic moduli of British Standard rolled steel joists. *Brit. Weld. Res. Assn. Report*, FE1/33 (1953).
32. J. F. Baker and M. R. Horne. The effect of internal stresses on the behaviour of members in the plastic range. *Engineering*, **171,** 212 (1951).
33. F. Panlilio. The theory of limit design applied to magnesium alloy and aluminium alloy structures. *J. Roy. Aero. Soc.*, **51,** 534 (1947).

Examples *

1. Verify the results given in Table 1.1 for the plastic moduli of an I-section bent about its minor axis and for a solid circular tube.

2. The bending moment at a particular section of a beam of rectangular cross-section, depth h, is $0{\cdot}88M_p$. Find the depth of the elastic core, assuming no upper yield stress. If the bending moment is reduced to zero, find the greatest residual stress, assuming elastic behaviour on unloading. Verify that the bending moment could then vary between the values $0{\cdot}88M_p$ and $0{\cdot}453M_p$ in the opposite sense without further yield taking place. (This result shows that the elastic range of bending moment remains at the value $1{\cdot}333M_p = 2M_y$, which is the elastic range for the section when initially unstressed.)

3. A 6 in. × 6 in. × ½ in. Tee-section may be regarded as composed of two rectangles, the flange being 6 in. × ½ in. and the web 5½ in. × ½ in. For bending about an axis perpendicular to the web the elastic section modulus is 4·61 in³. Find the corresponding value of the shape factor.

4. A beam of square cross-section, side b, is composed of a material whose yield stress in compression is $1{\cdot}5\, f_L$, the tensile yield stress being f_L. Find the position of the neutral axis in the fully plastic stress distribution, and the value of the fully plastic moment.

5. Show that a beam of the cross-section of example 3 and the material of example 4 has two different fully plastic moments, depending on whether the tip of the web is in tension or compression, and find the ratio of these fully plastic moments.

* *Answers to examples are given on p. 345.*

CHAPTER

2

Simple Cases of Plastic Collapse

2.1. Introduction

REFERENCE has already been made to the fact that the object of the plastic theory is the calculation of collapse loads, at which structures continue to deform while the loads remain constant, and that the plastic hinge hypothesis forms the basis of the calculations. It has also been shown that if the loads applied to a statically determinate structure, such as a simply supported beam, are steadily increased, collapse occurs as soon as a single plastic hinge forms, for the structure is thereby transformed into a mechanism which can continue to deform under constant load. In contrast, a statically indeterminate structure, such as a fixed-ended beam, does not in general collapse when the first plastic hinge forms at the most highly stressed cross-section, for this hinge will not be sufficient to permit a mechanism motion. Instead, further increases in the loads can be borne by the structure. During this further loading, rotation takes place at the first plastic hinge, and the bending moment at this hinge stays constant at the fully plastic value. Eventually, another plastic hinge will form ; if this hinge in conjunction with the first hinge enables the structure to deform as a mechanism, collapse will then occur. If not, the loads can increase still further, while rotations take place at both the plastic hinges. Finally, this process of the successive formation and rotation of plastic hinges will result in the formation of a sufficient number of hinges to transform the structure into a mechanism, and collapse will then occur while the loads remain constant.

As will be seen in Chapter 3, it is possible to determine the plastic collapse load and the corresponding collapse mechanism for a given structure and loading by direct methods, in which no consideration is given to the sequence in which the plastic hinges form as the loads are brought up to the values which cause collapse. Indeed, the simplicity of the plastic methods is due to the fact that direct calculations of this kind can be made. Nevertheless, a thorough understanding of the process by which those plastic hinges which participate in the collapse mechanism for a structure are successively formed under steadily increasing loads forms an essential preliminary to the study of the plastic methods themselves. Accordingly, this chapter is devoted to a study of this process.

The behaviour of three particular structures which are subjected to steadily increasing loads will be examined. These structures are a simply supported beam subjected to a central concentrated load, a fixed-ended beam subjected to a uniformly distributed load, and also to a central concentrated load, and a rectangular portal frame carrying both horizontal and vertical concentrated loads. Certain simplifying assumptions are made regarding the relations between bending moment and curvature for the members of these structures. It is then possible to trace their behaviour as the loads increase by means of a step-by-step method of calculation. As will be seen, the step-by-step calculations are somewhat tedious, so that although a value for the plastic collapse load for a given structure could always be found by calculations of this kind, such a process would have little value as a basis for a method of plastic design. It must be emphasized at the outset that these step-by-step calculations bear no relation to those based on the plastic theory; instead, their purpose is merely to illustrate the manner in which a few simple structures act under steadily increasing loads.

The step-by-step technique is also used to demonstrate two important facts concerning plastic collapse loads, namely, that within wide limits the value of the plastic collapse load is unaffected by the presence of residual stresses due to any cause whatsoever in the unloaded structure, or by the order in which the various load components are brought up to the values which cause collapse. These results are shown to be due to the fact that once the collapse mechanism is known, the value of the plastic collapse load can be calculated simply from a consideration of the requirements of statical equilibrium coupled with the knowledge of the fully plastic moments at the plastic hinge positions. It is shown how such calculations can be made for the simple structures which are considered. Alternative derivations of the plastic collapse loads based on a consideration of the kinematics of the collapse mechanisms are also given. In conclusion, it is pointed out that these direct calculations of plastic collapse loads assume a knowledge of the actual collapse mechanism, whereas for all but the simplest of structures there will be several possible mechanisms. Thus it is necessary to have available some principles by means of which the actual collapse mechanism can be singled out from all the possible mechanisms; these principles and their applications form the subject of Chapters 3 and 4.

2.2. Simply supported beam

Throughout this chapter it will be assumed that each member of the structure under consideration has an elastic flexural rigidity EI and behaves elastically until the fully plastic moment of magnitude M_p is attained, so that the relation between bending moment and curvature is of the form illustrated in Fig. 2.1. This *ideal* bending moment-curvature relation is assumed so that the illustrative calculations which follow can be made by reasonably simple methods. The qualitative character of the results which are obtained would be unchanged if the more general type of bending moment-curvature relation of Fig. 1.2, in which a restricted amount of plastic flow occurs before the fully plastic moment is attained, had been assumed.

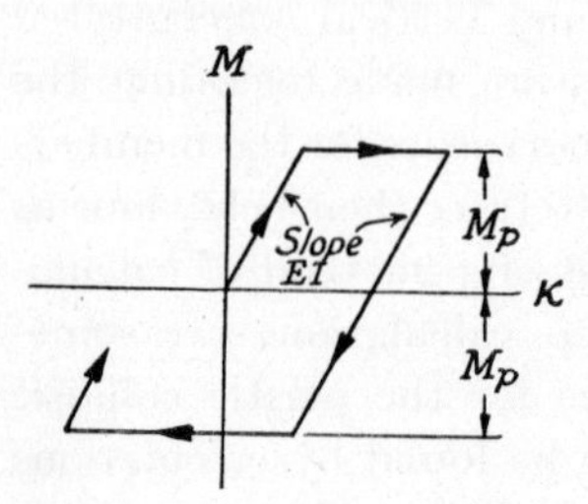

Fig. 2.1. *Ideal bending moment-curvature relation.*

The first structure to be considered is a simply supported beam of uniform cross-section, which has a span l and is subjected to a central concentrated load W, as shown in Fig. 2.2(a). The bending moment diagram for this beam is shown in Fig. 2.2(b), the maximum sagging bending moment at the centre of the beam being $\frac{1}{4}Wl$. Since the beam is statically determinate, the form of this diagram is independent of the properties of the beam, and in particular of the bending moment-curvature relation.

If the load W is increased steadily from zero the beam at first behaves elastically. Eventually, at a certain value of the load the central bending moment reaches the value M_p, and a plastic hinge then forms beneath the load. No further increase of load is possible if equilibrium is to be maintained, for the bending moment cannot rise above the value M_p. However, the plastic hinge can by hypothesis undergo rotation through any angle while the bending moment, and therefore the load, remains constant. The beam can thus continue to deflect at constant load due to this hinge action, and so fails by plastic collapse.

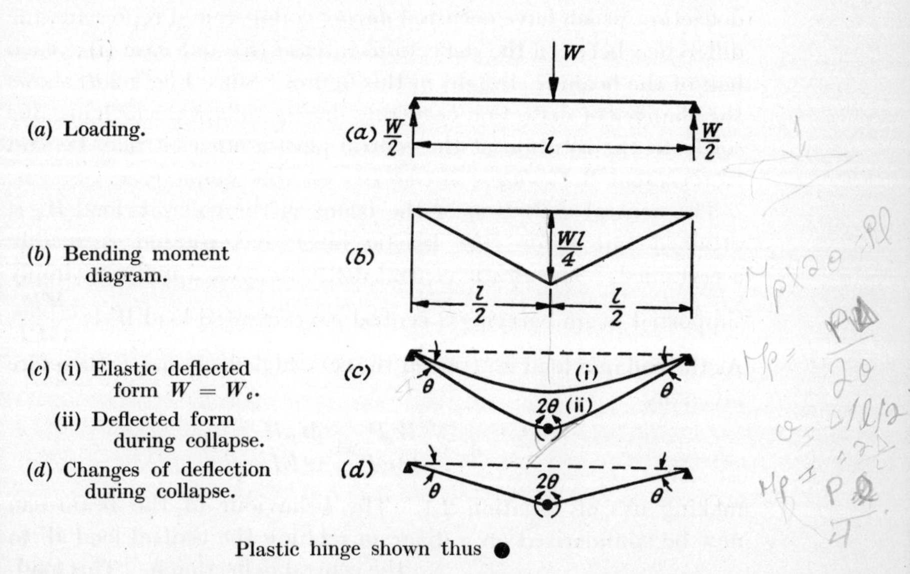

Fig. 2.2. Simply supported beam with central concentrated load.

The load at which this occurs is termed the plastic collapse load, and is denoted by W_c; its value is determined by equating the magnitude of the central bending moment to the fully plastic moment, giving

$$\tfrac{1}{4}W_c l = M_p$$

$$W_c = \frac{4M_p}{l} \quad . \quad . \quad . \quad . \quad 2.1$$

Since the bending moment at every cross-section except the central section is less than M_p, the beam remains elastic everywhere except at the centre, and the constancy of the load and therefore of the bending moments during collapse implies constancy of the curvatures. The *increases* of deflection *during* collapse are therefore due solely to the rotation at the central plastic hinge. This effect is illustrated in Fig. 2.2(*c*). Curve (i) is the deflected form of the beam just as the collapse load W_c is attained, but before any rotation has occurred at the central plastic hinge. Curve (ii) is the deflected form of the beam after the central hinge has undergone rotation through an arbitrary angle 2θ. The curved *shape* of each half of the beam is the same in case (ii) as in case (i). Thus Fig. 2.2(*d*) shows the *changes* of deflection which have occurred *during* collapse, and represents the difference between the deflections in case (ii) and case (i); each half of the beam is straight in this figure. Since Fig. 2.2(*d*) shows the changes of deflection occurring during collapse, which are due solely to the rotation at the central plastic hinge, it may be said to represent the *collapse mechanism* for this simple case.

The central deflection of the beam as the collapse load W_c is attained but before the plastic hinge has rotated is readily ascertained. The elastic central deflection δ of a uniform simply supported beam carrying a central concentrated load W is $\dfrac{Wl^3}{48EI}$. As the collapse load is attained the central deflection δ_c is therefore given by

$$\delta_c = \frac{W_c l^3}{48EI} = \frac{M_p l^2}{12EI},$$

making use of equation 2.1. The behaviour of the beam can now be summarized on a diagram relating the central load W to the central deflection δ. This load-deflection relation is shown as *Oab* in Fig. 2.3. In this figure *Oa* is the behaviour in the elastic range, and *ab* represents the plastic collapse under constant load, the increase of deflection from *a* to *b* being shown as $\frac{l}{2}\theta$, as in the mechanism of Fig. 2.2(*d*).

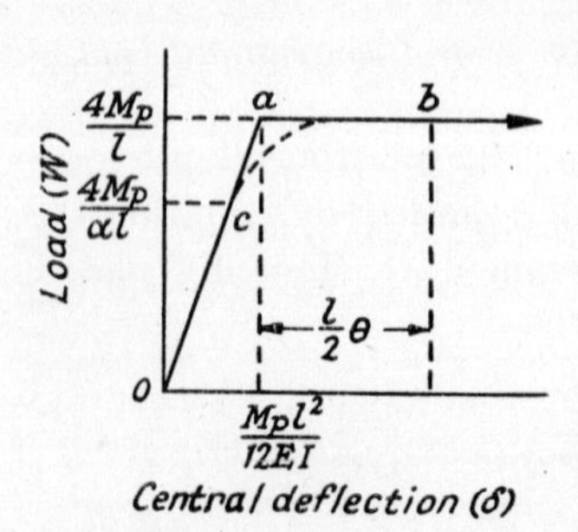

Fig. 2.3. Load-deflection relation for simply supported beam.

When the deflections are very large, the load will be supported partly by direct tension in the two halves of the beam. However, effects of this kind must be regarded as secondary, in that they would not in practice be appreciable until the deflections were quite excessive.

The dotted curve commencing at c in Fig. 2.3 shows qualitatively how the load-deflection diagram would be modified if the bending moment-curvature relation was not assumed to be of the ideal type shown in Fig. 2.1. If the yield moment M_y were less than the fully plastic moment M_p, so that the bending moment-curvature relation was of the more general type of Fig. 1.2, elastic behaviour would cease at c, when the central bending moment was M_y. Thus if $M_y = \frac{M_p}{\alpha}$, α being the shape factor, the load W_y at which elastic behaviour ceased and yield first occurred would be $\frac{W_c}{\alpha}$, or $\frac{4M_p}{\alpha l}$. At this value of the load the yield stress would just be attained in the outermost fibres at the central cross-section. If the load were then increased yield would spread inwards at this section towards the neutral axis and also along the beam, for the yield moment would then be attained at cross-sections at some distance from the centre of the beam. The central deflection would increase more rapidly per unit increase of load, until finally the fully plastic moment was attained at the centre of the beam at the same value of W as before, and collapse would then occur. Thus it is seen that the value of the collapse load is unaffected by the assumption of the ideal bending moment-curvature relation, but the deflections developed prior to collapse do depend on the particular form of bending moment-curvature relation which is assumed. It should also be pointed out that in practical rolled steel members strain-hardening causes a small rise of load with increase of deflection during collapse [see Fig. 1.1(a)].

For this simple example the ratio of the collapse load W_c to the yield load W_y is equal to α, the shape factor. The ratio of W_c to W_y is always α for any statically determinate beam or frame, for the bending moment distribution in such a structure is determined solely by statical considerations. Thus the greatest bending moment for a given statically determinate structure and loading is proportional to the load and occurs at the same position regardless of the value of the load. Yield occurs when this greatest bending moment is equal to M_y, and collapse occurs when it is equal to M_p, for the introduction of a single hinge is always sufficient to reduce such a structure to a mechanism. It follows that the ratio of W_c to W_y is the same as the ratio of

M_p to M_y, which is by definition the shape factor α. Moreover, it is clear that the plastic collapse load for any statically determinate structure can be obtained by constructing the bending moment diagram and then equating the maximum bending moment to the fully plastic moment. This procedure is based essentially on statical considerations. It is of considerable importance in the plastic theory to realise that the plastic collapse load can also be found by a *kinematical* procedure, as first pointed out by Horne.[1] Thus the *changes* of deflection and hinge rotation during any arbitrary motion of the collapse mechanism are as shown in Fig. 2.2(d). During collapse there is no change in the elastic strain energy stored in the beam, since the bending moment distribution remains unaltered. Thus the work done by the load during a small motion of the collapse mechanism must be equal to the work absorbed in the plastic hinge, since the motion is quasi-statical. In the mechanism motion of Fig. 2.2(d) the load W_c moves through a distance $\frac{l}{2}\theta$ and so does work $\frac{1}{2}W_c l\theta$. The rotation at the plastic hinge, where the bending moment is M_p, is 2θ, so that the work absorbed in the hinge is $2M_p\theta$. Equating the work done to the work absorbed, it is found that

$$\tfrac{1}{2}W_c l\theta = 2M_p\theta,$$

$$W_c = \frac{4M_p}{l},$$

which agrees with equation 2.1.

2.3. Fixed-ended beam

The behaviour of a fixed-ended beam when subjected to a uniformly distributed load will now be considered. The beam will be assumed to be of uniform cross-section and of length l, the total load being W, as shown in Fig. 2.4(a). This problem differs from the problem just considered in that there is now a single *redundancy* in the system. This redundancy may be thought of as the equal hogging fixing moments at each end of the beam, shown as M_E in Fig. 2.4(a).

If W is increased steadily from zero the behaviour is at first wholly elastic, and the bending moment diagram is as shown in Fig. 2.4(b). A convenient method of construction for this diagram is to draw first the *free bending moment* diagram, representing the distribution of bending moment $M^{(F)}$ which would result if the redundant end moments M_E were zero. This is the bending moment diagram for the beam if assumed to be simply supported, and is a parabola whose mid-ordinate is $\frac{1}{8}Wl$ measured below a datum line. The sign convention which is adopted here is that sagging bending moments are regarded as positive, and represented by ordinates below the datum line. The actual bending moment M at any section of the beam must be the sum of the free bending moment $M^{(F)}$ and the *reactant bending moment* $M^{(R)}$, where $M^{(R)}$ is defined as the bending moment due to the redundancy alone. Thus if the reactant moment diagram is drawn with the sign of the reactant moment reversed, the difference in ordinate between the free and reactant diagrams will represent the actual bending moment, since

$$M = M^{(F)} + M^{(R)} = M^{(F)} - (-M^{(R)}).$$

In the elastic range the redundant end moments M_E are both hogging and of equal magnitude $\frac{1}{12}Wl$, so that the reactant moment $M^{(R)}$ at any section of the beam is equal to $-\frac{1}{12}Wl$. The reactant moment diagram with reversed sign thus represents

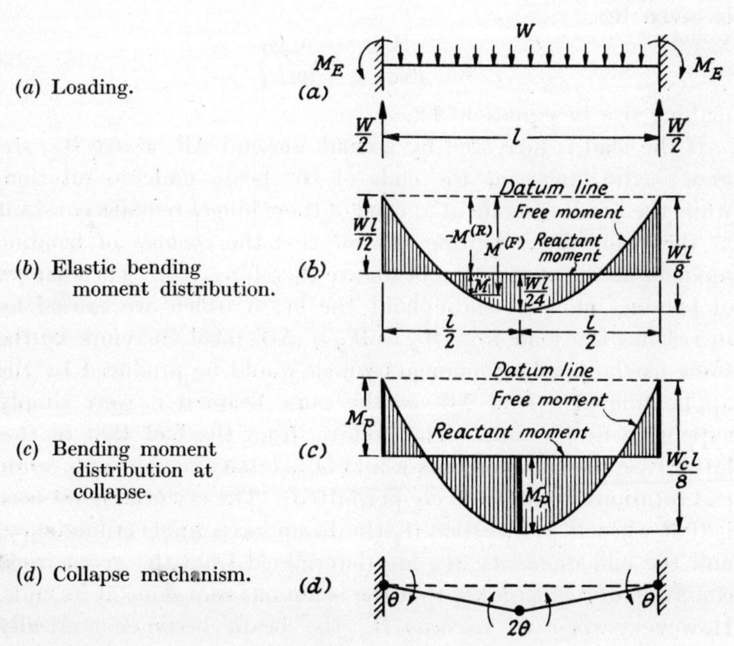

Fig. 2.4. Fixed-ended beam with uniformly distributed load.

a uniform sagging moment $\frac{1}{12}Wl$, as shown in Fig. 2.4(*b*). This reactant moment diagram can be regarded as the base-line from which the actual bending moment is measured as the ordinate to the free moment diagram. It is seen that the central bending moment is $\frac{1}{24}Wl$ sagging.

Elastic behaviour ceases when the end bending moments, which are the largest occurring at any section of the beam, attain the value $-M_p$, so that plastic hinges form at the ends of the beam. The corresponding value of the load W, which is termed the yield load and denoted by W_y, is given by

$$-\frac{W_y l}{12} = -M_p$$

$$W_y = 12\frac{M_p}{l} \quad . \quad . \quad . \quad . \quad 2.2$$

At the yield load the sagging bending moment at the centre of the beam is $\frac{W_y l}{24}$, or $\frac{1}{2}M_p$. The central deflection δ in the elastic range is $\frac{Wl^3}{384EI}$, so that the central deflection δ_y at the yield load is given by

$$\delta_y = \frac{W_y l^3}{384EI} = \frac{M_p l^2}{32EI}, \quad . \quad . \quad . \quad 2.3$$

making use of equation 2.2.

If the load is increased by a small amount ΔW above W_y, the two plastic hinges at the ends of the beam undergo rotation, while the bending moment at each of these hinges remains constant at the value M_p. During this process the *changes* of bending moment at the ends of the beam are therefore zero. The changes of bending moment throughout the beam which are caused by increasing the load from W_y to $W_y + \Delta W$ must therefore be the same as the bending moments which would be produced by the application of a load ΔW to the same beam if it were simply supported at its ends. This follows from the fact that in this latter case the end moments would of necessity remain zero, while end rotations would be freely permitted. The essential point here is that when W is less than W_y the beam has a single redundancy, and the end moments M_E are determined from the geometrical condition of *compatibility* that the beam has zero slope at its ends. However, when W exceeds W_y the beam becomes statically determinate, for the end moments are then known to be $-M_p$.

This also follows from the fact that the reactant moment line can in this case be drawn at a distance M_p below the datum line, so that the bending moment distribution throughout the beam is known. The known values of the end moments could then be used to find the rotations at the plastic hinges at the ends of the beam if required. Thus through the formation of the end hinges the end moments become known but the direct knowledge of the slope at the ends of the beam is lost.

For a simply supported beam of length l subjected to a uniformly distributed load ΔW the distribution of bending moment is parabolic, the maximum sagging bending moment being $\frac{1}{8}\Delta Wl$ at the centre of the beam. For the fixed-ended beam the central sagging bending moment at the yield load W_y was $\frac{1}{2}M_p$, so that an increase of load of ΔW above W_y would cause this bending moment to become $\frac{1}{2}M_p + \frac{1}{8}\Delta Wl$. This *régime*, in which rotations at the plastic hinges at the ends of the beam continue while the remainder of the beam behaves elastically, ceases when the central sagging bending moment becomes equal to M_p. The increment of load ΔW required to bring this moment up to the value M_p is given by

$$\frac{1}{2}M_p + \frac{1}{8}\Delta Wl = M_p$$

$$\Delta W = \frac{4M_p}{l} \qquad . \quad . \quad . \quad 2.4$$

Making use of equation 2.2, it is seen that when this condition is reached the total load on the beam is $W_y + \Delta W = \frac{16M_p}{l}$, and the distribution of bending moment is as shown in Fig. 2.4(c). At this load a plastic hinge forms at the centre of the beam, and it is evident that no further increase of load is possible without the fully plastic moment M_p being exceeded in magnitude at either the ends or the centre of the beam. In fact, at this load the beam collapses, for the formation of the central hinge reduces the structure to a mechanism. This mechanism is indicated in Fig. 2.4(d), and is similar to that encountered for the simply supported beam which has just been considered, as shown in Fig. 2.2(d). The only difference between these two mechanisms is that the plastic hinges which occur at the ends of the fixed-ended beam replace the simple supports which permit rotation under zero bending moment in the former example. Since the beam behaves elastically except where the bending moment is of magnitude M_p, the two portions of the beam between the plastic hinges do not undergo any changes of curvature while collapse is occurring, for the bending moment distribution remains unchanged. Thus the mechanism shown in Fig. 2.4(d) represents the *changes* of deformation which occur due to additional rotations θ at the end hinges during plastic collapse.

The collapse load for this beam, which is denoted by W_c, thus has the value $16\frac{M_p}{l}$, this result being due originally to Kazinczy.[2] It is of interest to determine the value of the central deflection δ_c which has developed as W just attains the value W_c but before any rotation of the central plastic hinge has occurred. This is done by noting that the central deflection of a simply supported beam of length l carrying a uniformly distributed load W is $\frac{5Wl^3}{384EI}$. Thus the change in the central deflection $(\delta_c - \delta_y)$ as the load increases from W_y to W_c, an increment of $\frac{4M_p}{l}$, is given by

$$\delta_c - \delta_y = \frac{5(W_c - W_y)l^3}{384EI} = \frac{5M_pl^2}{96EI}$$

It follows from equation 2.3 that

$$\delta_c = \frac{M_pl^2}{32EI} + \frac{5M_pl^2}{96EI} = \frac{M_pl^2}{12EI}$$

The behaviour of the beam can be summarized conveniently on a diagram relating the central deflection to the load, as in Fig. 2.5. In this figure the load-deflection relation for the beam, assuming the ideal bending moment-curvature relation, is $Oabc$.

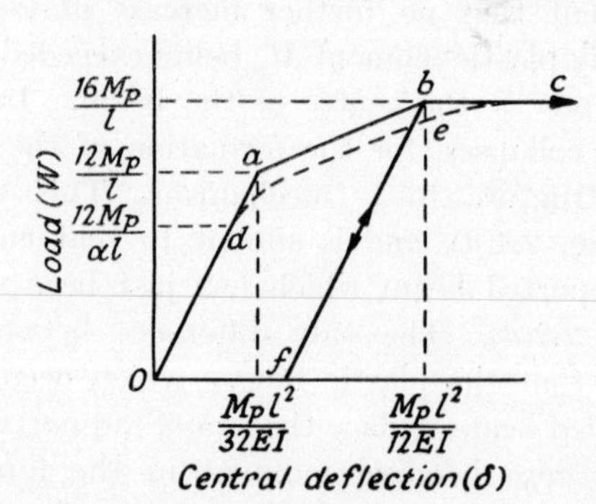

Fig. 2.5. *Load-deflection relation for fixed-ended beam.*

Oa represents the elastic behaviour, ab refers to the condition in which plastic hinges are rotating at the ends of the beam, and plastic collapse occurs along bc. The dotted line $Odec$ represents the type of load-deflection relation which would be obtained if the yield moment M_y of the beam did not coincide with the fully plastic moment M_p. Elastic behaviour would cease at d, the load then being $\frac{12M_p}{\alpha l}$, where α is the shape factor of the beam. At this load the outermost fibres of the beam at its ends would just attain the yield stress, but plastic hinges would not form until the load had been raised somewhat so that the hogging end bending moments were raised from M_y to M_p. Thus the load-deflection relation would curve over gradually before becoming more or less parallel to ab. The yield moment would be attained at the centre of the beam at e, and ultimately a central plastic hinge would be formed at the same load as before.

This load-deflection relation is typical for a beam or frame with one redundancy. When the first plastic hinge forms at the yield load (in this case a symmetrical *pair* of hinges), the structure is rendered statically determinate for further increases of the load, and the plastic hinge rotations which then occur cause a reduction in the slope of the load-deflection relation. Collapse does not occur until a further plastic hinge forms, thus reducing the structure to a mechanism. In general a finite increase in the load above the yield value will be required to bring the moment at the final plastic hinge position up to its fully plastic value.

The behaviour of the fixed-ended beam is thus fundamentally different from the behaviour of a simply supported beam, for which a typical load-deflection relation is shown in Fig. 2.3. Here the formation of a single plastic hinge causes collapse, and the ratio of the collapse load W_c to the yield load W_y is the same as the ratio of the fully plastic moment M_p to the yield moment M_y, which is equal to the shape factor α. However, for the fixed-ended beam just considered the yield load W_y, taking into account the effect of the shape factor, was $\frac{12M_p}{\alpha l}$, while the collapse load W_c was $\frac{16M_p}{l}$, so that the ratio of W_c to W_y was $\frac{4}{3}\alpha$. The greater margin between the yield and collapse loads for the fixed-ended beam is seen to be a consequence of the redundancy which exists in this case.

Behaviour on unloading

If the load on the beam was removed after some rotations at the end hinges had occurred, rotation at these hinges would cease and the behaviour during unloading would be wholly elastic, according to the assumed bending moment-curvature relation of Fig. 2.1. Suppose for instance that the load was increased to the collapse value $16\frac{M_p}{l}$ and was then removed before the central hinge had undergone any rotation. The unloading line would be *bf* in Fig. 2.5, this line being parallel to the original elastic line *Oa*, and the deflection at *f* is readily shown to be $\frac{M_p l^2}{24EI}$. The reason for the existence of this residual deflection is that the unloaded beam would no longer be free from bending moment. The plastic hinge rotations at the ends of the beam stay constant during the unloading, so that if the beam were freed at its ends after unloading there would be an abrupt change of slope at the ends in the hogging sense. The zero end slope required at the ends is thus achieved by the existence of equal sagging bending moments at the ends of the beam. These residual moments are readily calculated by noting that in the elastic range the end bending moment is $\frac{1}{12}Wl$ hogging. Thus the removal of the load $16\frac{M_p}{l}$ causes an elastic *change* in this moment of $\frac{1}{12}\left(16\frac{M_p}{l}\right)l$, or $\frac{4}{3}M_p$ sagging, and since at the point of collapse this moment was M_p hogging the residual moment is $\frac{1}{3}M_p$ sagging. Similarly, the central bending moment in the elastic range is $\frac{1}{24}Wl$ sagging, so that the removal of the load $16\frac{M_p}{l}$ changes this moment by $\frac{2}{3}M_p$ in the hogging sense. At the point of collapse this moment was M_p sagging, and the residual moment is thus $\frac{1}{3}M_p$ sagging, agreeing with the previous result.

The fact that residual moments can be induced in a structure by previous loading into the elastic-plastic range shows that the Principle of Superposition cannot be applied in such cases. For instance, if the beam were reloaded from the point *f* in Fig. 2.5 the beam would of necessity behave elastically along *fb* until the collapse load was reached at *b*, since the elastic behaviour during unloading is reversible. During such a reloading the bending moments and deflections produced by a given load would be different from those arising during the first loading, whereas if the Principle of Superposition held true, all bending moments and deflections would be unique linear functions of the applied load. The reason for the failure of the Principle of Superposition in such cases is that Hooke's law is not obeyed, for this demands linear proportionality between bending moment and curvature.

At first sight this would seem to be a grave disadvantage of the plastic methods of structural analysis, for the Principle of Superposition is of considerable utility and importance in conventional elastic methods. For instance, if instability effects are not involved, so that the Principle of Superposition is applicable, an elastic analysis of the stresses in a structure can be made for the effect of each load acting separately, and the results can then be superposed to determine the worst possible combination of loads as far as the stresses at any particular section are concerned. This is no longer possible in an elastic-plastic analysis, even when instability effects need not be considered. However, it will appear in Chapters 3 and 4 that this disadvantage is more than offset by the fact that the calculation of the plastic collapse load for a given structure and loading is far easier than the corresponding elastic calculation.

Direct calculation of collapse load

For the fixed-ended beam considered above it can be seen that there is only one possible collapse mechanism, which is the mechanism of Fig. 2.4(*d*). It is therefore simple to determine the collapse load directly, without having recourse to the step-by-step procedure just outlined. Two methods can be used, based on statical and kinematical considerations, respectively. The statical method is at once evident from Fig. 2.4(*c*). The free bending moment diagram is drawn with a maximum central ordinate of $\frac{1}{8}W_c l$ below the datum line. Knowing that at collapse the end moments are $-M_p$ and the central bending moment is M_p, the reactant moment line must be drawn as indicated. The change of bending moment from end to centre of the beam is then seen to be $2M_p$; equating this to $\frac{1}{8}W_c l$ it follows immediately that $W_c = 16\frac{M_p}{l}$. This example reveals the advantage of rapidity possessed by the plastic methods, for the computations necessary to determine the end bending moment $\frac{1}{12}Wl$ for the elastic case are quite lengthy in comparison with this trivial calculation.

The plastic collapse load can also be found by considering the kinematics of the collapse mechanism of Fig. 2.4(d). The vertical displacement of the centre of the beam during this small motion of the collapse mechanism is $\frac{1}{2}l\theta$, and so the *average* vertical displacement of the uniformly distributed load W_c is half this value, or $\frac{1}{4}l\theta$. The work done by the load is therefore $\frac{1}{4}W_c l\theta$. The central plastic hinge undergoes a rotation of magnitude 2θ in the sagging sense, and the bending moment at this hinge is sagging and of magnitude M_p, so that the work absorbed in this hinge is $2M_p\theta$. At each end hinge the rotation is of magnitude θ in the hogging sense, and the bending moment is hogging and of magnitude M_p, so that the work absorbed in each of these hinges is $M_p\theta$. Thus the total work absorbed in the hinges is $4M_p\theta$. Equating the work done to the work absorbed,

$$\frac{1}{4}W_c l\theta = 4M_p\theta$$

$$W_c = \frac{16M_p}{l}$$

as obtained previously. It should be noted that the work absorbed in each of the plastic hinges is positive, regardless of the sense of its rotation. It will be clear on physical grounds that this must always be the case ; it is also evident from Fig. 1.2, which shows that when the fully plastic moment is attained the sign of the curvature, which then becomes infinite, is the same as the sign of the fully plastic moment.

Effect of central concentrated load

In exceptional cases the yield and collapse loads may coincide even when the structure is redundant. This will happen whenever a structure remains elastic until a number of plastic hinges sufficient to transform the structure into a mechanism are formed simultaneously. A simple example is afforded by the case of a uniform fixed-ended beam of length l subjected to a central concentrated load, as shown in Fig. 2.6(a). The elastic bending moment distribution is as shown in Fig. 2.6(b), the end hogging bending moments and the central sagging bending moment both being of magnitude $\frac{1}{8}Wl$. It follows that this beam would behave elastically until these moments were brought up to the value M_p,

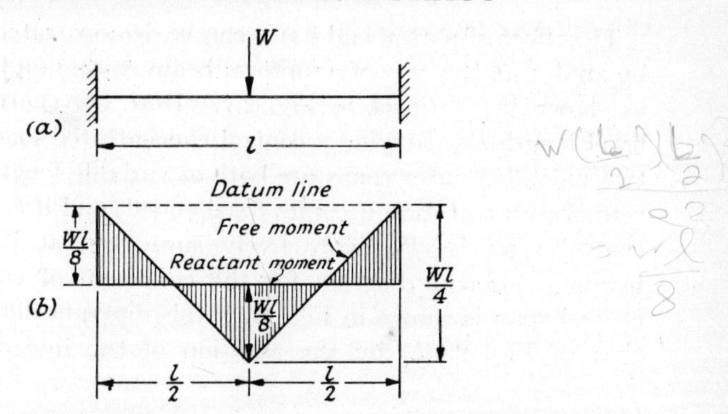

Fig. 2.6. Fixed-ended beam with central concentrated load.

and then plastic hinges would form simultaneously at the ends and at the centre, thus causing plastic collapse. The yield load W_y and the collapse load W_c would therefore both be equal to $\frac{8M_p}{l}$. If the effect of the shape factor were taken into account the yield load would be reduced to $\frac{8M_p}{\alpha l}$, and the ratio of W_c to W_y would therefore be α, as is the case for a simply supported beam.

2.4. Effect of partial end-fixity

One of the great advantages of the plastic theory is that the values of plastic collapse loads do not depend on the actual rigidity of joints or supports. It is impossible in practice to ensure perfect end-fixity of the kind assumed in the foregoing examples, and of course the elastic bending moment distribution will depend on the actual degree of end-fixity. However, considering the example of Fig. 2.6(*a*) it is seen that collapse can only occur in one way, with plastic hinges at the ends and centre of the beam. Thus, provided that the end connections are capable of developing the fully plastic moment, the collapse load is uniquely determined by statical considerations, the bending moment diagram at collapse being as shown in Fig. 2.6(*b*). However, the load-deflection relation prior to collapse is obviously dependent on the degree of end-fixity. By the same argument it is evident that the sinking of one or more of the supports during the loading cannot affect the value of the collapse load, as pointed out by Maier-Leibnitz.[3]

The effect of partial end-fixity can be demonstrated most simply by analysing the case of a uniform beam resting on four supports, as shown in the inset to Fig. 2.7. Here the central span is of fixed length l_2, carrying a central concentrated load W, and the two unloaded outer spans are both of variable length l_1. If l_1 is zero the central span becomes fixed-ended, and if l_1 is infinite the central span is effectively freely supported at its ends. The bending moment diagram for the condition of collapse of the central span is shown in Fig. 2.7; this diagram differs only from that of Fig. 2.6(b) by the addition of the linear variation of

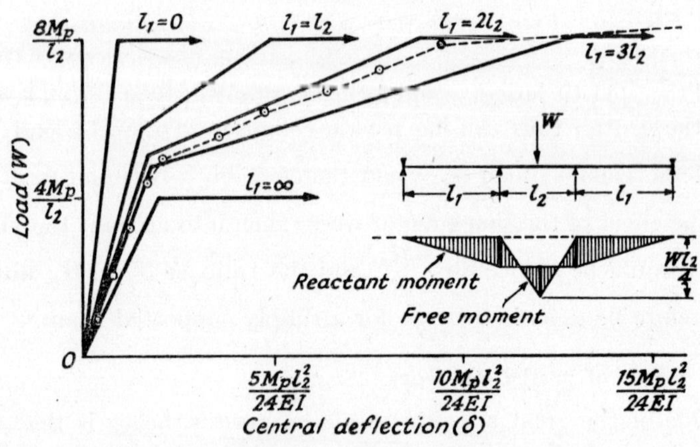

Fig. 2.7. Continuous beam on four supports.

moment in the two outer spans. The collapse load W_c is at once seen to be $8\frac{M_p}{l_2}$, which is independent of the value of l_1.

In the elastic range the greatest bending moment occurs beneath the load, and the yield load W_y and the central deflection δ_y at yield are readily shown to be

$$W_y = \frac{8M_p}{l_2}\left[\frac{2l_1 + 3l_2}{4l_1 + 3l_2}\right]$$

$$\delta_y = \frac{M_p{l_2}^2}{24EI}\left[\frac{8l_1 + 3l_2}{4l_1 + 3l_2}\right]$$

Above the yield load rotation occurs at the central hinge. Tracing the subsequent behaviour by the step-by-step process, it can be confirmed that the moments at the two middle supports reach the value $-M_p$ when $W = 8\frac{M_p}{l_2}$, and the corresponding central deflection δ_c at the point of collapse is

$$\delta_c = \frac{M_p l_2}{24EI}(4l_1 + l_2)$$

Load-deflection relations derived from these results are shown in Fig. 2.7. It will be seen that as l_1 increases, so that the degree of end-fixity is reduced, the slope of the load deflection relation between yield and collapse is progressively reduced, and the deflection δ_c at the point of collapse becomes larger. For values of l_1 in excess of about $3l_2$, the load-deflection relation becomes so flat after the load W_y is attained that unacceptably large deflections would develop before the calculated collapse load was reached. In such cases the theoretical collapse load would be of little interest, for the normal purpose of calculating the collapse load is to determine the load at which large deflections are *imminent.* This observation has been made by Kazinczy,[4] who discussed load-deflection relations similar to those of Fig. 2.7 for the case of a beam with partial end-fixity carrying a uniformly distributed load. The statement that the value of the collapse load is independent of the actual rigidity of joints, while correct from the analytical point of view, must therefore be interpreted with caution when the joint flexibility is very large. In fact, when it is suspected that large deflections may develop before the collapse load is reached, the calculation of the plastic collapse load should be supplemented by an estimate of the deflection at the point of collapse. This question is considered in Chapter 5.

In the extreme case when l_1 is infinite, the deflection δ_c at the point of collapse also becomes infinite, and so the slope of the load-deflection relation between yield and collapse becomes zero. The load-deflection relation for this case thus appears to correspond to a collapse load of only $\frac{4M_p}{l_2}$, which would be the collapse load for a simply supported beam of span l_2, whereas the calculated plastic collapse load is still $8\frac{M_p}{l_2}$. However, this apparent paradox, which was pointed out by Stüssi and Kollbrunner,[5] is resolved when it is realized that the horizontal load-deflection relation which occurs in this case when $W = W_y$ is merely the limiting case in which the slope of the load-deflection relation between yield and collapse tends to zero as l_1 tends to infinity.

Stüssi and Kollbrunner carried out tests of this kind on small beams of I-section, 4·7 cm. × 3·6 cm. In these tests l_2 was 60 cm., and values of l_1 of $0{\cdot}5l_2$, l_2, $2l_2$ and $3l_2$ were used. Comparative tests on simply supported beams spanning 60 cm. were also made. The only load-deflection curves given in their paper were for $l_1 = 2l_2$. The average of their observations from two tests of this kind (beams 532/6 and 534/8) are shown in Fig. 2.7; to reduce these observations to non-dimensional form the following values were assumed:

$$E = 2{,}100 \text{ tonnes per sq. cm.}$$
$$I = 16{\cdot}73 \text{ cm.}^4$$
$$M_p = 25{\cdot}7 \text{ tonne. cm.}$$

This value of M_p was observed in a test on a simply supported beam of the same section, the results of which were quoted by Maier-Leibnitz.[6] It will be seen that the comparison between the observations and the theoretical relation is good. The calculated collapse load for these tests is 3·43 tonnes, and Stüssi and Kollbrunner stated that "collapse" occurred at a load of 3·90 tonnes, presumably by buckling. For three tests on simply supported beams of the same span the average "collapse" load, again presumably the load at which buckling occurred, was 2·36 tonnes. Stüssi and Kollbrunner argued that the collapse load for the continuous beam tests should have been twice this value, or 4·72 tonnes, and cited the lower values observed as evidence against the validity of the simple plastic theory. However, the plastic theory does not predict buckling loads, but instead determines the loads at which large deflections are imminent, so that a comparison of buckling loads of the kind made by Stüssi and Kollbrunner is not valid.

Further tests of a similar nature were carried out by Maier-Leibnitz.[7] A type of test which is similar in principle is obtained by applying a central vertical load to a rectangular portal frame (as in Fig. 2.8 with $H = 0$), in which case the horizontal member functions as a partially fixed-ended beam. Tests of this kind have been described by Girkmann,[8] Baker and Roderick,[9] and also by Hendry,[10] who showed that increasing the height of the frame while leaving the span constant did not affect the collapse load but increased the deflections prior to collapse. Similar tests, but with symmetrical two-point loading, have been described by Rusek, Knudsen, Johnston and Beedle.[11] The effect of partial end-fixity on the design of beams subjected to uniformly distributed loads, whose ends are encased in reinforced concrete or masonry, was the subject of a theoretical and experimental investigation by Kazinczy.[4] The behaviour of a full scale portal frame whose feet were supported by short piled footings was investigated experimentally by Baker and Eickhoff,[12] who showed that for this frame the collapse load was not affected by the partial fixity of the feet to any appreciable extent.

2.5. Rectangular portal frame

The behaviour of the simple rectangular portal frame whose dimensions and loading are illustrated in Fig. 2.8 will now be discussed. All the members of this frame are assumed to be uniform and of the same cross-section and material, and the relation between bending moment and curvature for each member is the ideal relation of Fig. 2.1. The joints at sections 2 and 4 are assumed to be rigid, and the feet of the vertical members are assumed to be rigidly built in at sections 1 and 5. In the first place it will be assumed that the horizontal and vertical loads H and V remain equal to the same value W throughout the loading, W being increased steadily until collapse occurs. This type of loading is referred to as *proportional loading*. It will be appreciated that this frame has three redundancies, for if a cut is made at any section and the values of the shear force, axial thrust and bending moment are specified at this section, the frame will become statically determinate.

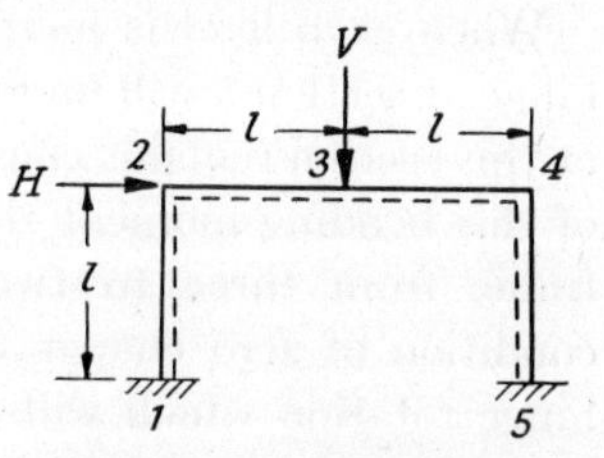

Fig. 2.8. *Rectangular portal frame.*

Along each of the four segments of the frame which are straight and free from external load, namely, 12, 23, 34 and 45, the shear force must be constant. The bending moment must therefore vary linearly along each of these segments. The values of the bending moments at the five cross-sections numbered from 1 to 5 in Fig. 2.8 therefore specify the bending moment distribution throughout the frame. Moreover, since the bending moment cannot exceed M_p in magnitude at any cross-section, it follows that plastic hinges can only occur at the ends of these segments. Thus the only possible locations of plastic hinges are the five cross-sections numbered from 1 to 5 in Fig. 2.8. This excludes the special case in which the shear force is zero in a segment, so that the bending moment is constant along the segment.

The elastic solution for this frame and loading shows that the greatest elastic bending moment occurs at the cross-section 5 in Fig. 2.8, and that the magnitude of this bending moment is $0{\cdot}413Wl$. Thus elastic behaviour will cease when the fully plastic moment is attained at this section. The corresponding value of W is the yield load W_y, and is given by

$$0{\cdot}413W_yl = M_p$$

$$W_y = 2{\cdot}424\frac{M_p}{l}.$$

When each load is increased above the value W_y, the plastic hinge at section 5 will undergo rotation while the bending moment at this section remains constant at the value M_p. The knowledge of this bending moment reduces the degree of redundancy of the frame from three to two. Correspondingly, the *compatibility* condition of zero change of slope at this section is lost, for the hinge rotation which will take place as the loads increase is not known *a priori*. To investigate the subsequent behaviour it is noted that if each load is increased from W_y to $W_y + \Delta W$, the increments of load ΔW will cause no change in the bending moment at section 5, while a hinge rotation of unknown magnitude will take place. Elsewhere the structure will still behave elastically, for the bending moment varies linearly along the member 45, so that the fully plastic moment is only attained at section 5. Thus the subsequent *increments* of bending moment will be the same as the bending moments which would be caused by loads ΔW applied to the frame if it were behaving elastically, but with the rigid joint at section 5 replaced by a pin-joint. These bending moments may be calculated for this two-redundancy problem by any of the conventional methods of elastic structural analysis.

The value of ΔW is adjusted so that when the corresponding increments of bending moment are added to the bending moments caused by the yield load W_y a second bending moment is just brought up to the fully plastic value, so that the frame then has only one redundancy. Beyond this load a new *régime* is established in which the changes of bending moment are calculated as for an elastic structure with pin-joints inserted at the two plastic hinge positions. This *régime* ends when a third plastic hinge forms, rendering the frame statically determinate. The calculations proceed in this way until a fourth hinge is formed, causing collapse.

The results of these calculations are set out in Table 2.1. The first row of this table shows the elastic solution when the loads are $W_y = 2{\cdot}424\frac{M_p}{l}$, the sign convention adopted for the bending moments being that a bending moment is positive when it causes tension in those fibres of the member which are adjacent to the dotted line in Fig. 2.8, and thus tends to open out the frame. The second row of this table shows the elastic solution for the frame when a pin-joint is placed at cross-section 5. The bending moment increments of this solution can be added to the bending moments of the wholly elastic solution so long as no further plastic hinges are formed. It is found that an increment of W of $0{\cdot}143\frac{M_p}{l}$ causes the fully plastic moment to be reached at cross-section 4, and the elastic solution in the second line of the table corresponds to this value of the increment ΔW in W. The third row of the table shows the result of adding the bending moments given in the first and second rows, and thus gives the bending moment distribution for $W = W_y + \Delta W = 2{\cdot}567\frac{M_p}{l}$.

TABLE 2.1

Step-by-step calculations for proportional loading to collapse

$\frac{\Delta Wl}{M_p}$	$\frac{Wl}{M_p}$	$\frac{M_1}{M_p}$	$\frac{M_2}{M_p}$	$\frac{M_3}{M_p}$	$\frac{M_4}{M_p}$	$\frac{M_5}{M_p}$
	2·424	−0·515	−0·030	0·727	−0·939	1
0·143		−0·067	0·015	0·049	−0·061	0
	2·567	−0·582	−0·015	0·776	−1	1
0·390		−0·331	0·058	0·224	0	0
	2·957	−0·913	0·043	1	−1	1
0·043		−0·087	−0·043	0	0	0
	3	−1	0	1	−1	1

If W is increased still further, rotations at the plastic hinges at the cross-sections 4 and 5 will take place, so that the subsequent behaviour is investigated by means of an elastic solution in which pin-joints are placed at the cross-sections 4 and 5. This solution is given in the fourth row of Table 2.1 for an increment of $0{\cdot}390\frac{M_p}{l}$ in W, sufficient to bring the bending moment at the cross-section 3 to the fully plastic value. By adding these changes of bending moment to the bending moments found to occur when $W = 2{\cdot}567\frac{M_p}{l}$, the bending moment distribution for $W = 2{\cdot}957\frac{M_p}{l}$ is obtained, with fully plastic moments at the cross-sections 3, 4 and 5. This distribution is given in the fifth row of the table. The sixth row of the table gives the elastic solution in which pin-joints are placed at these three cross-sections, and it is found that an increment of W of $0{\cdot}043\frac{M_p}{l}$ then brings the bending moment at cross-section 1 to the fully plastic value. The value of W is then $3\frac{M_p}{l}$, and the corresponding bending moment distribution is given in the last row of Table 2.1, with four plastic hinges at the cross-sections 1, 3, 4 and 5. These four plastic hinges are sufficient to transform the frame into a mechanism, so that collapse occurs while the loads and bending moments remain unchanged. The collapse load W_c for this frame is thus $3\frac{M_p}{l}$, whereas the yield load W_y was shown to be $2{\cdot}424\ \frac{M_p}{l}$, or $2{\cdot}424\frac{M_p}{\alpha l}$ allowing for the effect of the shape factor. Thus the ratio of W_c to W_y in this case is $\frac{3}{2{\cdot}424}\alpha$, or $1{\cdot}24\alpha$. The collapse mechanism is shown in Fig. 2.9.

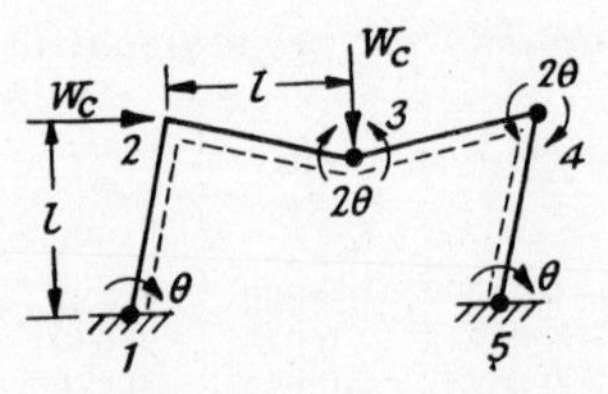

Fig. 2.9. *Collapse mechanism.*

At each stage of the loading programme it is possible to determine the deflections by means of a slope-deflection analysis, or other well-known methods. The results of such calculations are given in Fig. 2.10(a), in which the horizontal deflection h at the

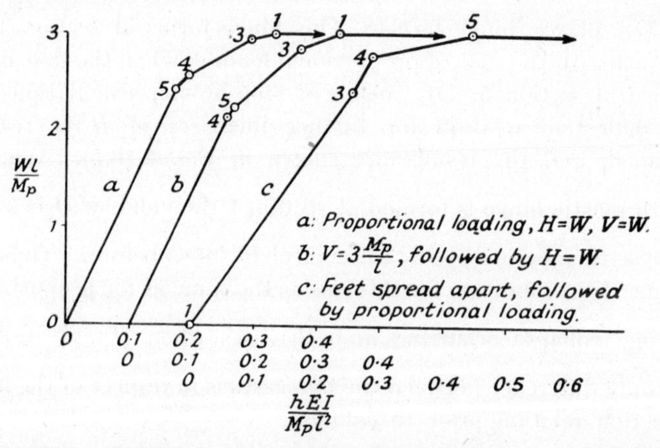

Fig. 2.10. *Load-deflection relations for rectangular portal frame.*

top of either vertical member, expressed non-dimensionally as $\frac{hEI}{M_p l^2}$, is shown plotted against the value of either load expressed non-dimensionally as $\frac{Wl}{M_p}$. Each point at which a fresh plastic hinge forms is indicated in the figure by the number of the cross-section at which the hinge forms, in accordance with Fig. 2.8.

2.6. Invariance of collapse loads

In an actual structure which is subjected to more than one load, the loads will rarely increase in proportion to one another, and it might be supposed that the order of application of the loads would have some influence on the value of the collapse load. Fortunately, however, this is not the case, and wide variations can occur in the manner in which the various loads are brought up to their collapse values without affecting the collapse load. Thus for the frame just considered it was seen that collapse occurs when $H = V = 3\frac{M_p}{l}$. To take an extreme case, suppose that the vertical load V is brought up to the value $3\frac{M_p}{l}$, and then held constant while H is steadily increased from zero. It can be shown that the load $V = 3\frac{M_p}{l}$ is borne by wholly elastic action, and that H must then be increased to the value $2{\cdot}133\frac{M_p}{l}$ before the first plastic hinge forms. This hinge forms at section 4, in contrast with the case of proportional loading when the first hinge formed at section 5. By means of step-by-step calculations the load-deflection relation for further increases of H is readily obtained, and the results are shown in Fig. 2.10(b). When a fourth plastic hinge is formed at section 1 the value of H is $3\frac{M_p}{l}$; collapse then occurs by the same mechanism as before. Thus the collapse load condition for this case is the same as for proportional loading, collapse occurring in both cases when $H = V = 3\frac{M_p}{l}$. The only difference between the two cases is in respect of the load-deflection relation prior to collapse.

A further general result of considerable importance is that the collapse load condition is always independent of the values of any residual stresses which may be present in the unloaded structure, whether these be due to welding, imperfect fit of members, the plastic hinge rotations which may have occurred during previous loading, or the movement of supports. To illustrate this point, suppose that in the frame of Fig. 2.8 the feet had spread apart, while the vertical members remained fixed in direction, the amount of this spread being just sufficient to cause the bending moments at the feet 1 and 5 to attain the value $-M_p$ in the unloaded condition. An elastic analysis then shows that the beam is then subjected to a uniform sagging bending moment $\frac{1}{3}M_p$. If the loads H and V are then increased proportionately, rotation at the plastic hinge at section 1 begins immediately, while the bending moment at section 5 is reduced below the fully plastic value. A step-by-step analysis then enables the resulting load-deflection relation, which is shown in Fig. 2.10(c), to be derived. It is again found that when $H = V = 3\frac{M_p}{l}$ collapse occurs as a fourth plastic hinge forms at section 5, although the load-deflection relation is different from the case of proportional loading for the initially stress-free frame shown in Fig. 2.10(a), and the sequence of formation of the hinges is also quite different.

The fact that residual stresses have no effect on the value of the collapse load for a given structure was pointed out by Kazinczy.[13] Apart from the work of Maier-Leibnitz [3] and of Horne [14] on continuous beams, in which the effect of initial lowering of supports on the collapse load was shown to be negligible, direct experimental confirmation of this point is lacking. However, it will be realized that any welded frame must contain residual stresses due to the welding process unless a stress relieving treatment is used, so that indirect confirmation has been supplied by the lack of any noticeable effect in the many tests which have been carried out on such structures. It is worth noting that residual stresses may have a considerable effect on the values of buckling loads, as discussed by Horne.[15]

This independence of the collapse load of the previous history of loading and the values of any residual stresses which may be present is obviously of prime importance in the plastic theory. Its explanation is comparatively simple, and stems from the fact that collapse can only occur when a sufficient number of plastic hinges have formed to transform the structure into a mechanism. Generally speaking, this number of hinges will be one more than the number which is required to render the whole structure or some part of it statically determinate. Thus in the rectangular frame just considered there were three redundancies. If three plastic hinges have formed at any stage of the loading, so that the bending moment at each of the three hinge positions is equal to the fully plastic moment, these three redundancies can be found by statics, so that the bending moment at any section of the frame can be written down in terms of the applied loads. If a fourth hinge is then formed, the structure must be reduced to a mechanism, and the knowledge of the bending moment at this fourth hinge enables the loads to be calculated. It follows that once the mechanism of collapse is known, the collapse load can be calculated by considering only the conditions of statical equilibrium coupled with the knowledge of the bending moments at the plastic hinges. Neither of these considerations is altered by the presence of residual stresses, the order of application of the loads, or the imperfect rigidity of joints, and so it is evident that the collapse load is unaffected by such factors.

To illustrate this point, imagine that the frame is cut at the centre of the beam, and that the three redundancies are therefore specified as the shear force V, the thrust H and the sagging bending moment M_3 at this section, as shown in Fig. 2.11. The central vertical load W_c is shown arbitrarily as acting on the right-hand half of the frame; if it was considered to act on the left-hand half the calculations would be unaffected except for the value which is found for V. At collapse plastic hinges are known to form at sections 1, 3, 4 and 5, and the signs of the bending moments at these plastic hinges are at once evident from a study of Fig. 2.9, which shows the collapse mechanism. Thus at section 1 the rotation at the hinge is such as to cause compression of the extreme fibres of the member adjacent to the dotted line, so that the bending moment at this hinge must be $-M_p$; similarly the

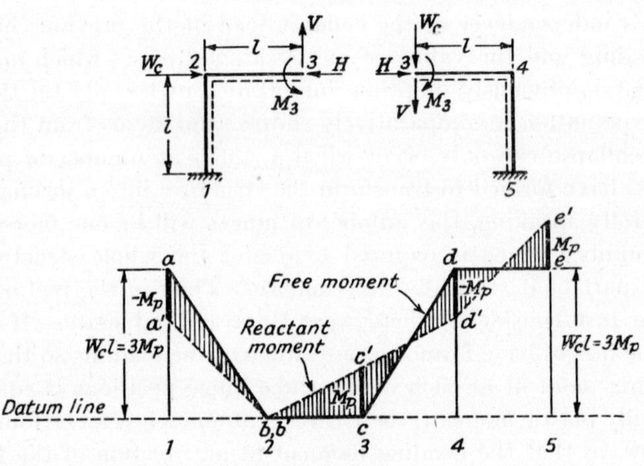

Fig. 2.11. *Bending moment diagram for rectangular portal frame at collapse.*

bending moments at the sections 3, 4 and 5 must be M_p, $-M_p$ and M_p respectively when collapse occurs. By taking moments about the four sections 1, 3, 4 and 5, it can be seen from Fig. 2.11 that

$$M_1 = M_3 + Hl + Vl - W_c l = -M_p$$
$$M_3 = M_p$$
$$M_4 = M_3 - Vl - W_c l = -M_p$$
$$M_5 = M_3 - Vl + Hl - W_c l = M_p,$$

and from these four equations of statics the three redundancies H, V and M_3, together with the collapse load W_c, can be determined. The values obtained are

$$H = 2\frac{M_p}{l}, \quad V = -\frac{M_p}{l}, \quad M_3 = M_p, \quad W_c = 3\frac{M_p}{l},$$

the value of W_c agreeing with the result obtained previously.

It is usually advisable when conducting a statical analysis of this kind to draw a bending moment diagram for the frame, and this diagram is shown in Fig. 2.11. The construction of this diagram was carried out by drawing separately the free and reactant bending moment diagrams for the frame, with the frame imagined to be unfolded to form a horizontal datum line. The free bending moment at any section is the moment due to the horizontal and vertical external loads W_c, with the redundancies H, V and M_3 all zero. It is readily seen that the free bending moments at sections 2 and 3 are both zero, while at sections 1, 4 and 5 the free bending moments each have the value $-W_c l$, or $-3M_p$. The corresponding free bending moment diagram is shown in Fig. 2.11 as the full line *abcde*. The reactant moment at any section is the moment due to the redundancies H, V and M_3 alone, with the loads zero. Thus the reactant moment at section 1 is given by

$$M_1^{(R)} = M_3 + Hl + Vl = M_p + 2M_p - M_p = 2M_p .$$

The reactant moments at sections 2, 3, 4 and 5 are similarly shown to be 0, M_p, $2M_p$ and $4M_p$, respectively. The dotted line $a'b'c'd'e'$ in Fig. 2.11 shows the corresponding reactant moment diagram in which the signs of the reactant moments have been changed. The difference in ordinate between the free and reactant moment diagrams then represents the actual bending moment. It will be seen that the bending moment at section 2 is zero at collapse, agreeing with the value given in Table 2.1.

This type of construction, while perhaps unnecessarily involved for this particular problem, is of value when the loading on a portal frame, and thus the free bending moment diagram, is more complicated. Its application to frames of this kind was first suggested by Partridge.[16]

An analysis of this kind constitutes an essentially statical procedure for the determination of the collapse load. As already pointed out, once the collapse mechanism is known it is also possible to determine the collapse load alternatively by a kinematical procedure. Thus an examination of the collapse mechanism of Fig. 2.9 shows that during the small motion of the mechanism which is indicated in this figure the distance between the knees 2 and 4 only changes by an amount which is of the second order of small quantities in θ. Both of the stanchions therefore rotate effectively through the same angle θ. The horizontal movement at the knee 2 is thus $l\theta$, and the horizontal load W_c does work $W_c l\theta$. Since there is no hinge at the knee 2, the left-hand half of the beam rotates clockwise through an angle θ, so that the vertical load W_c moves through a distance $l\theta$ and does work $W_c l\theta$. This neglects the second order vertical movement of $\frac{1}{2}l\theta^2$ due to the fact that the stanchion 12 is now inclined at an angle θ to the vertical. The total work done by the two applied loads is thus seen to be $2W_c l\theta$. Since the vertical movement at the centre of the beam is $l\theta$, the right-hand half of the beam of length l must rotate counter-clockwise through an angle θ, and the hinge rotations at sections 3 and 4 are therefore both of magnitude 2θ. The total hinge rotation is thus 6θ, and the total work absorbed in the plastic hinges is $6M_p\theta$. Equating the work done to the work absorbed,

$$2W_c l\theta = 6M_p\theta$$

$$W_c = 3\frac{M_p}{l}$$

as obtained previously.

As will be seen from the statical and kinematical calculations just given, the determination of the collapse load once the actual collapse mechanism is known is remarkably simple. However, the actual collapse mechanism cannot usually be foreseen unless the structure is extremely simple. As already remarked, in the case of the fixed-ended beam of Fig. 2.4 there was only one possible collapse mechanism, which must therefore be the actual collapse mechanism. In the example which has just been considered, it can be shown that there are in fact three possible collapse mechanisms. One of these is the mechanism of Fig. 2.9, which was shown by the step-by-step analysis to be the actual collapse mechanism. The other two possible collapse mechanisms are as illustrated in Fig. 2.12. The mechanism of Fig. 2.12(*a*) simply represents failure of the beam in a manner similar to that of the fixed-ended beam of Fig. 2.4(*d*), and the mechanism of Fig. 2.12(*b*) represents a *sidesway* failure in which both of the stanchions rotate through the same angle θ. If the step-by-step calculations had not been performed it would not be known *a priori* which of these three mechanisms was the actual collapse mechanism. In more complicated frames there is naturally a

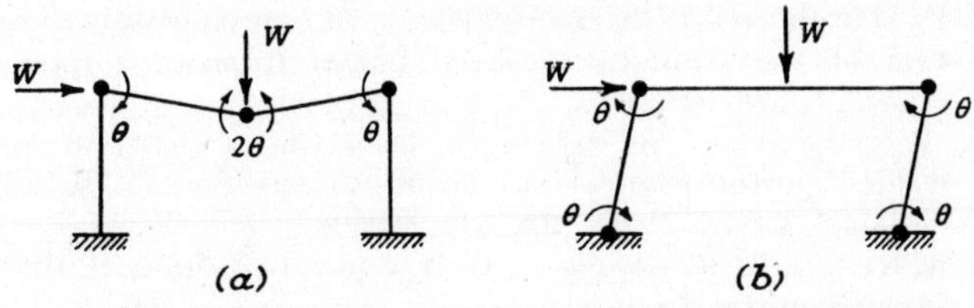

(*a*) Beam mechanism. (*b*) Sidesway mechanism.

Fig. 2.12. *Alternative collapse mechanisms.*

much wider choice of possible collapse mechanisms, and it is clearly necessary to have some guiding principles which enable the actual collapse mechanism to be found, for otherwise the simplicity of the calculation of the collapse load by these direct methods would be of no value. The available principles which enable the actual collapse mechanism to be singled out from amongst all the possible collapse mechanisms will be stated and discussed in Chapter 3.

References

1. J. F. Baker. The design of steel frames. *Struct. Engr.*, **27**, 397 (1949), see p. 421.
2. G. v. Kazinczy. Kisérletek befalazott tartókkal. *Betonszemle*, **2**, 68 (1914).
3. H. Maier-Leibnitz. Beitrag zur Frage der tatsächlichen Tragfähigkeit einfacher und durchlaufender Balkenträger aus Baustahl St. 37 und aus Holz. *Bautechnik*, **6**, 11 (1928).
4. G. v. Kazinczy. Die Bemessung unvollkommen eingespannter Stahl I-Deckenträger unter Berücksichtigung der plastischen Formänderungen. *Proc. Intern. Assn. Bridge and Struct. Engng.*, **2**, 249 (1934).
5. F. Stüssi and C. F. Kollbrunner. Beitrag zum Traglastverfahren. *Bautechnik*, **13**, 264 (1935).
6. H. Maier-Leibnitz. Versuche, Ausdeutung und Anwendung der Ergebnisse. *Prelim. Pubn. 2nd Congr. Intern. Assn. Bridge and Struct. Engng.*, 97. Berlin (1936).
7. H. Maier-Leibnitz. Versuche zur weiteren Klärung der Frage der tatsächlichen Tragfähigkeit durchlaufender Träger aus Baustahl. *Stahlbau*, **9**, 153 (1936).
8. K. Girkmann. Über die Auswirkung der " Selbsthilfe " des Baustahls in Rahmenartigen Stabwerken. *Stahlbau*, **5**, 121 (1932).

9. J. F. BAKER and J. W. RODERICK. An experimental investigation of the strength of seven portal frames. *Trans. Inst. Weld.*, **1**, 206 (1938).

10. A. W. HENDRY. An investigation of the strength of certain welded portal frames in relation to the plastic method of design. *Struct. Engr.*, **28**, 311 (1950).

11. J. M. RUSEK, K. E. KNUDSEN, E. R. JOHNSTON and L. S. BEEDLE. Welded portal frames tested to collapse. *Weld. J.*, Easton Pa., **33**, 469-s (1954).

12. J. F. BAKER and K. G. EICKHOFF. The behaviour of saw tooth portal frames. Prelim. Vol., Conference on the correlation between calculated and observed stresses and displacements in structures, *Instn. Civ. Engrs.*, 107 (1955).

13. G. v. KAZINCZY. Versuche mit innerlich statisch unbestimmten Fachwerken. *Bauingenieur*, **19**, 236 (1938).

14. M. R. HORNE. Experimental investigations into the behaviour of continuous and fixed-ended beams. *Prelim. Pubn. 4th Congr. Intern. Assn. Bridge and Struct. Engng.*, 147. Cambridge (1952).

15. M. R. HORNE. The influence of residual stresses on the behaviour of ductile structures. *Residual Stresses in Metals and Metal Construction.* Ed. W. R. Osgood. Reinhold (N.Y.) 139 (1954).

16. J. F. BAKER, J. W. RODERICK and M. R. HORNE. Plastic design of single bay portal frames. *Brit. Weld. Res. Assn. Report FE*1/2 (1947), see p. 17.

Examples

1. A uniform beam whose fully plastic moment is M_p is simply supported over a span l. Calculate the collapse load by the kinematical method for the following two loading cases

(i) Uniformly distributed load W,

(ii) Concentrated load W at a distance $\frac{1}{3}l$ from a support,

and verify the result in each case from the bending moment diagram.

2. Find the collapse load for the beam of example 1, loading case (ii), if the ends of the beam are fixed instead of simply supported.

3. Show that the test carried out by Maier-Leibnitz on a simply supported beam, Fig. 1.1(a), was consistent with a fully plastic moment of 5·88 tonne. metres. Assuming this value of the fully plastic moment, verify that the value of W which would cause collapse of the fixed-ended beam with two-point loading, as shown in Fig. 1.1(b), is 14·7 tonnes.

4. The frame of Fig. 2.8 is subjected to proportional loading, $H = V = W$, until the value of W is $2{\cdot}957\dfrac{M_p}{l}$, and the loads are then removed. Using the results in Table 2.1, calculate the values of the residual moments at the sections numbered from 1 to 5 in Fig. 2.8. From the values of the residual moments at sections 1, 2 and 3 calculate the redundancies H and V defined in Fig. 2.11, and verify that the residual moments at sections 4 and 5, as calculated from these values of H and V and the value of the residual moment at section 3, are in agreement with the values already obtained.

CHAPTER

3

Plastic Collapse—Basic Theorems and Simple Examples

3.1. Introduction

As was shown in Chapter 2, the value of the plastic collapse load for a given structure and loading can be calculated very simply by either a statical or a kinematical procedure, once the actual collapse mechanism is known. For certain very simple structures it is obvious that there is only one possible collapse mechanism, and in such cases the calculation of the collapse load presents no difficulty. However, when there is more than one possible collapse mechanism it becomes necessary to be able to distinguish the actual collapse mechanism from among the various possibilities, for otherwise the simplicity of the direct methods for calculating plastic collapse loads would be valueless. The main purpose of this chapter is to state those theorems concerned with the values of plastic collapse loads which are available for this purpose, and to discuss the application of each theorem to a few simple examples so that their significance can be understood. Formal proofs of these theorems are not given in the text but have been placed in an appendix; however, some indication of the reasons for the validity of the theorems is given when their applications are discussed. Two simple methods of analysis for determining plastic collapse loads are indicated, but it is pointed out that these methods are only of limited application. More general methods for the determination of plastic collapse loads are given in Chapter 4.

The validity of the theorems depends essentially on the assumption that the bending moment-curvature relation for each member of the frame is of the general form indicated in Fig. 1.2. The two fundamentally important features of this type of relation are that an increment of bending moment always causes an increment of curvature of the same sign, and that the magnitude of the curvature always tends to become indefinitely large whenever the magnitude of the bending moment tends to its limiting fully plastic value M_p. As a corollary, it is assumed that whenever the fully plastic moment is attained at any cross-section, a plastic hinge forms which can undergo rotation of any magnitude so long as the bending moment stays constant at the fully plastic value. For the present it is assumed that the value of the fully plastic moment is a definite constant for a given member, regardless of the values of the axial and shear forces which the member may be called upon to sustain. In fact, the value of the fully plastic moment is affected by axial and shear forces, and also by such factors as the local stress concentrations which occur beneath the points of application of concentrated loads. However, these effects are often negligibly small, and their discussion is deferred until Chapter 6.

The foregoing assumptions are the special requirements of the simple plastic theory. In addition, it will be assumed that the deflections of the frames under consideration are small enough for the equations of statical equilibrium to be sensibly the same as those for the undistorted frames. It will be appreciated that this assumption also underlies all the conventional methods of elastic frame analysis such as the slope-deflection method and the moment distribution method. By virtue of this assumption any consideration of failure by buckling before the attainment of the theoretical collapse load is excluded.

3.2. Statement of theorems

For each of the simple structures which were considered as examples in Chapter 2, it was evident that when the collapse load was reached a sufficient number of plastic hinges had formed to transform the structure into a mechanism. The deflections could then increase under constant load due to rotations occurring at these hinges, while the bending moments at the hinges remained constant at their fully plastic values. From a consideration of the requirements of statical equilibrium it followed that the bending moment distribution throughout the structure stayed unchanged during collapse. These results were obviously true for the simple cases considered, but it is desirable that they should be established for the general case. Formal proofs were supplied by Greenberg [1] using the terminology of truss-type structures; an adaptation of his argument to the case of framed structures is given in Appendix C. The first step is to define a state of plastic collapse as one in which the deflections of a frame can continue to increase while the external loads remain constant. From this definition it can be proved that during plastic collapse the distribution of bending moment in a frame remains unaltered as the deflections increase. It follows from the properties of the assumed bending moment-curvature relation that during plastic collapse there can be no changes of curvature at any cross-sections of the frame except at those cross-sections where the bending moment has the fully plastic value. The increases of the deflections during plastic collapse must therefore be due solely to rotations at plastic hinges, and these plastic hinges must be formed at a sufficient number of cross-sections to transform the structure into a mechanism. As a corollary, it is evident that during plastic collapse the work done by the external loads must be equal to the work absorbed in the plastic hinges, as was assumed in Chapter 2 when deriving the values of the plastic collapse loads by the kinematical procedure. This follows from the fact that during collapse there are no changes of curvature at any sections other than those at which the plastic hinges are rotating, so that no internal work can be done by the bending moments except at the hinges.

Static theorem

The first theorem to be stated is based on a consideration of the requirements of statical equilibrium for a frame. In general, there will exist many distributions of bending moment throughout a given redundant frame which satisfy all the conditions of statical equilibrium with a prescribed set of external loads. Greenberg and Prager [2] have termed distributions of this kind *statically admissible*. In addition, a distribution of bending moment in which the fully plastic moment is not exceeded anywhere in the frame is described as *safe*. It is clear that a frame could not conceivably carry a set of loads for which it is impossible to find any statically admissible distribution of bending moment which is also safe, for the requirements of equilibrium must of necessity be fulfilled if a frame is to carry a given set of loads, and it is impossible for the bending moment at any cross-section to exceed the fully plastic value. It follows that a *neces-*

sary condition which must be fulfilled if a frame is to be capable of carrying a given set of loads is that there must exist at least one safe distribution of bending moment throughout the frame which is statically admissible with the given loads. The static theorem states that this condition is also *sufficient* to ensure that the frame is capable of carrying the given set of loads.

A formal statement of this result can be given as follows.

Suppose that a given frame is subjected to several loads, each load being applied at a given point in a specified direction. Let one of the loads be denoted by W, and let each of the other loads be some given multiple of W. Then the set of loads is specified completely by the value of W, and so can be referred to collectively as the set of loads W. The set of loads which would cause plastic collapse if applied to the frame is denoted by W_c, this value of W being termed the collapse load. The static theorem can now be stated as follows.

Static theorem. For a given frame and loading, if there exists any distribution of bending moment throughout the frame which is both safe and statically admissible with a set of loads W, the value of W must be less than or equal to the collapse load W_c.

An obvious corollary of this theorem is that if for a given set of loads W it can be shown that no distribution of bending moment exists which is both safe and statically admissible, this value of W must be greater than the collapse load W_c. The static theorem therefore expresses the fact that any frame can actually carry the highest loads which could conceivably be carried without collapse, for the collapse load W_c is seen to be the highest value of W at which the two necessary requirements that statical equilibrium should be maintained and that the bending moment shall not exceed the fully plastic moment at any section can be fulfilled.

The static theorem was first suggested by Kist [3] as an intuitive axiom; its proof was supplied by Greenberg and Prager [2] and also by Horne,[4] and is given in Appendix C.

Another interesting corollary of this theorem relates to the effect of strengthening a frame by increasing the fully plastic moment of one or more of the members. It is easily shown that this cannot result in a decrease of the collapse load. Thus if a frame subjected to a given loading will collapse under a set of loads W_c, there must be at least one distribution of bending moment which is safe and statically admissible with these loads. This same distribution of bending moment must remain safe and statically admissible with these loads if the fully plastic moment is increased at one or more cross-sections, for the requirements of statical equilibrium remain unchanged, and if the fully plastic moment was not exceeded anywhere in the original frame it will certainly not be exceeded in the strengthened frame. This result has been stated by Feinberg[5] as an axiom, no proof being offered.

In case this result should appear to be obvious, it may be pointed out that when considering the elastic behaviour of a frame it is possible that by increasing the section of one member the maximum stress in some other member, which may be the most highly stressed member in the frame, might be increased.

Kinematic theorem

As shown by various examples in Chapter 2, if the actual collapse mechanism is known for a given frame and loading, the value of the collapse load can be found by equating the work done by the loads during a small motion of the collapse mechanism to the work absorbed in the plastic hinges. When the actual collapse mechanism is not known, it is possible to write down a work equation of this kind for any assumed collapse mechanism. A value of W will then be obtained which "corresponds" to the assumed mechanism. The kinematic theorem is concerned with such corresponding values of W, and can be stated as follows.

Kinematic theorem. For a given frame subjected to a set of loads W, the value of W which is found to correspond to any assumed mechanism must be either greater than or equal to the collapse load W_c.

The importance of this result is obvious, for it follows that if the values of W corresponding to all the possible collapse mechanisms are found, the actual collapse load W_c will be the smallest of these values. By proceeding in this way the collapse load could be determined in a given case from purely kinematical considerations.

A formal proof of this theorem is given in Appendix C. It was first established by Greenberg and Prager,[2] using an interesting physical argument based on Feinberg's axiom, which has been shown to be a direct consequence of the static theorem. Any assumed mechanism of collapse for a given frame and loading would undoubtedly be the correct collapse mechanism for this frame if the fully plastic moments were left unchanged at those cross-sections where plastic hinges occur in the assumed collapse mechanism, but increased indefinitely at all other cross-sections. Since this process would either increase the collapse load of the original frame or leave it unaltered, it follows that the collapse load of the original frame could not be greater than the value of W found to correspond to the assumed mechanism.

Uniqueness theorem

The static and kinematic theorems can be combined to form a uniqueness theorem. Thus it is known from the static theorem that for any value of W above W_c it is impossible to find any distribution of bending moment which would be both safe and statically admissible. Moreover, it is known from the kinematic theorem that it is impossible to find any mechanism for which the corresponding load is less than W_c. Combining these results, the following theorem can be stated.

Uniqueness theorem. If for a given frame and loading at least one safe and statically admissible bending moment distribution can be found, and in this distribution the bending moment is equal to the fully plastic moment at enough cross-sections to cause failure of the frame as a mechanism due to rotations of plastic hinges at these sections, the corresponding load will be equal to the collapse load W_c.

This theorem was proved by Horne.[4] Its value is that if it is thought that the actual collapse mechanism is known, confirmation can at once be obtained by the construction of the corresponding bending moment diagram.

Before turning to the applications of these theorems it is of interest to note that their counterparts have been established by Drucker, Prager and Greenberg[6] for the general case of solid bodies composed of material whose stress-strain relations are appropriate generalizations of the ideal-plastic relation of Fig. 1.4(*b*).

3.3. Simple illustrative example

The significance of the theorems stated in general terms in Section 3.2 will now be explained in detail in connection with a particular problem. The frame selected for this purpose is the simple rectangular portal frame whose dimensions are shown in

Fig. 3.1(*a*). Each member of the frame is supposed to be of uniform cross-section and material, with a fully plastic moment of magnitude M_p. The frame is subjected to a horizontal load $3W$ and a vertical load $2W$ as shown, and it is required to find the value W_c of W which would cause plastic collapse. This problem is similar to the rectangular portal frame problem discussed in Section 2.5 (see Fig. 2.8); the only difference is in the ratio of the horizontal to the vertical load. As was pointed out in Section 2.6, for this type of frame and loading there are three possible collapse mechanisms, which were shown in Figs. 2.9 and 2.12. The problem is therefore to determine which of these is the actual mechanism of collapse, and to find the corresponding value of the collapse load.

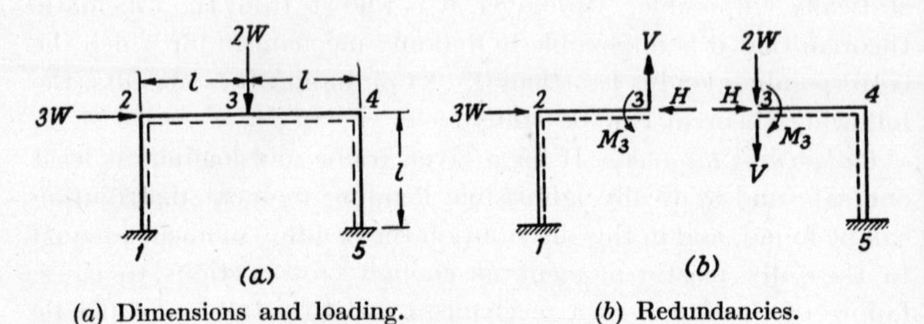

(*a*) Dimensions and loading. (*b*) Redundancies.

Fig. 3.1. *Rectangular portal frame.*

In the following discussion extensive use will be made of the equations of equilibrium, so that the first step will be to derive these equations. The sign convention for the bending moments will be that a positive bending moment in a member causes tension in those fibres which are adjacent to the dotted line in Fig. 3.1, thus tending to open out the frame.

The bending moment diagram for the frame will consist of a series of straight line segments between the sections numbered 1, 2, 3, 4 and 5 in Fig. 3.1(*a*), so that if the values of the bending moments at these five cross-sections are known the distribution of bending moment throughout the entire frame is determined. Owing to the linear variation of bending moment between these five cross-sections, it is impossible for plastic hinges to form at any other cross-sections, except in the special case in which the bending moment remains constant between two adjacent numbered cross-sections. Thus the equations of equilibrium need only be concerned with the values of the bending moments at the five numbered cross-sections.

For this frame there are three redundancies, which are conveniently specified as the thrust H, shear V and bending moment M_3 at the cross-section 3, as shown in Fig. 3.1(b). Since there are five unknown bending moments, there must be two equations of equilibrium relating these bending moments to the applied loads. These two equations of equilibrium are found conveniently by writing down expressions for the bending moments at sections 1, 2, 4 and 5 in terms of the applied loads and the three redundancies, and then eliminating the values of H and V between these equations. Thus by taking moments about these four cross-sections, it is readily shown that

$$M_1 = M_3 + Hl + Vl - 3Wl \quad . \quad . \quad 3.1$$

$$M_2 = M_3 + Vl \quad . \quad . \quad . \quad . \quad 3.2$$

$$M_4 = M_3 - Vl - 2Wl \quad . \quad . \quad . \quad 3.3$$

$$M_5 = M_3 + Hl - Vl - 2Wl \quad . \quad . \quad 3.4$$

Eliminating H and V from these equations, it is found that

$$3Wl = M_2 - M_1 + M_5 - M_4 \quad . \quad . \quad 3.5$$

$$2Wl = 2M_3 - M_2 - M_4 \quad . \quad . \quad . \quad 3.6$$

Equations 3.5 and 3.6 constitute the two equations of equilibrium for the frame. It should be noted that these two equations express merely the conditions of statical equilibrium, and must be obeyed whether or not the subsequent analysis is made according to the plastic theory.

Analysis of sidesway mechanism

With the aid of these two equations of equilibrium various deductions can be made if a particular mechanism, not necessarily the actual collapse mechanism, is analysed. In the first place the sidesway mechanism shown in Fig. 3.2 will be selected arbitrarily for discussion. This figure shows the additional deformations of the frame due to a small motion of the mechanism. It will be seen that there are four plastic hinges at the

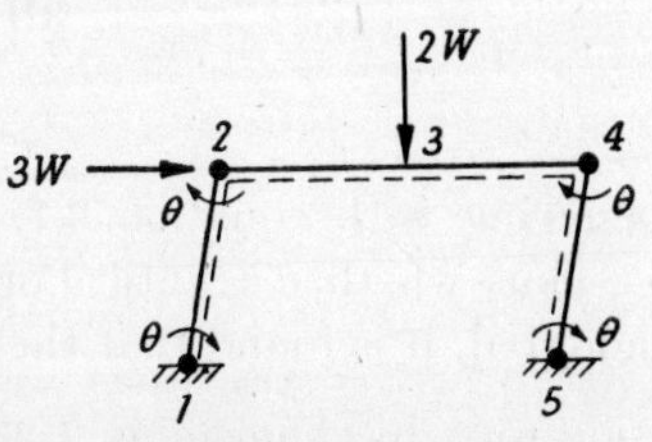

Fig. 3.2. *Sidesway mechanism.*

cross-sections 1, 2, 4 and 5. During this mechanism motion the beam remains horizontal, and the stanchions 12 and 54 each rotate through the same angle θ, so that the rotation at each of the four hinges is θ in magnitude. The rotation at hinge 1 is such as to cause compression in those fibres of the member adjacent to the dotted line in Fig. 3.1(a), so that the bending moment at this hinge is $-M_p$. The signs of the bending moments at the other three hinges can be ascertained similarly, and it is found that

$$M_1 = -M_p\,,\quad M_2 = M_p\,,\quad M_4 = -M_p\,,\quad M_5 = M_p\,.$$

It will be seen that these bending moments at the four assumed plastic hinges are the four bending moments appearing on the right-hand side of the equilibrium equation 3.5. Substituting their values in this equation, it is found that

$$\begin{aligned} 3Wl &= M_2 - M_1 + M_5 - M_4 \\ &= M_p - (-M_p) + M_p - (-M_p) = 4M_p \end{aligned}$$

$$W = 1{\cdot}333\frac{M_p}{l} \qquad \ldots \qquad 3.7$$

This value of W, which corresponds to the sidesway mechanism, can also be deduced from a consideration of the kinematics of this mechanism. Thus from Fig. 3.2 it is seen that since each stanchion is of length l and undergoes a rotation θ, the horizontal movement of the horizontal load $3W$ is $l\theta$, so that this load does work $3Wl\theta$. To the first order of small quantities the beam does not move vertically, so that the work done by the vertical load is negligible by comparison. Since there are four plastic hinges, each of which undergoes a rotation θ while the bending moment remains constant and of magnitude M_p, the total work absorbed in the hinges is $4M_p\theta$. Equating the work done by the loads to the work absorbed in the hinges, it is found that

$$3Wl\theta = 4M_p\theta$$

$$W = 1{\cdot}333\frac{M_p}{l}$$

agreeing with equation 3.7.

Thus whether a statical or a kinematical method of analysis is adopted, it is found that the value of W which corresponds to the sidesway mechanism is $1{\cdot}333\frac{M_p}{l}$. According to the kinematic theorem the value of W which corresponds to any assumed mechanism is always greater than or equal to the collapse load W_c. As a direct consequence of this theorem, it follows that

$$W_c \leqslant 1{\cdot}333\frac{M_p}{l}.$$

This analysis has therefore established an *upper bound* on the value of W_c.

The fact that this value of W is an upper bound on W_c can be deduced without reference to the kinematic theorem. The equation of equilibrium 3.5 from which this value of W was obtained is

$$3Wl = M_2 - M_1 + M_5 - M_4$$

Each of the four bending moments which appear on the right-hand side of this equation must lie between the limits $\pm M_p$. It is at once evident that the greatest possible value of the right-hand side of this equation is $4M_p$, obtained by setting

$$M_1 = -M_p, \quad M_2 = M_p, \quad M_4 = -M_p, \quad M_5 = M_p.$$

From the point of view of this equation alone, it follows that $3Wl$ cannot exceed $4M_p$, or that W cannot exceed $1{\cdot}333\frac{M_p}{l}$. The formal proof of the kinematic theorem can be regarded as a generalization of this type of argument.

It will be noticed that the values of the four bending moments M_1, M_2, M_4 and M_5 which maximize the right-hand side of equation 3.5 are precisely the values of these bending moments which were found to correspond to the sense of the hinge rotations in the sidesway mechanism. Thus the sidesway mechanism can be said to correspond to the *breakdown* of the equilibrium equation 3.5, for at the limiting value of W above which this equation cannot be satisfied the four bending moments involved in this equation would have the fully plastic values demanded by the sidesway mechanism.

The analysis of the sidesway mechanism may be completed by determining the value of M_3, which is the only remaining unknown bending moment. In equation 3.6, M_2 and M_4 are now known to be M_p and $-M_p$, respectively, while W has the value $1{\cdot}333\frac{M_p}{l}$. Substituting these values in equation 3.6, it is found that

$$2Wl = 2M_3 - M_2 - M_4$$
$$2{\cdot}666M_p = 2M_3 - M_p + M_p$$
$$M_3 = 1{\cdot}333M_p$$

Since M_3 cannot by hypothesis exceed the fully plastic moment M_p, it follows that the sidesway mechanism cannot be the actual collapse mechanism.

As pointed out by Greenberg and Prager,[2] a *lower bound* on the collapse load can also be obtained very simply from the complete statical analysis of any assumed mechanism. Thus for the sidesway mechanism which has just been analysed, the complete bending moment distribution was found to be:

$$M_1 = -M_p, \quad M_2 = M_p, \quad M_3 = 1{\cdot}333M_p,$$
$$M_4 = -M_p, \quad M_5 - M_p,$$

and these bending moments satisfied the equations of equilibrium with $W = 1{\cdot}333\frac{M_p}{l}$, thus being statically admissible. Since the equations of equilibrium are linear in the bending moments and loads, it follows that the bending moments obtained by multiplying each of the above bending moments by any positive factor k would be statically admissible with the loads corresponding to $W = k\left(1{\cdot}333\frac{M_p}{l}\right)$. If k is chosen as $0{\cdot}75$, the corresponding set of bending moments also becomes safe, since M_3 is thereby reduced to M_p. Thus the bending moments

$$M_1 = -0{\cdot}75M_p, \quad M_2 = 0{\cdot}75M_p, \quad M_3 = M_p,$$
$$M_4 = -0{\cdot}75M_p, \quad M_5 = 0{\cdot}75M_p$$

are both safe and statically admissible with the loads corresponding to $W = 0{\cdot}75\left(1{\cdot}333\frac{M_p}{l}\right) = \frac{M_p}{l}$. It follows at once from the static theorem that the collapse load W_c must be greater than or equal to $\frac{M_p}{l}$. In this bending moment distribution there is only one bending moment which is equal to the fully plastic moment, at section 3, and it is clear that the presence of a single plastic hinge at this section would not transform the frame into a mechanism. This bending moment distribution does not therefore satisfy the requirements of the uniqueness theorem, so that the corresponding load $W = \frac{M_p}{l}$ cannot be the actual collapse load W_c. It follows that W_c must be greater than $\frac{M_p}{l}$. Hence combining this lower bound with the upper bound already found, it is seen that

$$\frac{M_p}{l} < W_c < 1{\cdot}333\frac{M_p}{l}$$

Analysis of beam mechanism

The second possible collapse mechanism to be analysed is the beam mechanism illustrated in Fig. 3.3. From a consideration of the sense of the hinge rotations at each of the three plastic hinges, it is seen that the bending moments at these hinges must be

$$M_2 = -M_p, \quad M_3 = M_p, \quad M_4 = -M_p$$

These bending moments are the three bending moments occurring

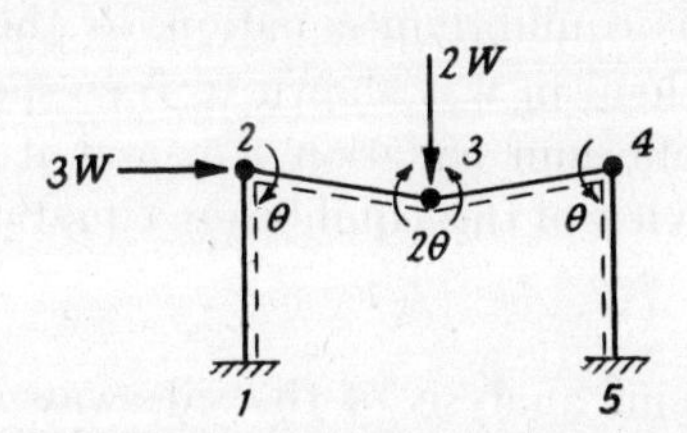

Fig. 3.3. Beam mechanism.

in the equilibrium equation 3.6, and when their values are substituted in this equation it is found that

$$\begin{aligned} 2Wl &= 2M_3 - M_2 - M_4 \\ &= 2M_p - (-M_p) - (-M_p) = 4M_p \end{aligned}$$

$$W = 2\frac{M_p}{l} \quad . \quad . \quad . \quad . \quad . \quad 3.8$$

This value of W which corresponds to the beam mechanism can also be deduced by considering the kinematics of this mechanism. Due to the rotations θ at each of the hinges at sections 2 and 4, the vertical load $2W$ moves through a distance $l\theta$, and so does work $2Wl\theta$. The distance between the knees 2 and 4 only changes by an amount which is of the second order of small quantities in θ, so that the vertical members do not undergo any appreciable rotation, and the work done by the horizontal load $3W$ is negligible. Since the rotations at the three plastic hinges are of magnitude θ, 2θ and θ, totalling 4θ, the total work absorbed in these hinges must be $4M_p\theta$. Equating the work done by the loads to the work absorbed in the hinges, it follows that

$$2Wl\theta = 4M_p\theta$$
$$W = 2\frac{M_p}{l}$$

agreeing with equation 3.8.

This value of W which corresponds to the beam mechanism is known to be an upper bound on the value of W_c, by the kinematic theorem. Alternatively, it could be argued that the values of the three bending moments M_2, M_3 and M_4 which occur at the plastic hinges in this mechanism are precisely the values which maximize the right-hand side of the equilibrium equation 3.6. It therefore follows that the beam mechanism corresponds to the breakdown of this equilibrium equation, in the same sense that the sidesway mechanism was shown to correspond to the breakdown of the equilibrium equation 3.5, and it also follows that from the point of view of this equilibrium equation alone W cannot exceed $2\frac{M_p}{l}$.

Since the previous analysis of the sidesway mechanism established an upper bound of $1{\cdot}333\frac{M_p}{l}$ on W_c, it is evident that the beam mechanism, for which the corresponding value of W is $2\frac{M_p}{l}$, cannot be the actual collapse mechanism. Despite this, however, the analysis of the beam mechanism will be completed in order to furnish another illustration of the application of the static theorem. There are two unknown bending moments, M_1 and M_5, and the only available equation of equilibrium is equation 3.5. Substituting $M_2 = -M_p$, $M_4 = -M_p$ and $W = 2\frac{M_p}{l}$ in this equation, it is found that

$$\begin{aligned} 3Wl &= M_2 - M_1 + M_5 - M_4 \\ 6M_p &= -M_p - M_1 + M_5 + M_p \\ &= M_5 - M_1 \qquad . \quad . \quad . \quad . \quad 3.9 \end{aligned}$$

Thus M_1 and M_5 are not determined uniquely in this case. This is due to the fact that the total number of unknowns in the problem of determining the collapse load for this frame is four, namely, the three redundancies and the value of the collapse load itself. Since the beam mechanism has only three hinges, at which the bending moment is known, it is evidently impossible to determine these four unknowns uniquely. This is in contrast with the case of the sidesway mechanism which has four hinges, thus enabling the collapse load and the complete bending moment distribution at collapse to be found.

Despite the fact that M_5 and M_1 are not known, it can be seen that there is no way in which equation 3.9 can be satisfied by values of these two moments within the limits $\pm M_p$, and it could therefore be concluded on this basis also that the beam mechanism is not the actual collapse mechanism.

To find a lower bound on the value of W_c it is noted that if any values are assigned to M_1 and M_5 which are in accordance with equation 3.9, a statically admissible bending moment distribution will result. Thus if the values $M_1 = -3M_p$, $M_5 = 3M_p$ are chosen, it follows that the bending moments

$$M_1 = -3M_p, \quad M_2 = -M_p, \quad M_3 = M_p, \\ M_4 = -M_p, \quad M_5 = 3M_p$$

are statically admissible with $W = 2\frac{M_p}{l}$. A safe and statically admissible set of bending moments is obtained by multiplying each of the above bending moments and the value of W by the factor 0·333, and it then follows from the static theorem that $W = 0{\cdot}333\left(2\frac{M_p}{l}\right) = 0{\cdot}666\frac{M_p}{l}$ is a lower bound on W_c. This can be seen to be the highest value of the lower bound which can be obtained, for any other choice of the values of M_5 and M_1 consistent with equation 3.9 would increase the magnitude of one of these moments and thus decrease the lower bound. This value of the lower bound is less than the lower bound $\frac{M_p}{l}$ obtained in a similar manner from the sidesway mechanism.

Analysis of combined mechanism

The final mechanism to be analysed is shown in Fig. 3.4. This mechanism is termed the combined mechanism because the deflections and hinge rotations which occur during a small motion of this mechanism may be regarded as the sum of the deflections and hinge rotations occurring in small motions of the sidesway and beam mechanisms of Figs. 3.2 and 3.3, respectively. The bending moments at the four plastic hinges in this mechanism are seen to be

$$M_1 = -M_p, \quad M_3 = M_p, \quad M_4 = -M_p, \quad M_5 = M_p$$

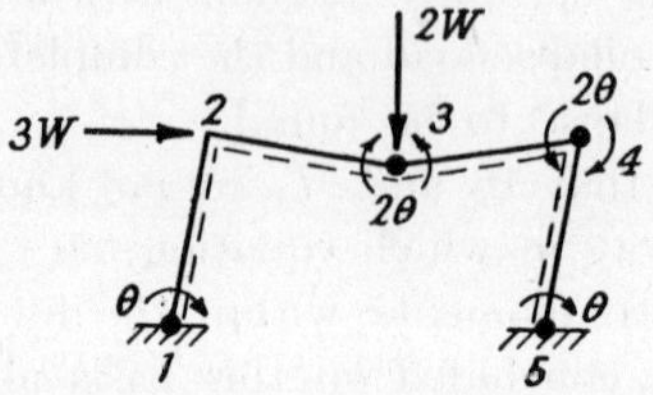

Fig. 3.4. *Combined mechanism.*

In analysing the sidesway and beam mechanisms, it was found that the fully plastic moments occurring at the plastic hinges were the only moments involved in the equations of equilibrium 3.5 and 3.6, respectively. For the combined mechanism there is no plastic hinge at section 2, whereas the moment M_2 occurs in both these equations of equilibrium. However, an equilibrium equation which does not involve M_2 is found at once by adding equations 3.5 and 3.6, giving

$$5Wl = 2M_3 - M_1 + M_5 - 2M_4 \quad . \quad . \quad 3.10$$

When the above values of the bending moments are substituted in this equation, it is found that

$$5Wl = 2M_p - (-M_p) + M_p - 2(-M_p) = 6M_p$$

$$W = 1{\cdot}2\frac{M_p}{l} \quad . \quad . \quad . \quad . \quad . \quad . \quad . \quad 3.11$$

This value of W corresponding to the combined mechanism can also be deduced kinematically. The kinematics of this type of mechanism has already been considered in Section 2.6 (see Fig. 2.9). Thus the horizontal and vertical loads both move through the same distance $l\theta$, doing work $3Wl\theta$ and $2Wl\theta$, respectively, so that the total work done by the loads is $5Wl\theta$. The hinge rotations at sections 1, 3, 4 and 5 are respectively of magnitude θ, 2θ, 2θ and θ. Thus the total hinge rotation is 6θ and the work absorbed in the plastic hinges is $6M_p\theta$. Equating the work done to the work absorbed, it follows that

$$5Wl\theta = 6M_p\theta$$

$$W = 1{\cdot}2\frac{M_p}{l}$$

agreeing with equation 3.11.

This value of W is an upper bound on W_c, as can be seen from the kinematic theorem or by noting that those fully plastic moments appearing in the collapse mechanism are such as to maximize the right-hand side of equation 3.10. To obtain a lower bound the statical analysis is completed by determining the value of M_2 from either of the equations 3.5 or 3.6. Thus in equation 3.6, substituting $M_3 = M_p$, $M_4 = -M_p$ and $W = 1{\cdot}2\dfrac{M_p}{l}$, it is found that

$$2Wl = 2M_3 - M_2 - M_4$$
$$2{\cdot}4M_p = 2M_p - M_2 + M_p$$
$$M_2 = 0{\cdot}6M_p$$

Since this value of M_2 is less than the fully plastic moment M_p, the bending moment distribution corresponding to the combined mechanism is not only statically admissible with a value of W of $1{\cdot}2\dfrac{M_p}{l}$, but also safe, since the fully plastic moment is not exceeded anywhere in the frame. It follows at once from the static theorem that $1{\cdot}2\dfrac{M_p}{l}$ is a lower bound on W_c; since this has also been shown to be an upper bound on W_c it can be concluded that the actual collapse load W_c must be $1{\cdot}2\dfrac{M_p}{l}$. This result is also evident from the uniqueness theorem, for the bending moment distribution corresponding to the combined mechanism is both safe and statically admissible with $W = 1{\cdot}2\dfrac{M_p}{l}$, while the fully plastic moment is developed at a sufficient number of cross-sections to transform the frame into a mechanism.

To confirm the conclusion that the collapse mechanism is the combined mechanism of Fig. 3.4, and the corresponding collapse load is $1{\cdot}2\dfrac{M_p}{l}$, the bending moment diagram for the frame may be drawn by constructing the free and reactant moment diagrams. Referring to Fig. 3.1(*b*), it can be seen that with $W = 1{\cdot}2\dfrac{M_p}{l}$ the free bending moments at sections 1, 2, 3, 4 and 5 are $-3{\cdot}6M_p$, 0, 0, $-2{\cdot}4M_p$ and $-2{\cdot}4M_p$, respectively. Substituting the known values of the bending moments at the four plastic hinges in equations 3.1–3.4, it is found that the three redundancies have the values

$$H = 2\frac{M_p}{l}, \quad V = -0{\cdot}4\frac{M_p}{l}, \quad M_3 = M_p$$

The reactant moments at sections 1, 2, 3, 4 and 5 are thus seen to be $2{\cdot}6M_p$, $0{\cdot}6M_p$, M_p, $1{\cdot}4M_p$ and $3{\cdot}4M_p$, respectively. Fig. 3.5 shows the corresponding free and reactant moment diagrams, and it is seen that the actual bending moment does not exceed the fully plastic moment at any section, thus confirming that this bending moment distribution corresponds to the actual collapse load.

In the discussion of this example two possible methods for the determination of collapse loads have been revealed. The first method is to assume a mechanism of collapse and to carry out a

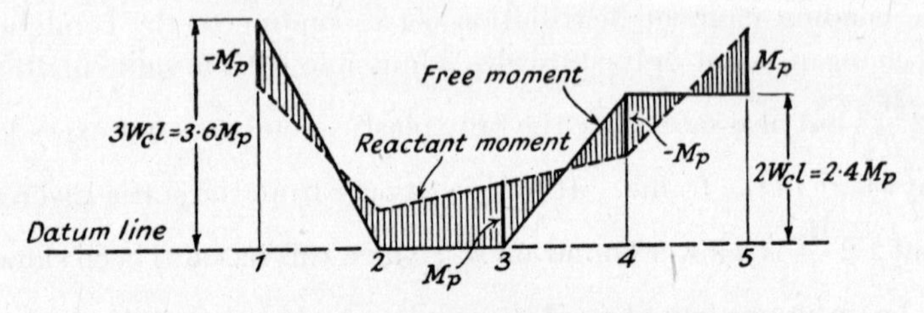

Fig. 3.5. Bending moment diagram at collapse.

complete statical analysis for this mechanism, thus obtaining the complete bending moment distribution. If this distribution is such that the fully plastic moment is not exceeded anywhere in the frame, the corresponding load must be the actual collapse load, by the uniqueness theorem. If not, other mechanisms of collapse are analysed similarly until the correct collapse mechanism is found. This method, which may be termed the *trial and error* method, was first proposed by Baker,[7] and examples of its use are given in Chapter 4. The second method is to examine all the possible collapse mechanisms, writing down a work equation for each mechanism and thus deriving the corresponding value of the collapse load. The actual value of the collapse load will then be the smallest value thus obtained, by the kinematic theorem. For simple frames it is comparatively easy to carry out the necessary computations, but in fairly complicated frames there would clearly be a great number of possible mechanisms, and the procedure would become very tedious. A method for circumventing the necessity for analysing all the possible mechanisms is described in Chapter 4.

A great deal of experimental work has been carried out on rectangular portal frames subjected to horizontal and vertical loading, and it has been found that the collapse loads predicted by the plastic theory are usually in excellent agreement with the observed loads at which large deflections are imminent. The most comprehensive series of tests were those reported by Baker and Heyman [8] on miniature frames, and confirmation of these results was provided by some full-scale tests described by Baker and Roderick.[9] Further full-scale tests have been reported by Schilling, Schutz and Beedle.[10]

3.4. Application of the Principle of Virtual Work

In the example just considered it was shown how the value of W corresponding to any assumed collapse mechanism could be found either from an equation of equilibrium or by writing down a work equation for the mechanism. For each of the three mechanisms it was found that these two methods led to the same value of W. At first sight there would appear to be little connection between these two techniques. However, it can be shown that they are much more closely related than would appear at first sight. The reason for this is that any equation of equilibrium for a given structure and loading can always be derived by an application of the Principle of Virtual Work to a small motion of the appropriate mechanism, in which the moments acting at the hinges are merely assumed to be in statical equilibrium with the loads. The hinges are thus not regarded as plastic hinges, but are merely supposed to occur in order to make possible a small imaginary movement of the structure while equilibrium is maintained. If the hinge positions are chosen to be the same as those occurring in the possible collapse mechanisms which would be analysed by the plastic theory, the equations of equilibrium obtained will be precisely those which correspond to these collapse mechanisms.

Thus consider for example the sidesway mechanism of Fig. 3.2. In a small motion of this mechanism the rotations at each of the four hinges have been shown to be of magnitude θ. It is now necessary to determine the signs of each of these rotations, for the hinges are not to be regarded as plastic hinges, and so it is no longer possible to dispense with a knowledge of their signs on the grounds that the work absorbed at a plastic hinge is always positive, regardless of the sense of rotation at the hinge. The sign convention for bending moments has already been stated to be that a positive bending moment causes tension in the fibres of the member adjacent to the dotted line in Fig. 3.2 ; for consistency a positive hinge rotation must be defined as causing extension in the same fibres. The rotations at these four hinges are thus seen to be as follows :

Cross-section	*Hinge rotation*
1	$-\theta$
2	θ
4	$-\theta$
5	θ

These hinge rotations cause a horizontal displacement $l\theta$ of the horizontal load and a negligible vertical displacement of the vertical load. Now suppose that this small motion of the sidesway mechanism is imagined to occur while the horizontal load $3W$ and the vertical load $2W$ are acting and the frame is in statical equilibrium. The virtual work done by the horizontal load $3W$ would be $3Wl\theta$, while the virtual work done by the vertical load would be zero. If the bending moment acting at section 1 is M_1, the virtual work absorbed in this hinge would be $-M_1\theta$, the rotation at this hinge being $-\theta$. In a similar manner it is seen that the virtual work absorbed in the hinges 2, 4 and 5 would be $M_2\theta$, $-M_4\theta$ and $M_5\theta$, respectively. The bending moments M_1, M_2, M_4 and M_5 must of course fulfil the requirements of equilibrium with the given loads. The virtual work done by the loads can now be equated to the virtual work absorbed in the hinges, giving

$$3Wl\theta = -M_1\theta + M_2\theta - M_4\theta + M_5\theta$$
$$3Wl = M_2 - M_1 + M_5 - M_4$$

Since the moments appearing in this equation are required to be in equilibrium with the applied loads, this equation must represent a condition of equilibrium, and is in fact identical with equation 3.5, which was derived from purely statical considerations.

It can be shown similarly that the equilibrium equations 3.6 and 3.10 can be derived by applying the Principle of Virtual Work to the beam mechanism of Fig. 3.3 and the combined mechanism of Fig. 3.4, respectively. In general, it can be seen that an equation of equilibrium can always be derived by applying the Principle of Virtual Work to a mechanism in this way.

As already pointed out, once an equation of equilibrium such as equation 3.5 has been derived, an upper bound on the value of W_c is at once obtained by noting that the right-hand side of the equation cannot exceed a certain limit set by the fact that each bending moment cannot exceed the fully plastic moment in magnitude. Thus in equation 3.5 the right-hand side is maximized by putting

$$M_1 = -M_p, \quad M_2 = M_p, \quad M_4 = -M_p, \quad M_5 = M_p,$$

and these are precisely the bending moments which would occur at the plastic hinges in the sidesway mechanism.

It follows that the procedure of writing down a work equation to determine the value of W corresponding to an assumed mechanism can be regarded as combining two steps into one. The first of these steps is to establish an equilibrium equation by the Principle of Virtual Work, and the second is to insert in this equation those values of the bending moments which maximize the load. Interpreted in this way, it is clear that by writing down a work equation a value of the load must be obtained which is identical to that found by a statical analysis. This fact is of obvious importance for a full understanding of the close relationship existing between the statical and kinematical methods of analysis, which will be developed further in Chapter 4.

A final point is that whereas for the establishment of an equilibrium equation by the virtual work method careful attention must be paid to the signs of the hinge rotations, these signs need not be known when a work equation is written down for an assumed collapse mechanism to determine the corresponding value of the load. Physically this is because the work absorbed at a plastic hinge must always be positive; analytically this follows from the fact that the fully plastic moments inserted in an equation of equilibrium such as equation 3.5 are given signs which make each term on the right-hand side of the equation positive.

3.5. Distributed loads

When a member in a frame is subjected to a uniformly distributed load, the position where the maximum bending moment occurs in this member is not known *a priori*, since the bending moment distribution in the member is parabolic. It follows that a plastic hinge might occur at any section along the member. Thus before the collapse load can be found for a frame in which some of the members are subjected to uniformly distributed loads, it is necessary to determine the precise locations of the plastic hinges in these members, unless it can be shown that the actual collapse mechanism does not involve plastic hinges in these members. The determination of the collapse load is thus more lengthy in such cases, although good approximations can be found by employing the technique of upper and lower bounds.

The example chosen for illustrative purposes is the simple portal frame whose dimensions and loading are as indicated in Fig. 3.6(*a*).

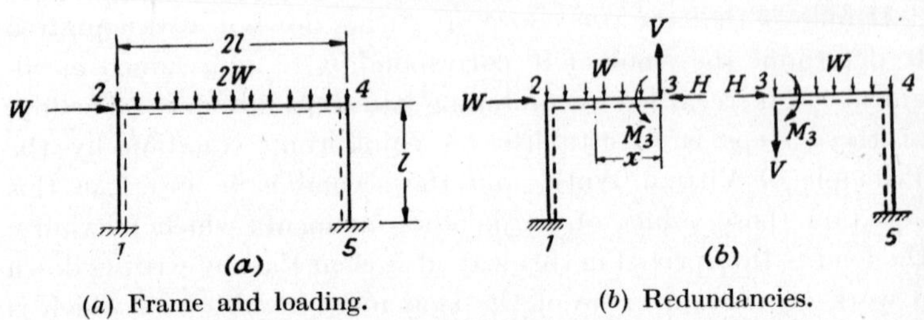

(*a*) Frame and loading. (*b*) Redundancies.

Fig. 3.6. *Frame with uniformly distributed vertical load.*

In this frame the members are all of the same cross-section and material, with fully plastic moment M_p.

The dimensions of this frame are identical with those of the frame of Fig. 3.1, and the kinematics of each of the three basic possible collapse mechanisms, the sidesway, beam and combined mechanisms, is thus unchanged. However, the latter two mechanisms involve plastic hinges in the beam, and whereas in the former example these hinges were known to occur at the centre of the beam, their positions are now not known *a priori*. Nevertheless, it will first be assumed that they still occur at the centre of the beam. On this basis, corresponding values of the load can be obtained by either the statical or the kinematical method ; here the kinematical method will be adopted, since the kinematics of each of the mechanisms is already known.

Consider first the sidesway mechanism of Fig. 3.2. Here the horizontal load W, in moving through a distance $l\theta$, does work $Wl\theta$, while the vertical load does not move through an appreciable vertical distance. Each of the four hinges rotates through the same angle θ, so that the total work absorbed is $4M_p\theta$. Equating the work done to the work absorbed,

$$Wl\theta = 4M_p\theta$$

$$W = 4\frac{M_p}{l}$$

For the beam mechanism of Fig. 3.3 the centre of the beam moves vertically through a distance $l\theta$. Thus the *average* vertical displacement of the uniformly distributed vertical load $2W$ is $\frac{1}{2}l\theta$, so that the work done by this load is $Wl\theta$. The total of the rotations at the three plastic hinges is 4θ, so that the total work absorbed in these hinges is $4M_p\theta$. The corresponding value of W is thus given by

$$Wl\theta = 4M_p\theta$$

$$W = 4\frac{M_p}{l},$$

an identical result with that obtained for the sidesway mechanism.

Finally, in the combined mechanism of Fig. 3.4 the horizontal load W moves through a distance $l\theta$ and the uniformly distributed vertical load $2W$ moves through an average distance $\frac{1}{2}l\theta$, so that the total work done by both these loads is $2Wl\theta$. The total of the rotations at the four plastic hinges is 6θ, so that the total work absorbed is $6M_p\theta$. Equating the work done to the work absorbed,

$$2Wl\theta = 6M_p\theta$$

$$W = 3\frac{M_p}{l}$$

This is the smallest value of W corresponding to any of the three mechanisms. From the kinematic theorem it is therefore concluded that this is the value of the collapse load, and that collapse occurs in the combined mechanism, subject only to the proviso that some adjustment may be necessary to allow for the fact that the plastic hinge in the beam may not occur at mid-span. The necessity for an adjustment of this kind can best be seen by constructing the bending moment diagram for the frame corresponding to the combined mechanism with a plastic hinge at the centre of the beam. For this purpose the frame is imagined to be cut at the centre of the beam, and the three redundancies H,

V and M_3 introduced as in Fig. 3.6(b). With M_3 assumed to be M_p, it is found by taking moments about the three other plastic hinge positions that

$$M_1 = M_p + Hl + Vl - 1{\cdot}5Wl = -M_p$$
$$M_4 = M_p - Vl - 0{\cdot}5Wl = -M_p$$
$$M_5 = M_p + Hl - Vl - 0{\cdot}5Wl = M_p$$

Solving these equations it is found that

$$H = 2\frac{M_p}{l}, \quad V = 0{\cdot}5\frac{M_p}{l}, \quad W = 3\frac{M_p}{l}$$

With these values of the redundancies and of W the free and reactant moments are readily seen to be as follows:

Cross-section	1	2	3	4	5
Free moment	$-4{\cdot}5M_p$	$-1{\cdot}5M_p$	0	$-1{\cdot}5M_p$	$-1{\cdot}5M_p$
Reactant moment	$3{\cdot}5M_p$	$1{\cdot}5M_p$	M_p	$0{\cdot}5M_p$	$2{\cdot}5M_p$

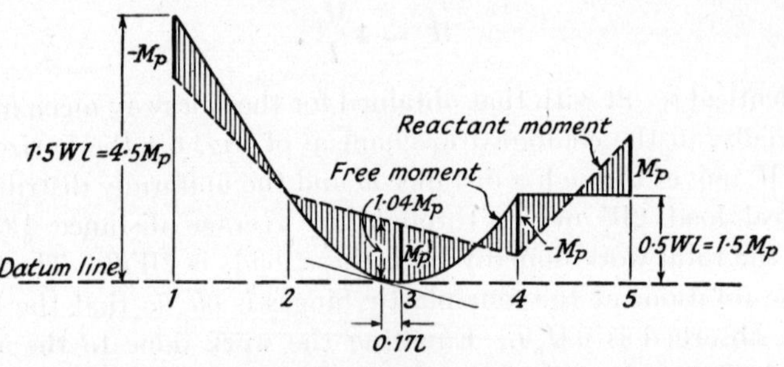

Fig. 3.7. ***Bending moment distribution for combined mechanism with hinge at centre of beam.***

Bearing in mind the fact that the free bending moment varies parabolically along the beam, the free and reactant moment diagrams are readily constructed from these values, as shown in Fig. 3.7.

It will be seen from this figure that in the corresponding bending moment distribution the fully plastic moment is slightly exceeded in the left-hand half of the beam. In fact the maximum bending moment in the beam will occur where the shear force is zero. The shear force at the centre of the beam is $V = 0{\cdot}5\frac{M_p}{l}$ acting upwards on the left-hand half of the beam, and the intensity of downward loading on the beam is $\frac{W}{l} = 3\frac{M_p}{l^2}$ per unit length.

Thus at a distance x to the left of the centre of the beam, as defined in Fig. 3.6(b), the shear force will be $\left(0{\cdot}5\,\frac{M_p}{l} - 3x\frac{M_p}{l^2}\right)$, so that the shear force becomes zero when $x = 0{\cdot}17l$. This position of zero shear force and thus of maximum bending moment is indicated in Fig. 3.7; it can also be obtained graphically as the position where the tangent to the free moment diagram is parallel to the reactant moment line. The value $M^{\max}$ of the bending moment at this cross-section is immediately found by taking moments in accordance with Fig. 3.6(b) to be

$$M^{\max} = M_3 + 0{\cdot}17Vl - \frac{1}{2}\left(\frac{W}{l}\right)(0{\cdot}17l)^2$$
$$= 1{\cdot}04M_p$$

It follows that the assumed mechanism of collapse was not quite correct, for the plastic hinge in the beam should not have been located in the centre. Thus the corresponding value of W, namely $3\frac{M_p}{l}$, is an upper bound on the value of W_c. To obtain a lower bound, it is only necessary to note that the maximum bending moment was $1{\cdot}04M_p$, so that the distribution of bending moment obtained by multiplying all the bending moments by the factor 0·96 would be safe. This distribution would be statically admissible with a value of W of $0{\cdot}96 \times 3\frac{M_p}{l}$, or $2{\cdot}88\frac{M_p}{l}$, and by the static theorem this value of W is a lower bound on W_c. Combining these upper and lower bounds, it follows that

$$2{\cdot}88\frac{M_p}{l} < W_c < 3\frac{M_p}{l}$$

This result, which defines the value of W_c to within $\pm 2\%$, would be accurate enough for many purposes. If the precise value of W_c is required, it is only necessary to find the value of W corresponding to the combined mechanism shown in Fig. 3.8, with the hinge in the beam at an arbitrary distance x from the centre of the beam, and to minimize this value of W with respect to x. This follows from the kinematic theorem, for any value of x defines a mechanism, for which the corresponding value of W is an upper bound on W_c. To determine the corresponding value of W the kinematic procedure will be used, but it will be appreciated that the same result could be derived on a statical basis. It will be seen from Fig. 3.8, in which the loads are omitted for the sake of clarity, that the vertical deflection at the plastic hinge in the beam is $(l - x)\theta$, where θ is the magnitude of the hinge rotation at section 1. The *average* vertical deflection of the distributed load $2W$ on the beam is thus $\frac{1}{2}(l - x)\theta$, so that the work done by this load is $W(l - x)\theta$. The horizontal deflection at section 2 is $l\theta$, so that the work done by the horizontal load W is $Wl\theta$. Thus the total work done by the loads is $W(2l - x)\theta$. The horizontal deflection at section 4 is also $l\theta$ to the first order, so that the rotation of the hinge at section 5 is θ. Since the vertical deflection at the plastic hinge in the beam was shown to be $(l - x)\theta$, the counter-clockwise rotation of the right-hand portion of the beam of length $(l + x)$ must be $\left(\frac{l - x}{l + x}\right)\theta$. The left-hand portion of the beam rotates clockwise through an angle θ, so that the rotation at the hinge in the beam must be of magnitude $\theta + \left(\frac{l - x}{l + x}\right)\theta$, or $\left(\frac{2l}{l + x}\right)\theta$.

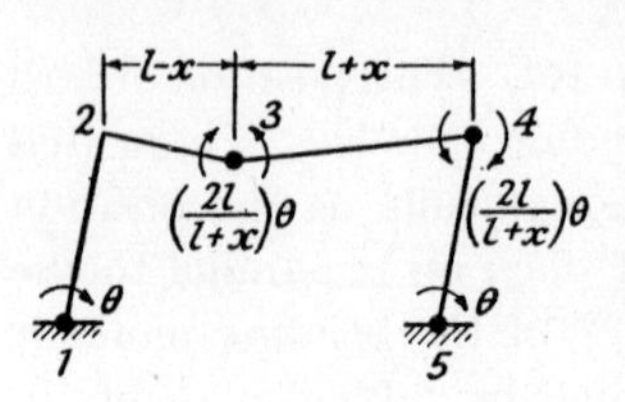

Fig. 3.8. *Combined mechanism.*

It can also be shown that this is the magnitude of the hinge rotation at section 4. The total of the hinge rotations in this mechanism is thus $2\theta + 2\left(\frac{2l}{l + x}\right)\theta$, or $2\left(\frac{3l + x}{l + x}\right)\theta$, and the total work absorbed in the hinges is $2\left(\frac{3l + x}{l + x}\right)M_p\theta$. Equating the work done to the work absorbed,

$$W(2l - x)\theta = 2\left(\frac{3l + x}{l + x}\right)M_p\theta$$

$$W = 2M_p\left[\frac{3l + x}{(2l - x)(l + x)}\right]$$

The value of x which is found to minimize W in this expression is $0{\cdot}16l$, and the corresponding value of W, which is the actual collapse load W_c, is $2{\cdot}96\frac{M_p}{l}$. The correct value of x differs only slightly from the value $0{\cdot}17l$ which defined the position of maximum bending moment in the distribution of Fig. 3.7, and it is worth noting that if instead of determining the precise value of x a fresh statical analysis had been carried out assuming the hinge to be at a distance $0{\cdot}17l$ from the centre of the beam, the corresponding value of W would have been equal to the correct value of W_c to three significant figures.

3.6. Partial and over-complete collapse

In the examples which were discussed in Sections 3.3 and 3.5, the simple portal frame had three redundancies, and the collapse mechanisms were both of the combined type and involved four plastic hinges. Since at any plastic hinge the value of the bending moment is known to be the fully plastic moment, the values of four bending moments at collapse were known in these cases, and the knowledge of these four bending moments enabled the three redundancies, together with the collapse value W_c of the load, to be found. Thus in these cases the bending moment distribution throughout the entire frame at collapse was uniquely determined by statics.

The generalization has often been made that if a frame has r redundancies, there will be $r + 1$ plastic hinges involved in the collapse mechanism, so that at collapse not only can the collapse value W_c of the load be found, but also the entire frame will be statically determinate. That this generalization is incorrect will be demonstrated by means of the specific examples which follow.

In the first place a simple case of what is termed *partial* collapse is described. A collapse is said to be partial if the plastic hinges which are formed in the collapse mechanism are not such as to render the entire frame statically determinate at collapse. In the case which is considered the frame has a number of redundancies r equal to three, and there are only three plastic hinges in the collapse mechanism, thus providing insufficient information for the determination of the three redundancies together with the collapse load.

Following the discussion of this case of partial collapse, three cases of what is termed *over-complete* collapse will be given. Over-complete collapse is said to occur when there are two or more mechanisms for which the corresponding value of the load is the same, this load value being the actual collapse load. It will be shown that in such cases a single collapse mechanism which has more than one degree of freedom can be constructed by amalgamating these mechanisms.

The term *complete* collapse is reserved for those cases in which there are $r + 1$ plastic hinges in the collapse mechanism, and this mechanism has only one degree of freedom. The cases described in Sections 3.3 and 3.5 were examples of complete collapse.

Example of partial collapse

A simple case of partial collapse is shown in Figs. 3.9(*a*) and (*b*). The members of this portal frame are of uniform cross-section throughout, the fully plastic moment being M_p. The equations of equilibrium for this frame and loading are readily derived either

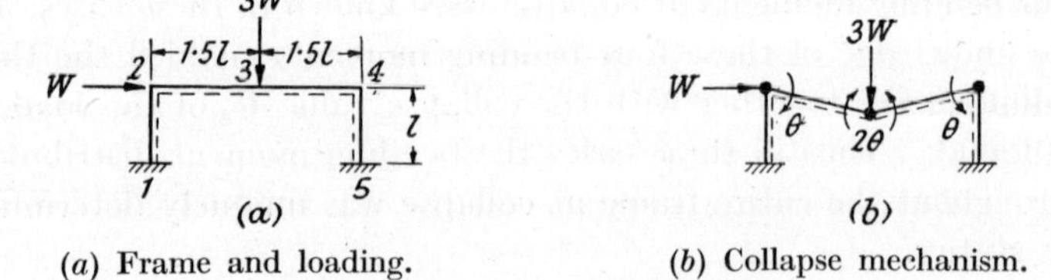

(*a*) Frame and loading. (*b*) Collapse mechanism.

Fig. 3.9. *Case of partial collapse.*

by the statical method of Section 3.3 or the virtual work method of Section 3.4. These equations are

$$4{\cdot}5Wl = 2M_3 - M_2 - M_4 \qquad . \qquad . \qquad 3.12$$

$$Wl = M_2 - M_1 + M_5 - M_4 \qquad . \qquad . \qquad 3.13$$

If the collapse mechanism is assumed to be the beam mechanism of Fig. 3.9(*b*), the bending moments at the three plastic hinges are

$$M_2 = -M_p, \quad M_3 = M_p, \quad M_4 = -M_p$$

Substituting these values in equation 3.12, the corresponding value of W is found to be $0{\cdot}89\frac{M_p}{l}$. Using this value of W in equation 3.13, it is then found that

$$M_5 - M_1 = 0{\cdot}89M_p \qquad . \qquad . \qquad . \qquad 3.14$$

It thus appears that the values of M_1 and M_5 are not uniquely determined for this mechanism. This is obvious from the fact that for this frame the number of redundancies r is three, and there are only three plastic hinges in the assumed mechanism.

It is clear that there are many possible pairs of values of M_1 and M_5 which satisfy equation 3.14 and in addition do not exceed

M_p in magnitude, for instance, $M_1 = -0{\cdot}89M_p$, $M_5 = 0$, or $M_1 = -0{\cdot}4M_p$, $M_5 = 0{\cdot}49M_p$. Any such pair of moments would be safe and statically admissible with $W = 0{\cdot}89\dfrac{M_p}{l}$. It follows from the uniqueness theorem that this value of W is the collapse load W_c, and that the collapse mechanism is the beam mechanism of Fig. 3.9(b).

This example is a typical case of partial collapse, in which the entire frame is not statically determinate at collapse. In general, a part of the frame must become statically determinate in a case of partial collapse, owing to the formation of a mechanism. Thus it is always possible to calculate the collapse load corresponding to an assumed partial mechanism by writing down the work equation for this mechanism, or, of course, by a statical procedure. It is not then necessary to determine the actual bending moment distribution in the rest of the structure at collapse. So long as it can be shown that there exists *any* bending moment distribution consistent with the equations of equilibrium in which the fully plastic moment is not exceeded anywhere in the frame, it is known that collapse must occur in this partial mechanism. In the present example it can be shown that if the idealized bending moment-curvature relation of Fig. 2.1 is assumed, the bending moments at sections 1 and 5 at collapse under proportional loading would be $0{\cdot}05M_p$ and $0{\cdot}94M_p$ respectively, but a knowledge of these moments is not required to decide that the actual collapse mechanism is the beam mechanism.

It will be realized that for more complicated frames for which partial collapse mechanisms may occur with considerably fewer than $r + 1$ plastic hinges, the demonstration of the existence of at least one safe and statically admissible bending moment distribution throughout the entire frame corresponding to an assumed partial mechanism and its load may present considerable difficulties. The trial and error method for the determination of collapse loads, in which assumed collapse mechanisms are analysed statically, is thus of limited application in cases where partial collapse is likely to occur.

Example of over-complete collapse

In certain special cases it is possible that the actual collapse mechanism for a structure will be a mechanism with more than one degree of freedom. Consider for example the portal frame shown in Fig. 3.10(*a*), in which each vertical member has a fully plastic moment M_p, while the beam has a fully plastic moment $1{\cdot}5M_p$, as indicated in the figure. This is the first example which has been encountered of a frame whose members do not all have the same fully plastic moment. The only additional consideration introduced by this lack of uniformity is that if a plastic hinge forms at either of the knees 2 or 4, it will occur in the corresponding stanchion, with a fully plastic moment of magnitude M_p, rather than in the stronger beam.

A kinematical approach will be adopted in this case. Consider first the sidesway mechanism of Fig. 3.10(*b*). The horizontal load W moves through a distance $2l\theta$, and does work $2Wl\theta$, while the vertical movement of the vertical load is negligible. The four plastic hinges all form in the stanchions, whose fully plastic moments are M_p; their total rotation is 4θ and the total work absorbed in the plastic hinges is thus $4M_p\theta$. The work equation for this mechanism is thus

$$2Wl\theta = 4M_p\theta$$

$$W = 2\frac{M_p}{l} \quad . \quad . \quad . \quad . \quad 3.15$$

For the combined mechanism of Fig. 3.10(*c*) the horizontal load W moves through a distance $2l\phi$ while the vertical load $1{\cdot}5W$ moves through a distance $l\phi$, so that the total work done is $3{\cdot}5Wl\phi$. The plastic hinges at sections 1, 4 and 5 in the stanchions undergo a total rotation of 4ϕ, so that the work absorbed in these three hinges is $4M_p\phi$. The plastic hinge at section 3 in the beam undergoes a rotation 2ϕ, so that the work absorbed in this hinge is $3M_p\phi$. The total work absorbed in the plastic hinges is thus $7M_p\phi$, and the work equation is

$$3{\cdot}5Wl\phi = 7M_p\phi$$

$$W = 2\frac{M_p}{l} \quad . \quad . \quad . \quad . \quad 3.16$$

The same value of W, $2\frac{M_p}{l}$, thus corresponds to both the sidesway and the combined mechanisms, and it is readily verified that a higher value of W, $3{\cdot}33\frac{M_p}{l}$, corresponds to the beam mechanism. From the kinematic theorem it follows that the value of the collapse load W_c is $2\frac{M_p}{l}$, but it appears that the collapse mechanism could be either the sidesway or the combined mechanism.

These results imply that the correct description of the collapse mechanism in this case is as indicated in Fig. 3.10(d), in which the displacements and hinge rotations of the two mechanisms of Figs. 3.10(b) and (c) have been added to form a mechanism with

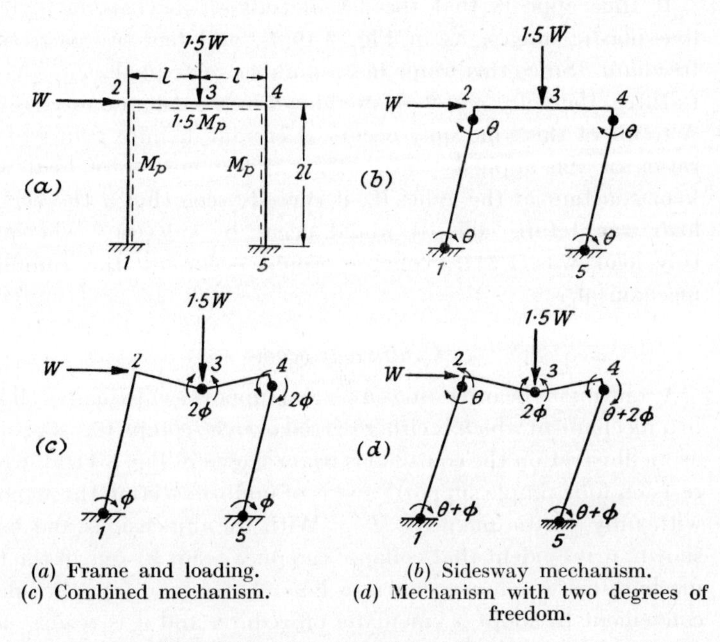

(a) Frame and loading.
(b) Sidesway mechanism.
(c) Combined mechanism.
(d) Mechanism with two degrees of freedom.

Fig. **3.10.** *Over-complete collapse.*

two degrees of freedom specified by the angles θ and ϕ. Here the only restrictions on θ and ϕ are that these angles should be positive, so that the hinge rotations at sections 2 and 3 are in the correct sense. The *relative* magnitude of θ and ϕ need not be specified. The corresponding work equation, as derived directly from a consideration of the kinematics of this mechanism, is

$$2Wl(\theta + \phi) + 1{\cdot}5Wl\phi = M_p(4\theta + 4\phi) + 1{\cdot}5M_p(2\phi)$$

$$2Wl\theta + 3{\cdot}5Wl\phi = 4M_p\theta + 7M_p\phi$$

$$W = \frac{M_p}{l}\left(\frac{4\theta + 7\phi}{2\theta + 3{\cdot}5\phi}\right) = 2\frac{M_p}{l}$$

It will be noticed that this work equation could have been obtained by adding the two work equations 3.15 and 3.16 for the sidesway and combined mechanisms, respectively, as is evident from the fact that the displacements and hinge rotations of these mechanisms were added to form the mechanism of Fig. 3.10(d). This explains the fact that the same corresponding value of W is found regardless of the relative magnitudes of θ and ϕ.

It thus appears that the actual collapse mechanism involves five plastic hinges, as in Fig. 3.10(d), and has two degrees of freedom. Since this frame has a number of redundancies r equal to three, there are $r + 2$ plastic hinges in the collapse mechanism. A result of this kind only occurs at certain definite values of the ratios of the applied loads. Thus if the horizontal load were kept constant at the value W, it is easily seen that if the vertical load was $1{\cdot}49W$ collapse would occur by sidesway, whereas if this load was $1{\cdot}51W$ collapse would occur by the combined mechanism.

Continuous beams

A continuous beam resting on several supports will usually collapse in a mechanism which is either partial or over-complete. Consider as an illustration the continuous beam shown in Fig. 3.11(a), which rests on four simple supports and is of uniform section throughout, with fully plastic moment M_p. With the dimensions and loads shown, it is evident that collapse can only occur by one of the two mechanisms which are shown in Figs. 3.11(b) and (c). It will be convenient to adopt a kinematic procedure, and it is readily seen that the work equations for these two mechanisms are as follows:

$$\text{Fig. 3.11}(b): \quad \tfrac{1}{2}Wl\theta = 4M_p\theta$$

$$W = 8\frac{M_p}{l}$$

$$\text{Fig. 3.11}(c): \quad \tfrac{1}{2}Wl\theta = 3M_p\theta$$

$$W = 6\frac{M_p}{l}$$

The lower of the two corresponding values of W is that obtained for the mechanism of Fig. 3.11(c). By the kinematic theorem, it follows that this is the actual collapse mechanism and that the collapse load W_c is $6\dfrac{M_p}{l}$.

It will be appreciated that this beam has two redundancies, for if the two intermediate supports were removed it would become statically determinate. Since there are only two plastic hinges in the collapse mechanism the collapse is therefore partial, and the beam will not be statically determinate at collapse. This can be seen by attempting to construct the bending moment diagram for the collapse condition, as illustrated in Fig. 3.11(*d*). In this figure the free bending moment diagram is drawn by imagining that each span is simply supported, with zero end moment. The central bending moment in the loaded spans would then be $\frac{1}{4}Wl$.

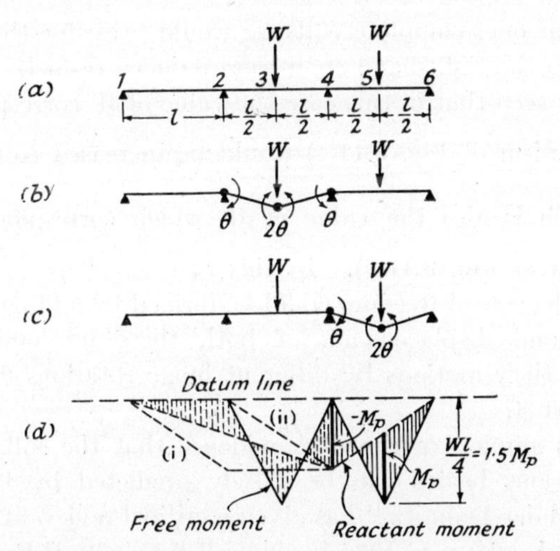

Fig. 3.11. *Continuous beam on four supports.*

The reactant bending moment diagram can be regarded as representing the bending moment distribution for the beam, now regarded as continuous over the intermediate supports, which is due to the reactions at these two supports. This diagram consists of linear variations within each of the three spans with zero bending moment at each outer support. In the right-hand span the reactant moment line is determined uniquely from the fact that the bending moments at sections 4 and 5 are $-M_p$ and M_p respectively, as shown in Fig. 3.11(*d*). However, these are the only known bending moments, so that the reactant moment line is not determined in the two remaining spans. Fig. 3.11(*d*) shows a possible bending moment distribution in these spans, indicated by broken lines, corresponding to a linear reactant moment line between sections 4 and 1. Since the fully plastic moment is not exceeded in this distribution, its existence confirms that the correct collapse mechanism has been found, by the uniqueness theorem. Two other reactant moment lines are indicated as (i) and (ii) in Fig. 3.11(*d*), representing the two extreme possible positions of this line. In (i) the bending moment at section 2 would be $-M_p$, while in (ii) the bending moment at section 3 would be M_p. With a reactant moment line lying outside these limits, the fully plastic moment would be exceeded at either section 2 or section 3.

A case of over-complete collapse would arise for this beam if the load on the right-hand span were reduced from W to $0{\cdot}75W$. It is easily seen that in this case the value of W corresponding to the mechanism of Fig. 3.11(*c*) would be increased from $6\frac{M_p}{l}$ to $8\frac{M_p}{l}$, which is also the value of W which corresponds to the mechanism of Fig. 3.11(*b*). In this case a collapse mechanism with two degrees of freedom could be formed by adding the hinge rotations and displacements of both these mechanisms, and specifying their motions by different hinge rotations θ and ϕ at the support 4.

There is ample experimental evidence that the collapse loads for continuous beams can be closely predicted by the plastic theory. Maier-Leibnitz [11] has given a critical review of the many tests carried out up to 1936 by himself,[12] Schaim,[13] Hartmann [14] and several other investigators. In some of the early tests by Maier-Leibnitz [15] intermediate supports were lowered before the test commenced, and tests of this kind were also carried out by Horne; [16] the results confirmed that the collapse load was not thereby affected. Horne [16] also showed that the collapse load was not affected by non-proportional loading.

Fixed-ended beam strengthened by flange plates

A final example of over-complete collapse is furnished by the case of a fixed-ended beam which is strengthened by flange plates welded on at its ends, as illustrated in Fig. 3.12. A full discussion of the economies which can be achieved by this type of reinforcement in fixed-ended beams has been given by Horne.[17] It will here be supposed that these plates each extend over a length $0{\cdot}1l$, as shown, the span of the beam being l. The beam, whose unreinforced cross-section has a fully plastic moment M_p, is subjected to a uniformly distributed load.

The optimum strengthening effect is achieved if the fully plastic moment of the reinforced section is adjusted so that at collapse the fully plastic moment is developed at the ends of the beam, and in addition the fully plastic moment of the unreinforced section is developed at the centre of the beam and also at those sections where the reinforcement begins. The corresponding bending moment diagram is shown in Fig. 3.12. A simple calculation then shows that the required value of the fully plastic

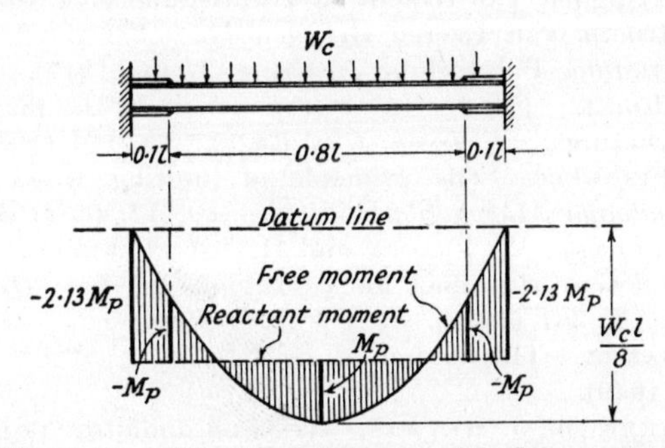

Fig.* 3.12. *Fixed-ended beam with flange plates.

moment at the ends of the beam is $2{\cdot}13M_p$, and from the figure it is then seen that the collapse load W_c is given by

$$\frac{W_c l}{8} = M_p + 2{\cdot}13M_p$$

$$W_c = 25{\cdot}0\frac{M_p}{l}$$

For the unreinforced beam the collapse load would be $16\frac{M_p}{l}$, as was shown in Section 2.3. Thus by strengthening the beam over 20% of its length in such a way as to raise its fully plastic moment from M_p to $2{\cdot}13M_p$, the collapse load is increased by 56%.

It will be clear that for this case there are two collapse mechanisms to which there corresponds the same value of the collapse load. One of these is with two end hinges and a central hinge, and the second is with two hinges at the termination of the reinforcement together with a central hinge. A collapse mechanism with two degrees of freedom could therefore be formed without difficulty.

References

1. H. J. GREENBERG. The principle of limiting stress for structures. *2nd Symposium on Plasticity*, Brown Univ., April 1949.
2. H. J. GREENBERG and W. PRAGER. On limit design of beams and frames. *Trans. Amer. Soc. Civ. Engrs.*, **117,** 447 (1952). [First published as Tech. Rep. A18-1, Brown Univ. (1949).]
3. N. C. KIST. Leidt een Sterkteberekening, die Uitgaat van de Evenredigheid van Kracht en Vormverandering, tot een goede Constructie van Ijzeren Bruggen en gebouwen? *Inaugural Dissertation*, Polytechnic Institute, Delft (1917).
4. M. R. HORNE. Fundamental propositions in the plastic theory of structures. *J. Instn. Civ. Engrs.*, **34,** 174 (1950).
5. S. M. FEINBERG. The principle of limiting stress (Russian). *Prikladnaya Matematika i Mekhanika*, **12,** 63 (1948).
6. D. C. DRUCKER, W. PRAGER and H. J. GREENBERG. Extended limit design theorems for continuous media. *Quart. Appl. Math.*, **9,** 381 (1952).
7. J. F. BAKER. The design of steel frames. *Struct. Engr.*, **27,** 397 (1949).
8. J. F. BAKER and J. HEYMAN. Tests on miniature portal frames. *Struct. Engr.*, **28,** 139 (1950).
9. J. F. BAKER and J. W. RODERICK. Tests on full-scale portal frames. *Proc. Instn. Civ. Engrs.*, **1,** (Part I), 71 (1952).
10. C. G. SCHILLING, F. W. SCHUTZ and L. S. BEEDLE. Behaviour of welded single-span frames under combined loading. *Weld. J.*, Easton, Pa., to be published.
11. H. MAIER-LEIBNITZ. Versuche, Ausdeutung und Anwendung der Ergebnisse. *Prelim. Pubn. 2nd Congr. Intern. Assn. Bridge and Struct. Engng.*, 97. Berlin (1936).
12. H. MAIER-LEIBNITZ. Versuche mit eingespannten und einfachen Balken von I-Form aus St 37. *Bautechnik*, **7,** 313 (1929).
13. J. H. SCHAIM. Der durchlaufende Träger unter Berücksichtigung der Plastizität. *Stahlbau*, **3,** 13 (1930).
14. F. HARTMANN. Die Formänderungen einfacher und durchlaufender Stahlträger. Mit einem Versuch. *Schweiz. Bauztg.*, **101,** 75 (1933).
15. H. MAIER-LEIBNITZ. Beitrag zur Frage der tatsächlichen Tragfähigkeit einfacher und durchlaufender Balkenträger aus Baustahl St. 37 und aus Holz. *Bautechnik*, **6,** 11 (1928).
16. M. R. HORNE. Experimental investigations into the behaviour of continuous and fixed-ended beams. *Prelim. Pubn. 4th*

Congr. Intern. Assn. Bridge and Struct. Engng., **147.** Cambridge (1952).

17. M. R. HORNE. Determination of the shape of fixed-ended beams for maximum economy according to the plastic theory. *Prelim. Pubn. 4th Congr. Intern. Assn. Bridge and Struct. Engng.*, **111.** Cambridge (1952).

Examples

1. A uniform continuous beam whose fully plastic moment is 7 tons ft. rests on five simple supports, A, B, C, D and E. $AB = 6$ ft., $BC = 8$ ft., $CD = 8$ ft., $DE = 10$ ft. Each span carries a concentrated load at its mid-point, these loads being $AB = W$, $BC = W$, $CD = 1{\cdot}4W$, $DE = 0{\cdot}5W$. Using the kinematical method, find the value of W which would just cause collapse, and construct a safe and statically admissible bending moment distribution. Find also the limits between which the reaction at A must lie.

2. A uniform continuous beam whose fully plastic moment is M_p rests on three simple supports at A, B and C, the two spans AB and BC each being of length l. The span AB is unloaded while the span BC is subjected to a uniformly distributed load W. Show that at collapse the plastic hinge which forms in the span BC is located at a distance $(\sqrt{2}-1)l$ from C, and find the value of W which would cause collapse. By means of an elastic analysis show that if W is steadily increased from zero yield first occurs at a distance $\frac{7}{16}l$ from C.

3. A continuous beam rests on four simple supports, A, B, C and D. $AB = BC = CD = 10$ ft. Each span carries a uniformly distributed load, as follows: $AB = 5$ tons, $BC = 10$ tons, $CD = 6$ tons. The beam is to be designed so that it is of uniform section in each span, but the fully plastic moments of the spans may all be different. Find the required value of the fully plastic moment for each span, assuming that the fully plastic moment of the centre span is greater than that of either outer span.

4. A uniform fixed-ended beam of length l and fully plastic moment M_p is subjected to a uniformly distributed load W together with a concentrated load P at a distance $\frac{1}{3}l$ from one end of the beam. Find the value of W which would cause collapse for the following three values of P: $0{\cdot}25W$, $0{\cdot}5W$ and W.

5. For the fixed-ended beam of Fig. 3.12 the collapse load W_c was increased from $16\frac{M_p}{l}$ to $25{\cdot}0\frac{M_p}{l}$ if the flange plates increased the fully plastic moment from M_p to $2{\cdot}13M_p$ over a distance $0{\cdot}1l$ at each end of the beam. If the ends of the beam were not strengthened in this way, find the central length which would require strengthening to the same increased value of the fully plastic moment to achieve the same increase in W_c.

6. Construct the bending moment diagram for the frame of Fig. 3.10(a) assuming that collapse occurs by the combined mechanism of Fig. 3.10(c), and hence verify that the corresponding value of the bending moment at section 2 is M_p.

7. A fixed-base rectangular portal frame is of height l and span $2l$, and is of uniform section throughout, with fully plastic moment M_p. The loading on this frame consists of a horizontal load H applied at the top of one of the stanchions together with a vertical load V at the centre of the beam. Find the value of W which would cause collapse for the following pairs of values of H and V, and in each case construct the bending moment diagram at collapse.

(i) $H = W, \quad V = 0$
(ii) $H = W, \quad V = 0{\cdot}5W$
(iii) $H = W, \quad V = W$
(iv) $H = W, \quad V = 2W$
(v) $H = W, \quad V = 3W$

Confirm that in cases (ii) and (iv) the collapse is over-complete. For case (iv) sketch the over-complete collapse mechanism with two degrees of freedom, and show that the value of W_c can be found from the work equation corresponding to this mechanism.

8. A pinned-base rectangular portal frame is of height l and span $3l$, and is of uniform section throughout, with fully plastic moment M_p. The frame is subjected to a horizontal load W applied at the top of one of the stanchions together with a vertical load W at a distance l from one end of the beam. Find the value of W which would cause collapse.

9. Construct the bending moment diagram for the frame of Fig. 3.6(a) assuming that collapse occurs by the sidesway mechanism. Find the greatest bending moment in this distribution, and hence determine a lower bound on the value of W_c.

10. For the frame of Fig. 3.6(a), find the value of W corresponding to the combined mechanism, with the hinge in the beam located at the position of maximum bending moment found in example 9. Find the maximum moment in the corresponding bending moment distribution, and hence determine a lower bound on the value of W_c.

11. A fixed-base rectangular portal frame is of height and span l. The stanchions each have a fully plastic moment $2M_p$, while the beam has a fully plastic moment M_p. One of the stanchions is subjected to a uniformly distributed horizontal load W. Use the trial and error method to find the value of W which would cause collapse. Find also the value of the collapse load if the fully plastic moment of the vertical members is reduced to M_p.

12. In a fixed-base rectangular portal frame $ABCD$ the stanchion AB is of height 10 ft. while the stanchion DC is of height 15 ft. The foot D is 5 ft. lower than the foot A, so that the beam BC, of length 10 ft., is horizontal. All the members of the frame have the same fully plastic moment M_p. The beam BC carries a central concentrated vertical load 5 tons, and a concentrated horizontal load 2 tons is applied at C in the direction BC. Find the value of M_p such that collapse would just occur. Find also the required value of M_p if the horizontal load 2 tons is reversed in direction.

13. A uniform beam of length l and fully plastic moment M_p is simply supported at one end and rigidly clamped at the other end. A concentrated load W may be applied anywhere within the span. Find the smallest value of M_p such that collapse would just occur when the load was in its most unfavourable position.

CHAPTER

4

General Methods for Plastic Design

4.1. Introduction

In this chapter three of the principal general methods which are available for designing frames so that plastic collapse would just occur under a prescribed loading are described in detail, and a brief outline of several other methods is given at the end of the chapter. In the first place, the *trial and error* method, due to Baker,[1] which was referred to briefly in Section 3.3, is discussed in more detail in Section 4.2. This method, which is based on the uniqueness theorem, consists of analysing statically an assumed collapse mechanism to see whether a corresponding safe and statically admissible bending moment distribution for the whole frame can be found. If such a distribution is shown to exist, it follows that the actual collapse mechanism has been found; if not, a fresh guess as to the collapse mechanism is made, guided by the results of the previous analysis, and the process repeated. This method is satisfactory provided that it is only necessary to investigate cases of complete or over-complete collapse, where the entire frame is statically determinate at collapse. If the actual collapse mechanism is partial, so that only a portion of the frame is statically determinate at collapse, its application is clearly more tedious. Such cases often occur in practice, so that there is a need for some other method for deciding upon the actual collapse mechanism. This is provided by the method of *combining mechanisms*, due to Neal and Symonds,[2, 3] which is described in Section 4.3.

As pointed out in Section 3.3, a valid method for determining the actual collapse mechanism and collapse load would be to write down a work equation for every possible mechanism of collapse so as to derive the corresponding value of the load. The collapse load would then be the lowest value of the load thus obtained, by the kinematic theorem. For a complicated frame where the number of possible mechanisms would be large this procedure would be impracticable. However, it can be shown that for a given frame and loading, all the possible mechanisms can be obtained as different combinations of a comparatively small number of *independent mechanisms*, which are readily identified for a given frame and loading. It is then found to be unnecessary to investigate all the possible combinations of the independent mechanisms, for the combination which is sought is the one to which there corresponds the lowest value of the load, and in a particular case it is at once evident that there are very few combinations which require investigation. It will be appreciated that this method is not more difficult to apply in cases of partial collapse, for there is no extra difficulty in writing down the work equation for a partial collapse mechanism.

Another method, which was developed by Horne,[4] is described in Section 4.4. This method is termed the *plastic moment distribution* method, because it bears some resemblance to the moment distribution method of elastic theory. In this case the approach is essentially statical, the object being to derive initially a distribution of bending moment which is in equilibrium with the given loads and is thus statically admissible. If the fully plastic moment of each member were then made equal to the largest bending moment occurring in the member, the bending moment distribution would also be safe, and so by the static theorem the loads would be less than or equal to the values necessary to cause collapse. Unless the fully plastic moment were attained at a sufficient number of cross-sections to transform the structure into a mechanism, the frame thus designed would be unnecessarily strong. Adjustments are therefore made to the bending moment distribution in such a way that equilibrium is preserved while this mechanism condition is fulfilled, so that an economical design is ultimately attained.

Finally, those other methods which have been developed for the determination of plastic collapse loads are described briefly in Section 4.5.

4.2. Trial and error method

The trial and error method will be illustrated by its application to the pitched roof portal frame whose dimensions and loading are shown in Fig. 4.1(*a*). All the loads on this frame are uniformly distributed along the members, and their magnitudes are indicated against the dotted arrows which show the directions in which the loads act. The uniformly distributed vertical loads of 2·6 tons on each rafter represent dead and snow loading. The uniformly distributed horizontal loads of 0·5 ton on each vertical stanchion member, and the uniformly distributed suctions of 0·4 ton and 0·8 ton on the rafters represent wind loading caused by wind blowing from left to right. All the joints, including those at the feet of the stanchions, are assumed to be capable of developing the fully plastic moment. The members of the frame are all

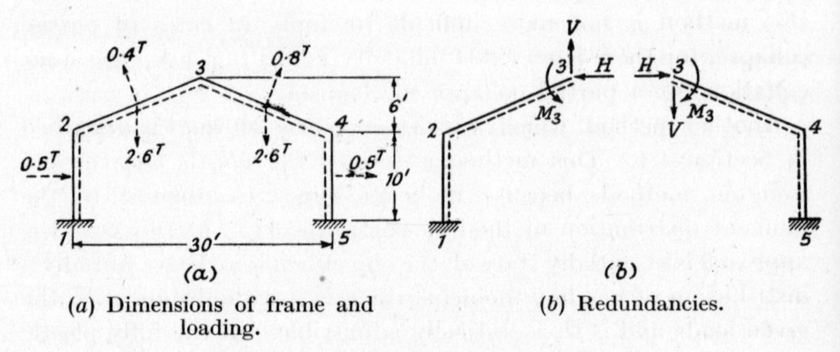

(*a*) Dimensions of frame and loading.

(*b*) Redundancies.

Fig. 4.1. *Pitched roof portal frame.*

supposed to have the same fully plastic moment M_p, and the value of M_p is required such that collapse would just occur under the given loads.

In a case of this kind in which the members are subjected to uniformly distributed loads, plastic hinges can occur anywhere along the members, and not simply at a few known positions as is the case when the loads are concentrated. The statical analysis of an assumed mechanism is thus best carried out by constructing the complete bending moment diagram for the frame, so that it is possible to see immediately whether the fully plastic moment is exceeded at any section. The most convenient method of construction is to draw free and reactant bending moment diagrams. For this purpose the frame is imagined to be cut at its apex, and the three redundancies are taken to be the horizontal and vertical reactions H and V at this section, together with the bending moment M_3, as shown in Fig. 4.1(*b*). The free bending moment diagram, representing the bending moments due to the applied loads with H, V and M_3 assumed to be zero, is then drawn. From the statical analysis of any assumed mechanism the values of the three redundancies and of M_p are found, and this enables the reactant moment diagram, representing the bending moments due to the three redundancies alone, to be drawn. This diagram is drawn with the signs of the reactant moments changed, so that the difference in ordinate between the free and reactant moment diagrams at any section represents the actual bending moment.

The free bending moment at any section depends only on the dimensions of the frame and its loading. Thus the free bending moment diagram is independent of any choice of assumed collapse mechanism, and can be constructed at the outset. By taking moments about the cross-sections 1, 2, 4 and 5 the free bending moments at these sections are found to have the values given in Table 4.1, the free bending moment at section 3 being zero.

TABLE 4.1

Free bending moments

Section	1	2	3	4	5
Free moment (tons ft.) .	−17·28	−16·27	0	−13·04	−7·57

The complete free bending moment diagram, consisting of four parabolic segments, is shown in Fig. 4.2(a), in which the frame is imagined to be opened out to form a horizontal datum from which the free moments are plotted as ordinates. The parabolic distributions in the rafters are readily constructed from a knowledge of the free moments given in Table 4.1; for the stanchions it is necessary to calculate in addition the shear forces at each end. The sign convention adopted for the bending moments is that a positive moment causes tension in those fibres of the member adjacent to the dotted line in Fig. 4.1(b).

The first mechanism to be analysed is as shown in Fig. 4.2(a), with plastic hinges at sections 1, 3, 4 and 5. It is not necessary to consider the kinematics of this mechanism beyond noting that the sense of the rotations at the hinges determines the signs of the fully plastic moments to be as follows:

$$M_1 = -M_p, \quad M_3 = M_p, \quad M_4 = -M_p, \quad M_5 = M_p$$

Thus in this mechanism one of the redundancies M_3 is already determined; the other two, H and V, together with the corresponding value of M_p, are found by taking moments about sections 1, 4 and 5. The moments due to the redundancies can be written down by inspection from Fig. 4.1(*b*), and the moments

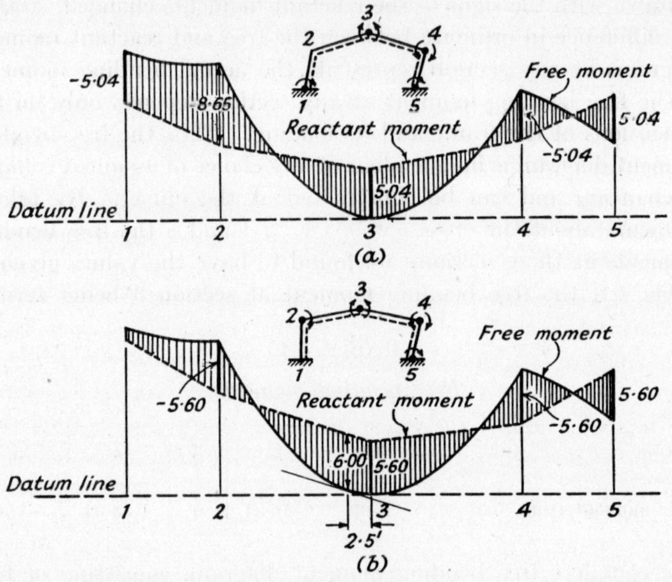

Fig. 4.2. *Bending moment diagrams for pitched roof portal frame.*

due to the loads are simply the free moments and are taken from Table 4.1. The relevant equations are

$$M_1 = M_3 + 16H + 15V - 17{\cdot}28 = -M_p$$
$$M_3 = M_p$$
$$M_4 = M_3 + 6H - 15V - 13{\cdot}04 = -M_p$$
$$M_5 = M_3 + 16H - 15V - 7{\cdot}57 = M_p$$

and the solution of these equations gives

$$H = 0{\cdot}461 \text{ tons}, \quad V = -0{\cdot}012 \text{ tons}, \quad M_3 = M_p = 5{\cdot}04 \text{ tons ft.}$$

The reactant moment diagram clearly consists of four straight line segments, one for each member, and it is therefore only necessary to know the reactant moment at each of the sections 1, 2, 3, 4 and 5 to construct this diagram. From Fig. 4.1(*b*), and using the values of the redundancies just obtained, it is seen that the reactant moment $M_1^{(R)}$ at section 1 has the value

$$M_1^{(R)} = M_3 + 16H + 15V = 12{\cdot}24 \text{ tons ft.}$$

The other reactant moments are found in a similar manner to have the values given below:

Section	1	2	3	4	5
Reactant moment (tons ft.) .	12·24	7·62	5·04	8·00	12·61

The corresponding reactant moment diagram is shown as the dotted line in Fig. 4.2(a). It can be seen from this figure that the greatest bending moment occurring in the resulting bending moment distribution is at section 2, where the bending moment is $-8{\cdot}65$ tons ft., thus exceeding in magnitude the value of M_p of 5·04 tons ft. It is at once evident that the assumed collapse mechanism was not correct. However, from the results of this first trial it is possible to establish upper and lower bounds on the required value of M_p. Thus if the fully plastic moment had the value 8·65 tons ft., the bending moment distribution of Fig. 4.2(a), as well as being statically admissible with the given loads, would also be safe. From the static theorem the given loads would then not cause collapse, and the frame would therefore be more than adequately strong to carry these loads. It follows that 8·65 tons ft. is an *upper bound* on the required value of M_p. If on the other hand the fully plastic moment had the value 5·04 tons ft., the analysis has shown that the given loads would be those found to correspond to the assumed mechanism, which is not the actual collapse mechanism. These loads would then be greater than the values which would cause collapse, by the kinematic theorem. It thus follows that with a fully plastic moment of 5·04 tons ft. the frame would not be strong enough to carry the given loads, and so this is a *lower bound* on the required value of M_p.

It will be recalled that in Section 3.3 the highest value of the bending moment obtained in a statical analysis of an assumed mechanism was used to derive a *lower bound* on the collapse load W_c by applying the static theorem, whereas the value of the load which was found to correspond to an assumed mechanism was shown to be an *upper bound* on W_c, by applying the kinematic theorem. In this case the problem was formulated as the *analysis* of a frame with given fully plastic moments, the object being to determine the collapse load. In contrast, the present problem is to *design* a frame which is subjected to given loads, so that the value of the fully plastic moment of the members is to be found such that collapse would just occur under these loads. As has been shown, the highest value of the bending moment obtained in a statical analysis is in this case an *upper bound* on the required value of M_p, whereas the value of M_p corresponding to an assumed mechanism is a *lower bound* on the required value of M_p.

The results of the statical analysis of the first assumed mechanism should be utilized when considering which mechanism should form the subject of the second trial. A study of the bending moment distribution of Fig. 4.2(a) suggests that the plastic hinge at section 1 should have been located at section 2, where a larger negative moment occurred. The second mechanism to be assumed is therefore as shown in Fig. 4.2(b). This mechanism is analysed as before by taking moments about those cross-sections where hinges occur to determine the values of the three redundancies and of M_p. In this way it is found that

$$\begin{aligned} M_2 &= M_3 + 6H + 15V - 16{\cdot}27 = -M_p \\ M_3 &= M_p \\ M_4 &= M_3 + 6H - 15V - 13{\cdot}04 = -M_p \\ M_5 &= M_3 + 16H - 15V - 7{\cdot}57 = M_p \end{aligned}$$

and the solution of these equations gives

$$H = 0{\cdot}574 \text{ tons}, \quad V = 0{\cdot}108 \text{ tons}, \quad M_3 = M_p = 5{\cdot}60 \text{ tons ft.}$$

The reactant moment diagram corresponding to these values of the redundancies is shown in Fig. 4.2(b), in which the free moment diagram is of course identical with that of Fig. 4.2(a).

Examination of the resulting bending moment diagram in Fig. 4.2(b) shows that the assumed mechanism was in this case substantially correct, for the calculated value of M_p is only exceeded to a small extent in the neighbourhood of the apex, section 3. In fact, the greatest bending moment occurs at about 2·5 ft. to the left of the apex, at the position where the tangent to the free moment diagram is parallel to the reactant moment diagram; this position may be determined from an accurate drawing or by calculating the position in this rafter member where the shear force is zero. The value of this maximum bending moment is 6·00 tons ft., so that this second trial has established upper and lower bounds on the required value of M_p of 6·00 tons ft. and 5·60 tons ft., respectively.

It is evident that the actual collapse mechanism is obtained from the mechanism of Fig. 4.2(b) by moving the plastic hinge from the apex to a position in the left-hand rafter which will be about 2·5 ft. from the apex. A value of M_p sufficiently accurate for all practical purposes can be obtained by assuming that this hinge should be precisely 2·5 ft. from the apex and carrying out a statical analysis to find the corresponding value of M_p. Details of this analysis will not be given; the result obtained is 5·73 tons ft.

The loads which are shown as acting on the frame are in fact representative of the *working loads* to which such a frame might be subjected. Thus if the frame were to be designed to a load factor of 2, it would be necessary to provide members whose fully plastic moments were at least 2 × 5·73 tons ft., or 11·46 tons ft.

Further examples of the application of the trial and error method need not be given here. Baker [1] has given an illustration of its application to a three-bay portal frame, and a comparison between the elastic designs of several pitched roof portal frames and plastic designs derived by the trial and error method has been given by Foulkes.[5] Several examples have also been given in a publication of the British Constructional Steelwork Association,[6] and Hendry [7] has applied the method to the design of Vierendeel girders, several models of this type of frame being tested to confirm the theory.

It is worth noting that for a single-bay portal frame of this type the design is often governed by the loading case in which the wind loads are absent, so that the frame is only subjected to a uniformly distributed vertical loading. In such a case it is evident that the bending moment distribution must be symmetrical, so that the redundancy V of Fig. 4.1(b) is zero. Another case of practical importance occurs when the feet of the stanchions are pinned, so that the bending moments at these positions are zero. In such a case the frame has only one redundancy, and the values of H, V and M_3 as defined in Fig. 4.1(b) are no longer independent but are connected by two equations stating that the bending moment at the foot of each stanchion is zero.

The trial and error method is admirably suited to those cases in which the collapse is complete, in the sense defined in Section 3.6, so that the entire frame is statically determinate at collapse. It is then only necessary to investigate assumed collapse mechanisms of the complete type, and at each trial the complete bending moment distribution is determined uniquely by statics, as in the example just considered. In fact, single-bay portal frames of this kind almost invariably fail in a complete manner, so that for such frames the trial and error method can be adopted with confidence. However, difficulties arise when the actual collapse mechanism is likely to be of the partial type, in which only a portion of the frame is statically determinate at collapse. When investigating an assumed collapse mechanism of this kind it is a simple matter to construct the uniquely determined bending moment diagram in that part of the frame which is statically determinate at collapse and to see whether the fully plastic moment is exceeded anywhere within this distribution. However, it is also necessary to examine the remainder of the frame, in which the distribution of bending moment is not uniquely determined, to see whether amongst all the various possible statically admissible bending moment distributions at least one safe distribution exists. Such an investigation may prove to be of great difficulty, particularly if, as is sometimes the case, the portion of the frame which is not rendered statically determinate is itself highly redundant. In fact an investigation of this kind would scarcely be carried out without some assurance that the assumed partial collapse mechanism was almost certainly the actual collapse mechanism.

Experience in the design of similar structures may afford the necessary guidance as to the likelihood of an assumed collapse mechanism being the correct one. However, in facing unusual problems there is a need for a method by means of which a close approximation to the actual collapse mechanism can be found quickly even though this mechanism may be of the partial type. The method of combining mechanisms, which will now be described, enables this to be done.

4.3. Method of combining mechanisms

The essential notion underlying the method of combining mechanisms is that for a given frame and loading every possible collapse mechanism can be regarded as some combination of a certain number of *independent mechanisms*. Once these independent mechanisms have been identified, a work equation can be written down and the corresponding value of M_p deduced for each of these mechanisms. Now the actual collapse mechanism is distinguished from amongst all the possible mechanisms by the fact that it has the highest corresponding value of M_p, by the kinematic theorem. Accordingly, those independent mechanisms with high corresponding values of M_p are examined to see whether they can be combined to form a mechanism which gives an even higher value of M_p. It is only necessary to examine a few of the more likely combinations in this way in order to arrive at a mechanism which is almost certainly the actual collapse mechanism or a close approximation to it. An independent statical check is then carried out which either verifies that the actual collapse mechanism has been found or indicates the minor adjustments that need to be made. By this procedure the apparent necessity for examining every possible collapse mechanism in order to determine which one gives the highest corresponding value of M_p is avoided.

The basic principles of the method will first be explained with reference to a simple rectangular portal frame problem, and its application to a two-bay rectangular portal frame subjected to concentrated loads will then be discussed. The design of a two-storey, single-bay rectangular frame is then given in order to illustrate a technique for dealing with cases in which the members are subjected to uniformly distributed loads. Finally, applications to single and two-bay pitched roof portal frames will be described, since the kinematics of some of the mechanisms for such frames are more complicated than for rectangular frames.

Simple rectangular portal frame

The technique of combining mechanisms will first be explained with reference to the simple rectangular portal frame whose dimensions and loading are given in Fig. 4.3(*a*). In this frame the beam is to have a fully plastic moment of M_p while the stanchions are each to have a fully plastic moment of $1{\cdot}5M_p$, as shown in the figure. The problem is to determine the value of M_p such that the frame would just collapse under the given loading.

From previous consideration of the collapse mechanisms for simple rectangular portal frames in Section 3.3, it is known that there are only three possible collapse mechanisms for such frames. These are as shown in Figs. 4.3(*b*), (*c*) and (*d*), in which the magnitudes of the hinge rotations for each of these mechanisms are given in terms of a single parameter θ. The mechanism of Fig. 4.3(*b*) represents collapse of the beam, and that of Fig. 4.3(*c*) represents sidesway collapse, the hinges at the corners 2 and 4 being located in the beam, which is weaker than the stanchions. In Fig. 4.3(*c*) the dotted lines represent the effect of superimposing the displacements and hinge rotations of the beam mechanism on

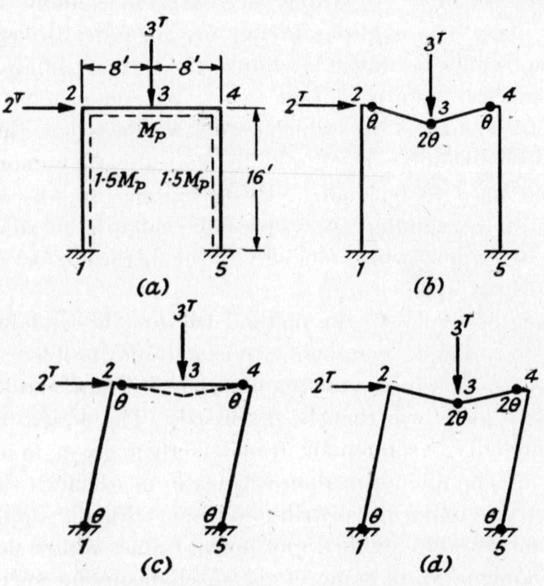

(*a*) Frame and loading. (*b*) Beam mechanism.
(*c*) Sidesway mechanism. (*d*) Combined mechanism.

Fig. 4.3. Rectangular portal frame.

those of the sidesway mechanism, and it is seen that by this superposition the mechanism of Fig. 4.3(*d*) is obtained. This mechanism is thus a combination of the beam and sidesway mechanisms, and for this reason is referred to as the combined mechanism. In fact, the beam and sidesway mechanisms may be regarded as the two independent mechanisms for this frame, from which the other possible collapse mechanism is obtained by combination. The important corollary to this statement is that the work equation for the combined mechanism can be derived by combining the work equations for the two independent mechanisms.

To show how the work equation for the combined mechanism is derived, the work equations for the two independent mechanisms must first be obtained. In the beam mechanism of Fig. 4.3(*b*) the horizontal load 2 tons does no work, and the vertical load 3 tons, moving through a distance 8θ ft., does work 24θ tons ft. The three plastic hinges all occur in the beam, whose fully plastic moment is M_p; their total rotation is 4θ and the total work absorbed is thus $4M_p\theta$. The work equation for the beam mechanism is thus

$$24\theta = 4M_p\theta \qquad . \quad . \quad . \quad . \qquad 4.1$$
$$M_p = 6 \text{ tons ft.}$$

In the sidesway mechanism of Fig. 4.3(*c*) the vertical load 3 tons does no work and the horizontal load 2 tons, moving through a distance 16θ ft., does work 32θ tons ft. The plastic hinges at sections 2 and 4 occur in the beam, whose fully plastic moment is M_p. Since their total rotation is 2θ the total work absorbed in these hinges is $2M_p\theta$. The plastic hinges at sections 1 and 5 occur in the stanchions, whose fully plastic moment is $1{\cdot}5M_p$; their total rotation is 2θ and the total work absorbed in these hinges is $3M_p\theta$. The total work absorbed in the four plastic hinges is thus $5M_p\theta$, and the work equation for the sidesway mechanism is

$$32\theta = 5M_p\theta \qquad . \quad . \quad . \quad . \qquad 4.2$$
$$M_p = 6{\cdot}4 \text{ tons ft.}$$

The derivation of the work equation for the combined mechanism from the work equations 4.1 and 4.2 will now be considered. The essential point involved is that the deflections and hinge rotations of this mechanism are obtained by adding the deflections and hinge rotations of the beam and sidesway mechanisms. It will be seen that in the beam mechanism the hinge at the knee 2 undergoes a rotation of magnitude θ in the hogging sense, whereas the rotation of the hinge at this section is of the same magnitude θ but in the sagging sense in the sidesway mechanism. Thus when the two mechanisms are combined these hinge rotations cancel. The only other cross-section at which hinges occur in both the beam and the sidesway mechanism is section 4. These hinges both undergo a hogging rotation of magnitude θ, and so may be added to give the rotation 2θ at this section in the combined mechanism. At sections 1, 3 and 5 a hinge rotation only occurs in one of the independent mechanisms, so that at any one of these three sections as well as at section 4, the magnitude of the hinge rotation in the combined mechanism is equal to the sum of the magnitudes of the hinge rotations in the two independent mechanisms.

To obtain the total work absorbed in the plastic hinges in the combined mechanism it is noted that the work absorbed in the hinges in the beam and sidesway mechanisms was $4M_p\theta$ and $5M_p\theta$, respectively, from equations 4.1 and 4.2. Each of these totals included a term $M_p\theta$ for the work absorbed in the hinge at section 2. Since this hinge is cancelled in forming the combined mechanism it follows that the total work absorbed in both of the independent mechanisms, $9M_p\theta$, exceeds the work absorbed in the combined mechanism by an amount $2M_p\theta$. Thus the total work absorbed in the combined mechanism must be $7M_p\theta$. Since the displacement at any point in the combined mechanism is obtained by adding the displacements at this point in the two independent mechanisms, the work done by the loads in the combined mechanism must be the sum of the work done by the loads for the two independent mechanisms. From equations 4.1 and 4.2 it is seen that the work done by the loads for these mechanisms was 24θ for the beam mechanism and 32θ for the sidesway mechanism, so that the total work done by the loads in the combined mechanism is 56θ.

The above argument can be summarized by noting that the work equation for the combined mechanism can be obtained from equations 4.1 and 4.2 by adding the terms on the left-hand sides to obtain the total work done by the applied loads, adding the terms on the right-hand sides to obtain the total work absorbed in the two combining mechanisms, and then subtracting $2M_p\theta$ to allow for the cancellation of the hinge at section 2, as follows:

Beam mechanism: $$24\theta = 4M_p\theta \qquad (4.1)$$

Sidesway mechanism: $$32\theta = 5M_p\theta \qquad (4.2)$$

Combined mechanism: $$56\theta = 9M_p\theta - 2M_p\theta = 7M_p\theta \qquad (4.3)$$

$$M_p = 8 \text{ tons ft.}$$

The value of M_p corresponding to the combined mechanism is thus seen to be 8 tons ft., which is higher than either of the values which were found to correspond to the beam and sidesway mechanisms. It follows from the kinematic theorem that this is the required value of M_p. This conclusion may be checked by constructing the corresponding bending moment diagram for the entire frame.

The most significant feature of this calculation is that by combining two mechanisms for which the corresponding values of M_p were 6 tons ft. and 6·4 tons ft., a mechanism was derived for which the corresponding value of M_p was 8 tons ft., this value being higher than the values for either of the mechanisms which were combined. This result is made possible by the cancellation of the hinge at section 2 which occurs when the mechanisms are combined, which causes the total work absorbed in the plastic hinges in the combined mechanism to be less than the total work absorbed in the hinges in the two mechanisms which were combined. In general, it may be said that in analysing a frame the object is to find the mechanism to which there corresponds the highest possible value of M_p. The independent mechanisms are therefore inspected to find two or more mechanisms each with high corresponding values of M_p, which can be combined so as to cause the cancellation of a hinge. In this way a combined mechanism may be obtained for which the corresponding value of M_p is higher than the value for either of the mechanisms which were combined. It will be clear that if there is no hinge cancellation the value of M_p corresponding to a combination of mechanisms will lie between the highest and lowest values of M_p for the combining mechanisms.

The procedure which has been described for deriving the work equation for the combined mechanism is little shorter for this particular example than the direct derivation from considering the kinematics of the combined mechanism. However, it is not difficult to see that in more complicated problems this procedure will usually result in a considerable reduction of computational work. This is particularly true if the kinematical calculations for one or more of the independent mechanisms involve some labour, for in such cases the work equation for any combined mechanism can be obtained without performing the further awkward kinematical calculations which would be demanded if this mechanism were treated directly.

The number of independent mechanisms

In applications of the method of combining mechanisms it is naturally of the utmost importance to determine at the outset the correct number of independent mechanisms. This number is in fact the same as the number of independent equations of equilibrium. The reason for this has already been touched upon in Section 3.3, where equations of equilibrium were established for the particular rectangular portal frame of Fig. 3.1(a). It was then shown that each of the three possible collapse mechanisms corresponded to the breakdown of a certain equation of equilibrium, in the sense that the fully plastic moments occurring at the plastic hinges in the mechanism were precisely the values which were found to maximize the load which occurred in the equilibrium equation. This correspondence of every mechanism to the breakdown of an equation of equilibrium, although only established in Section 3.3 in relation to a particular example, is quite general. Since all possible equilibrium equations for a given frame and loading can be obtained by combining a certain number of independent equations, it must follow that all possible mechanisms can be obtained by combining the same number of independent mechanisms. Thus to determine the correct number of independent mechanisms it is only necessary to determine the number of independent equations of equilibrium.

In view of the fundamental importance of this result a brief discussion of the correspondence of the relevant equations of equilibrium to the beam, sidesway and combined mechanisms for the above example will be given. To decide upon the number of independent equations of equilibrium it is noted that the frame has three redundancies, whereas the complete bending moment distribution is specified by the values of the bending moments at the five cross-sections numbered from 1 to 5 in Fig. 4.3(a). There must therefore be two independent equations of equilibrium relating these bending moments to the applied loads.

The derivation of these two equations can be made by a statical method, in which the frame is imagined to be cut at the centre of the beam, the redundant thrust, shear force and bending moment are introduced at this section, and expressions for each of the five unknown bending moments are written down in terms of these three redundancies and the loads. Elimination of the three redundancies from these five equations then leads to the two equations of equilibrium. Alternatively a kinematical method, in which the Principle of Virtual Work is applied to the beam and sidesway mechanisms in turn, can be employed. Details of the derivation of these equations will not be given, since both of the above methods were applied in Sections 3.3 and 3.4 to the similar case of the rectangular portal frame of Fig. 3.1(*a*), leading to the equilibrium equations 3.5 and 3.6. For the frame of Fig. 4.3(*a*) the two equations of equilibrium are

$$24 = 2M_3 - M_2 - M_4 \quad . \quad . \quad . \quad 4.4$$

$$32 = M_2 - M_1 + M_5 - M_4 \quad . \quad . \quad 4.5$$

with the usual sign convention for the bending moments.

In equation 4.4 each bending moment must lie within the limits $\pm M_p$, and the minimum possible value of M_p consistent with this equation being satisfied is obtained by putting

$$M_2 = -M_p, \quad M_3 = M_p, \quad M_4 = -M_p$$

so that $4M_p = 24$ and $M_p = 6$ tons ft., as obtained by the corresponding work equation 4.1. Thus from the point of view of this equation alone, M_p must be at least 6 tons ft., and this value is a lower bound on the required value of M_p. The above bending moment values are precisely those occurring at the plastic hinges in the beam mechanism of Fig. 4.3(*b*), so that this mechanism may be said to correspond to the breakdown of the equilibrium equation 4.4.

A similar argument in respect of equation 4.5 shows that a lower bound on the value of M_p is established by this equation when the following values of the bending moment are inserted:

$$M_1 = -1{\cdot}5M_p, \quad M_2 = M_p, \quad M_4 = -M_p, \quad M_5 = 1{\cdot}5M_p$$

giving $5M_p = 32$, or $M_p = 6{\cdot}4$ tons ft., as found by the work equation 4.2. These bending moments are those occurring at the plastic hinges in the sidesway mechanism, which therefore corresponds to the breakdown of the equilibrium equation 4.5.

The combined mechanism of Fig. 4.3(*d*) has no plastic hinge at section 2, and the corresponding equation of equilibrium is thus obtained by adding equations 4.4 and 4.5 to eliminate M_2, giving

$$56 = 2M_3 - M_1 + M_5 - 2M_4 \quad . \quad . \quad 4.6$$

This addition is the statical equivalent of combining the beam and sidesway mechanisms by adding their deflections and hinge rotations. From equation 4.6 it is seen that a lower bound on M_p is obtained by putting

$$M_1 = -1{\cdot}5M_p, \quad M_3 = M_p, \quad M_4 = -M_p, \quad M_5 = 1{\cdot}5M_p$$

which gives $7M_p = 56$, or $M_p = 8$ tons ft., as found by the work equation 4.3.

The above discussion emphasizes further the general point that each possible collapse mechanism for a given frame and loading corresponds to the breakdown of a particular equilibrium equation. Since all the possible equilibrium equations can be derived from a certain number of independent equations, it follows that all the possible collapse mechanisms can be derived from the same number of independent mechanisms. To determine the number of independent equilibrium equations and thus of independent mechanisms for the portal frame of Fig. 4.3(a), it was noted that the number of bending moments whose values were needed to specify the bending moment distribution for the whole frame was five, whereas the number of redundancies was three, so that there must be two independent equations of equilibrium relating the five unknown bending moments. In general, the number of independent mechanisms is $(n - r)$, where n is the number of bending moments necessary to specify the bending moment distribution in the frame, and r is the number of redundancies.

It will be appreciated that in the example of Fig. 4.3 any pair of the three mechanisms shown in this figure could have been selected as the two independent mechanisms. The main reason for choosing the beam and sidesway mechanisms as the two independent mechanisms is that if the combined mechanism is selected as one of the independent mechanisms, the combination of the independent mechanisms involves the subtraction of hinge rotations and displacements. For example, choosing the combined mechanism and the beam mechanism as the two independent mechanisms, the sidesway mechanism is derived by subtracting the displacements and hinge rotations of the beam mechanism from those of the combined mechanism. This would lead to a slight awkwardness in the corresponding calculations.

Two-bay rectangular portal frame

The technique of combining mechanisms will now be applied to the two-bay rectangular portal frame whose dimensions and loading are illustrated in Fig. 4.4(a). In this frame each of the three stanchions is to have a fully plastic moment M_p, and each of the two beams is to have a fully plastic moment $2M_p$. The

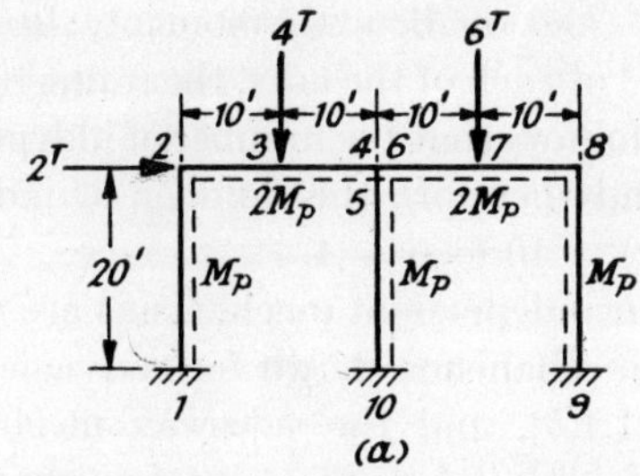

(*a*) Dimensions of frame and loading.

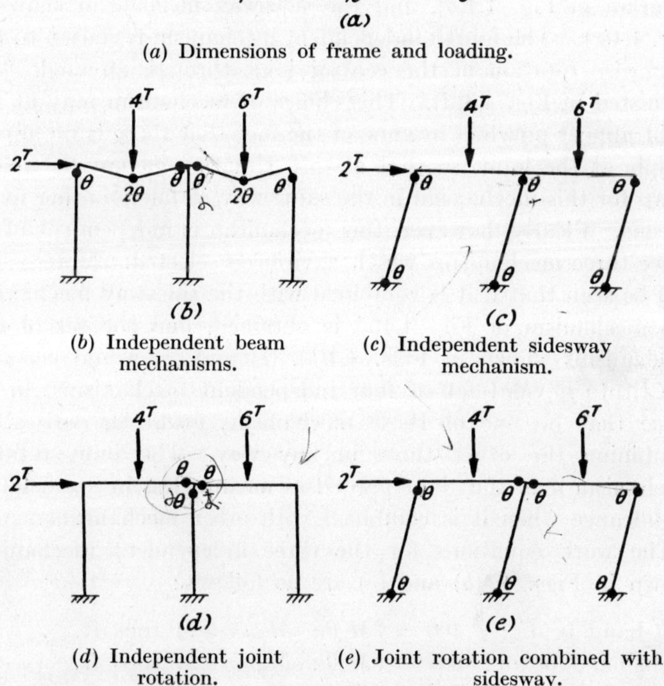

(*b*) Independent beam mechanisms.

(*c*) Independent sidesway mechanism.

(*d*) Independent joint rotation.

(*e*) Joint rotation combined with sidesway.

Fig. 4.4. Two-bay rectangular frame.

value of M_p is required such that the frame would just collapse under the given loads.

The first step is to decide upon the correct number of independent mechanisms. For this purpose it is noted that the number of bending moments n which is required to specify the bending moment distribution throughout the frame is ten, since the bending moment must vary linearly between those sections numbered from 1 to 10 in Fig. 4.4(a). Moreover, the number of redundancies r for this frame is six, since if two cuts are made, say at sections 3 and 7, and the bending moment, shear force and axial thrust are specified at each of the cuts, the frame becomes statically determinate. It follows that the number of independent equations of equilibrium, and therefore the number of independent mechanisms, is $(n - r) = 10 - 6 = 4$.

Three of the four independent mechanisms are readily identified as the two beam mechanisms shown for convenience on the same diagram in Fig. 4.4(*b*), and the sidesway mechanism shown in Fig. 4.4(*c*). The fourth independent mechanism is chosen to be a clockwise rotation of the central joint through an angle θ, as indicated in Fig. 4.4(*d*). This choice of mechanism may at first sight appear puzzling in view of the fact that there is no applied couple at the joint, so that a work equation cannot be written down for this mechanism in the same way as for the other mechanisms. Clearly, however, this mechanism is independent of the other three mechanisms which have been selected. Moreover, it will be seen that if it is combined with the sidesway mechanism, the mechanism of Fig. 4.4(*e*) is obtained, and the set of four mechanisms shown in Figs. 4.4(*b*), (*c*) and (*e*) would certainly constitute a valid set of four independent mechanisms, in the sense that no one of these mechanisms could be derived by combining the other three in any way. The joint rotation mechanism may thus be regarded as meaningless in itself, but of significance when it is combined with other mechanisms.

The work equations for the three independent mechanisms shown in Figs. 4.4(*b*) and (*c*) are as follows:

Left-hand beam:	$40\theta = 7M_p\theta$;	$M_p = 5{\cdot}71$ tons ft.	4.7
Right-hand beam:	$60\theta = 7M_p\theta$;	$M_p = 8{\cdot}57$ tons ft.	4.8
Sidesway:	$40\theta = 6M_p\theta$;	$M_p = 6{\cdot}67$ tons ft.	4.9

The derivation of these equations need not be discussed beyond noting that in the beam mechanisms the plastic hinges at sections 2 and 8 form in the stanchions rather than in the stronger beams. As already pointed out, no work equation can be written down for the joint rotation mechanism.

The independent mechanism which gives the highest corresponding value of M_p is the right-hand beam mechanism, for which M_p is 8·57 tons ft. In attempting to find a mechanism giving a higher value of M_p, it is natural to investigate the combination of this mechanism with the sidesway mechanism, which gives the next highest value of 6·67 tons ft. for M_p. A straightforward addition of the displacements and hinge rotations of these two mechanisms would be pointless, for they have no common hinge whose rotation would be cancelled by the addition and thus reduce the work absorbed in the plastic hinges in the combined mechanism. The mechanism resulting from this simple addition is shown in Fig. 4.5(*a*). However, it will be noticed that in this mechanism there are two plastic hinges at the central joint, one in the right-hand beam and one in the central stanchion. By rotating the joint clockwise through an angle θ, these two hinges can be replaced by a single hinge in the left-hand beam, thus producing the mechanism of Fig. 4.5(*b*). This reduces the work

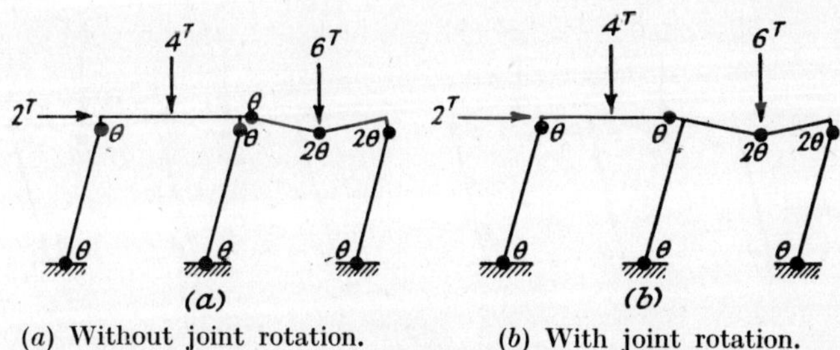

(*a*) Without joint rotation. (*b*) With joint rotation.

Fig. 4.5. *Combination of right-hand beam and sidesway mechanisms*

absorbed in the plastic hinges at this joint from $3M_p\theta$ to $2M_p\theta$, while leaving the other terms in the work equation unchanged.

The work equation for the mechanism of Fig. 4.5(*b*) is obtained by adding the work equations 4.8 and 4.9 for the right-hand beam and sidesway mechanisms which were combined, and then subtracting $M_p\theta$ from the resulting work absorbed in the plastic hinges to account for the effect of the joint rotation. The calculation is thus as follows:

Right-hand beam: $60\theta = 7M_p\theta$ 4.8

Sidesway: $40\theta = 6M_p\theta$ 4.9

Mechanism of Fig. 4.5(*b*): $100\theta = 13M_p\theta - M_p\theta = 12M_p\theta$ 4.10

$$M_p = 8{\cdot}33 \text{ tons ft.}$$

Despite the hinge cancellation achieved by the joint rotation, this value of M_p does not exceed the value of 8·57 tons ft. obtained for the collapse of the right-hand beam. It can therefore be concluded on the basis of the kinematic theorem that the mechanism of Fig. 4.5(*b*) is not the actual collapse mechanism.

The above calculation indicates the role which is played by the joint rotation mechanism. For every particular combination of mechanisms the joint is rotated into the position which minimizes the work absorbed in the plastic hinges at the central joint. In this way, the joint rotation mechanism features in the technique as an independent mechanism, although when considered by itself it has no physical meaning.

Another possible combination is obtained by adding the deflections and hinge rotations of the left-hand beam mechanism, Fig. 4.4(*b*), to those of the sideway mechanism of Fig. 4.4(*c*).

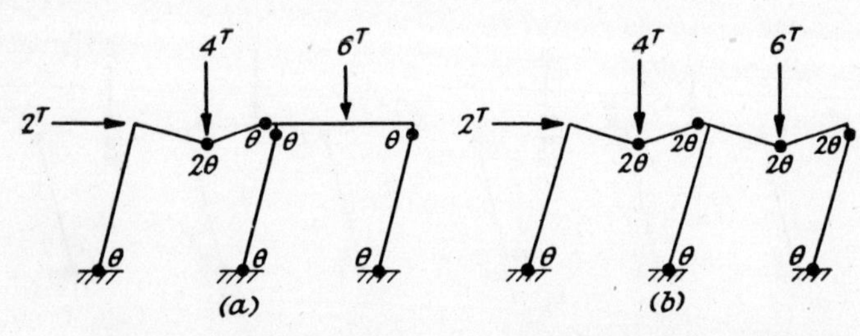

(*a*) Combination of left-hand beam and sidesway mechanisms.

(*b*) Combination of both beam and sidesway mechanisms with joint rotation.

Fig. 4.6. *Combined mechanisms.*

This results in the cancellation of the plastic hinge at section 2, producing the mechanism shown in Fig. 4.6(*a*). In this mechanism it can be seen that there is no advantage to be gained by rotating the central joint. For each of the combining mechanisms the total work absorbed in the plastic hinges included a term $M_p\theta$ for the hinge at section 2, so that a reduction of $2M_p\theta$ in the total work absorbed is achieved by this combination. The resulting work equation is obtained as follows:

Left-hand beam: $40\theta = 7M_p\theta$ . . . 4.7

Sidesway: $40\theta = 6M_p\theta$ . . . 4.9

Mechanism of Fig. 4.6(*a*): $80\theta = 13M_p\theta - 2M_p\theta = 11M_p\theta$ 4.11

$$M_p = 7{\cdot}27 \text{ tons ft.}$$

The corresponding value of M_p, while greater than the values of 5·71 tons ft. and 6·67 tons ft. for the two mechanisms which were combined, is less than the value of 8·57 tons ft. for the right-hand beam mechanism.

The only other possible combination is obtained by adding the deflections and hinge rotations of the right-hand beam mechanism to those of the mechanism of Fig. 4.6(*a*). A direct addition of these mechanisms would give three plastic hinges at the central joint, all with rotations of magnitude θ, so that the work absorbed at the joint would be $2M_p\theta$ in each beam and $M_p\theta$ in the stanchion, making a total of $5M_p\theta$. However, if the joint is rotated clockwise through an angle θ, the hinges in the stanchion and in the right-hand beam are cancelled and the hinge rotation in the left-hand beam becomes 2θ, so that the work absorbed at the joint is reduced to $4M_p\theta$. The resulting mechanism is shown in Fig. 4.6(*b*) and the work equation is obtained as follows:

Mechanism of Fig. 4.6(*a*): $80\theta = 11M_p\theta$ . . . 4.11

Right-hand beam: $60\theta = 7M_p\theta$ . . . 4.8

Mechanism of Fig. 4.6(*b*): $140\theta = 18M_p\theta - M_p\theta = 17M_p\theta$ 4.12

$$M_p = 8{\cdot}24 \text{ tons ft.}$$

Again the value of M_p is less than the value 8·57 tons ft. which corresponds to the right-hand beam mechanism. It is therefore concluded that the actual collapse mechanism is the right-hand beam mechanism, and that the required value of M_p is 8·57 tons ft. This conclusion will now be confirmed by a statical analysis.

Statical analysis

While it is possible to perform the statical calculations by constructing free and reactant bending moment diagrams in the manner adopted previously for single-bay portals, this method becomes progressively more difficult as the frame becomes more complicated. Even for a two-bay frame of the type under consideration, it may be simpler to establish the existence of a safe statically admissible bending moment distribution by using the equations of equilibrium, particularly if the frame is subjected to concentrated loads, so that the possible locations of plastic hinges within the spans of the members are known. The best method for obtaining the equilibrium equations is to apply the Principle of Virtual Work to the independent mechanisms, for the kinematics of these mechanisms has already been derived. The equations of equilibrium which correspond to the four independent mechanisms are found in this way to be:

Left-hand beam: $40 = 2M_3 - M_2 - M_4$ 4.13

Right-hand beam: $60 = 2M_7 - M_6 - M_8$ 4.14

Sidesway: $40 = M_2 - M_1 + M_5 - M_{10} + M_9 - M_8$ 4.15

Joint: $0 = M_4 + M_5 - M_6$ 4.16

The usual sign convention is adopted here for the bending moments. Equation 4.16, which corresponds to the joint mechanism of Fig. 4.4(*d*), merely expresses the fact that the central joint must be in rotational equilibrium.

The fully plastic moments occurring in the collapse mechanism for the right-hand beam are:

$$M_6 = -2M_p = -17{\cdot}14 \text{ tons ft.}$$
$$M_7 = 2M_p = 17{\cdot}14 \text{ tons ft.}$$
$$M_8 = -M_p = -8{\cdot}57 \text{ tons ft.}$$

These values satisfy equation 4.14, which corresponds to this beam mechanism. There are three remaining equations of equilibrium relating the other seven bending moments, and none of these bending moments is determined uniquely by these equations. To determine a statically admissible solution it is necessary to assign values to four of the unknown bending moments; the remaining three will then be determined from equations 4.13, 4.15 and 4.16. Guidance in making this choice is furnished by noting that the mechanism for which the corresponding value of M_p was nearest the value for the actual collapse mechanism was the combined mechanism of Fig. 4.5(*b*), which gave a value of 8·33 tons ft. This implies that the equilibrium equation which corresponds to this mechanism breaks down if M_p is less than 8·33 tons ft., and can therefore be satisfied by only a narrow range of bending moments not exceeding the fully plastic values when M_p is 8·57 tons ft. It follows that the four unknown bending moments which are to be selected must be chosen to have values close to those fully plastic moments occurring in the combined mechanism of Fig. 4.5(*b*) which do not occur in the right-hand beam mechanism.

In the combined mechanism of Fig. 4.5(*b*) plastic hinges occur at five cross-sections where hinges do not occur in the right-hand beam mechanism, namely, sections 1, 2, 4, 9 and 10. Since the values of only four bending moments need be selected, an arbitrary choice of any four of these sections is made, say, sections 1, 4, 9 and 10. The bending moments at these sections are then chosen to have the fully plastic values corresponding to the sense of the hinge rotations in the combined mechanism of Fig. 4.5(*b*), namely,

$$\begin{aligned} M_1 &= -\ M_p = -\ 8{\cdot}57 \text{ tons ft.} \\ M_4 &= -\ 2M_p = -\ 17{\cdot}14 \text{ tons ft.} \\ M_9 &= \ M_p = \ 8{\cdot}57 \text{ tons ft.} \\ M_{10} &= -\ M_p = -\ 8{\cdot}57 \text{ tons ft.} \end{aligned}$$

Substituting these values in equations 4.13, 4.15 and 4.16, the corresponding values of the remaining three unknown bending moments are found to be

$$\begin{aligned} M_2 &= 5{\cdot}72 \text{ tons ft.} \\ M_3 &= 14{\cdot}29 \text{ tons ft.} \\ M_5 &= 0 \end{aligned}$$

None of the seven bending moments found in this way exceeds the corresponding fully plastic value, so that a distribution of bending moment has been found for the entire frame which is not only statically admissible but also safe. This confirms that the right-hand beam mechanism is the actual collapse mechanism, by the uniqueness theorem. It will be realized that there are many other possible safe and statically admissible bending moment distributions corresponding to this partial collapse mechanism, but it is only necessary to establish the existence of one such distribution to verify the solution.

Two-storey, single-bay rectangular frame

The analysis of a two-storey, single-bay rectangular frame will now be outlined in order to illustrate a technique for dealing with distributed loads. The dimensions of the frame and its loading are given in Fig. 4.7(*a*). The load on each member is uniformly distributed over the length of the member, the magnitude of each load being indicated against the dotted arrow which shows the direction in which the load acts. The fully plastic moments of the members are shown in the figure as multiples of a single value M_p, and the problem is to determine the value of M_p such that the frame would just collapse under the given loading.

In order to establish the number of independent mechanisms

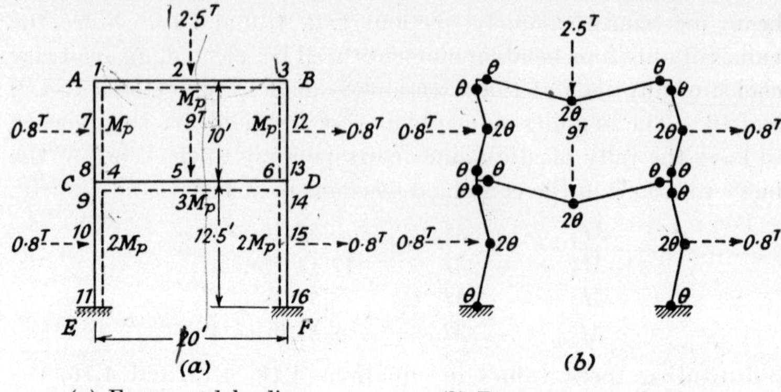

(*a*) Frame and loading.

(*b*) Beam type mechanisms.

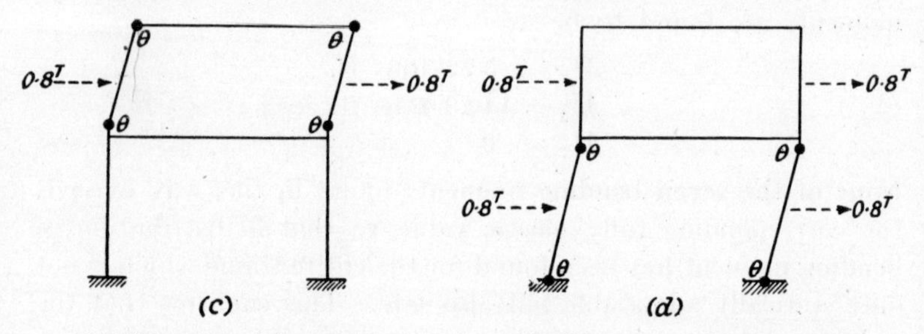

(*c*) Sidesway of top storey.

(*d*) Sidesway of bottom storey.

Fig. 4.7. *Two-storey, single-bay rectangular frame.*

for this frame and loading it is necessary to determine the number of bending moments which are required to specify the distribution of bending moment throughout the frame. In a member which is subjected to a uniformly distributed load, such as *AB* or *CE*, the distribution of bending moment will be parabolic. Since three points are necessary to define a unique parabola, the values of the bending moments at three cross-sections within the span of such a member, say at each end and at mid-span, will be sufficient to determine the complete bending moment distribution in the member. It follows that the number of bending moments n which must be known in order to specify the distribution of bending moment throughout the entire frame is 16, the relevant cross-sections being numbered from 1 to 16 inclusive in Fig. 4.7(*a*).

The number of redundancies r for this frame is 6, as can be seen by imagining the frame to be cut at two sections, for instance, the centres of the beams *AB* and *CD*. If the values of the shear force, thrust, and bending moment were known at each of these two sections the frame would become statically determinate, so that the values of these three quantities at the two sections could be taken as the 6 redundancies. The number of independent equations of equilibrium, and thus the number of independent mechanisms, is thus $(n - r) = 16 - 6 = 10$.

The independent mechanisms can be identified as follows. In the first place there is a beam type mechanism for each of the two beams, and also for each of the four stanchions. These six mechanisms are shown for convenience on the same diagram, Fig. 4.7(*b*). There are also two independent sidesway mechanisms, one for each storey, as shown in Figs. 4.7(*c*) and (*d*). To avoid confusion only those loads which do any work are shown on the figure for each mechanism. As well as the beam and sidesway mechanisms, there are two joint rotations to be counted as independent mechanisms, namely, at the joints C and D, where more than two members meet. The ten independent mechanisms are therefore made up as follows:

6 beam type mechanisms
2 sidesway mechanisms
2 joint rotations

Total = 10 independent mechanisms

In deriving the work equations for the beam type mechanisms it will be assumed in the first place that the plastic hinges within the spans occur at mid-span. In the actual collapse mechanism these plastic hinges may occur anywhere within the spans. However, in the initial stages of the analysis the beam type mechanisms will be combined with the other independent mechanisms while keeping the plastic hinges within the spans at mid-span, and only when it is considered that the actual collapse mechanism has been obtained will an adjustment be made to allow for the incorrect positioning of these hinges. A simple example of this procedure has already been given in Section 3.5 for the frame and loading of Fig. 3.6(*a*). On this basis, the work equations for the six beam type mechanisms are as follows:

$$AB: \quad 12{\cdot}5\theta = 4M_p\theta; \quad M_p = 3{\cdot}13 \text{ tons ft.} \qquad 4.17$$

$$CD: \quad 45\theta = 12M_p\theta; \quad M_p = 3{\cdot}75 \text{ tons ft.} \qquad 4.18$$

$$AC, BD: \quad 2\theta = 4M_p\theta; \quad M_p = 0{\cdot}5 \text{ tons ft.} \qquad 4.19$$

$$CE, DF: \quad 2{\cdot}5\theta = 8M_p\theta; \quad M_p = 0{\cdot}31 \text{ tons ft.} \qquad 4.20$$

The work equations for the two sidesway mechanisms are:

Top storey: $8\theta = 4M_p\theta$; $M_p = 2$ tons ft. . . 4.21

Bottom storey: $30\theta = 8M_p\theta$; $M_p = 3{\cdot}75$ tons ft. . 4.22

The highest value of M_p which is found to correspond to any of these independent mechanisms is 3·75 tons ft., for the beam *CD* and also for the sidesway of the bottom storey.

The beam type mechanisms cannot be combined with one another in such a way as to achieve the cancellation of any hinges, with a possible consequent increase in the corresponding value of M_p. The analysis therefore takes the form of investigating combinations of the beam type mechanisms with the sidesway mechanisms. Of the two independent sidesway mechanisms, the bottom storey mechanism, Fig. 4.7(*d*), gives a value of M_p of 3·75 tons ft., as compared with 2 tons ft. for the top storey mechanism, Fig. 4.7(*c*), and so it is natural to investigate first the possible combinations of the beam type mechanisms with the bottom storey mechanism. The beam type mechanisms for the stanchions have corresponding values of M_p not exceeding 0·5 tons ft., so that their combinations with the bottom storey mechanism are scarcely worth investigating. It is then seen from Fig. 4.7 that the only beam type mechanism which can be combined with the bottom storey mechanism so as to cause the cancellation of hinges is that for the beam *CD*, which has a corresponding value of M_p which is also 3·75 tons ft.

A direct addition of the hinge rotations and displacements of these two mechanisms does not eliminate any hinges, but at the joint *C* there are then hinge rotations of magnitude θ in both

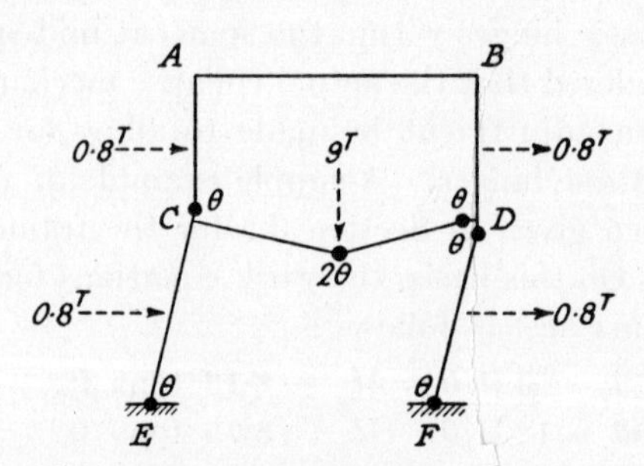

Fig. 4.8. Beam mechanism combined with bottom storey sidesway mechanism.

the beam CD, whose fully plastic moment is $3M_p$, and the stanchion CE whose fully plastic moment is $2M_p$, so that the total work absorbed at this joint is $5M_p\theta$. However, a clockwise rotation of the joint C through an angle θ eliminates these two hinges and replaces them by a single hinge with a rotation θ in the stanchion CA, whose fully plastic moment is M_p, so that the work absorbed at this joint becomes $M_p\theta$. This joint rotation thus reduces the work absorbed at the joint C by $4M_p\theta$. The work equation for this combination, which will be referred to as mechanism (i) and is illustrated in Fig. 4.8, is thus obtained as follows:

Beam CD: $\quad 45\theta = 12M_p\theta$ 4.18

Bottom storey: $\quad 30\theta = 8M_p\theta$ 4.22

Mechanism (i): $\quad 75\theta = 20M_p\theta - 4M_p\theta = 16M_p\theta \quad$ 4.23

$$M_p = 4{\cdot}69 \text{ tons ft.}$$

This value of M_p is higher than the value 3·75 tons ft. which corresponded to both of the mechanisms which were combined.

Another type of combination must now be investigated in which the two sidesway mechanisms are combined. Reference to Figs. 4.7(c) and (d) shows that the two sidesway mechanisms can be combined to form the sidesway mechanism illustrated in Fig. 4.9(a), which will be referred to as mechanism (ii). It will be seen from the figure that in combining these mechanisms the joint at C has been rotated through an angle θ clockwise. If this joint were not rotated, it can be seen from Figs. 4.7(c) and (d) that there would be hinge rotations of magnitude θ in

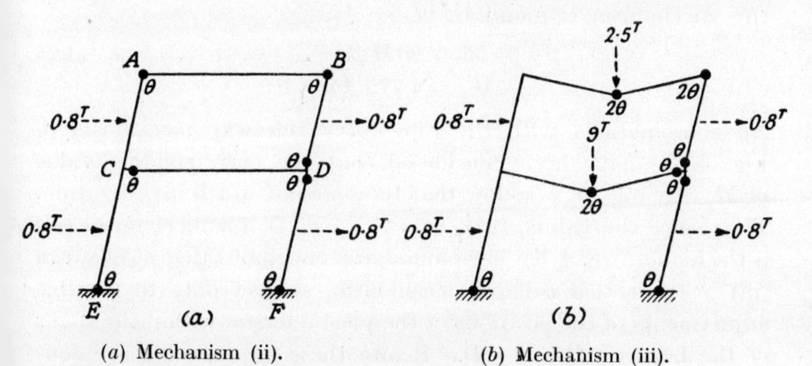

(a) Mechanism (ii). (b) Mechanism (iii).

Fig. 4.9. Combinations of top and bottom storey mechanisms.

the stanchions CA and CE, whose fully plastic moments are M_p and $2M_p$, respectively, so that the total work absorbed at the joint would be $3M_p\theta$. The joint rotation eliminates these hinges, but replaces them by a hinge with a rotation θ in the beam CD, whose fully plastic moment is $3M_p$. The work absorbed at this joint thus remains at the value $3M_p\theta$, so that there is no nett change in the work absorbed due to the joint rotation. The only purpose in rotating the joint is to anticipate the hinge cancellation which will result at the joint C when the beam type mechanism for the beam CD is combined. The joint at D is not rotated, for again a rotation would not alter the nett work absorbed, and a hinge in the beam CD at this joint could not be eliminated by combining the beam type mechanism for CD. Since there is no reduction in the work absorbed due to the combination, the work equation for mechanism (ii) is obtained as follows:

Top storey: $8\theta = 4M_p\theta$ 4.21

Bottom storey: $30\theta = 8M_p\theta$ 4.22

Mechanism (ii): $38\theta = 12M_p\theta$ . . 4.24

$M_p = 3{\cdot}17$ tons ft.

The combinations of the beam type mechanisms with mechanism (ii) can be investigated in the same way as described for the bottom storey sidesway mechanism. Details need not be given here; it is found that the corresponding value of M_p is increased by combining the beam type mechanisms for the beams AB and CD. The resulting mechanism, which is referred to as mechanism (iii), is shown in Fig. 4.9(b). The work equation for this mechanism is found to be

$$95{\cdot}5\theta = 20M_p\theta \quad . \quad . \quad . \quad . \quad 4.25$$

$$M_p = 4{\cdot}775 \text{ tons ft.}$$

No combinations with the top storey sidesway mechanism of Fig. 4.7(c) have been considered, but the corresponding value of M_p of 2 tons ft. is so low that they are not worth investigating. Thus since the value of M_p of $4{\cdot}775$ tons ft. for mechanism (iii) is the highest which has been found, it is concluded that mechanism (iii) is the actual collapse mechanism, subject only to possible adjustments of the positions of the plastic hinges within the spans of the beams AB and CD. Before these adjustments are considered, it is advisable to carry out a statical analysis to confirm that, provided attention is confined to the bending moments at the ends of the members, and at mid-span in the members subjected to uniformly distributed loads, at least one safe and statically admissible distribution of bending moment throughout the frame corresponding to mechanism (iii) can be found.

Statical analysis

The statical analysis can be carried out by writing down the equations of equilibrium corresponding to the independent mechanisms by the Principle of Virtual Work, substituting the values of the fully plastic moments occurring in mechanism (iii), and then solving these equations for the remaining unknown bending moments. Details of this procedure need not be given; the entire frame is statically determinate, and the computations therefore present no difficulty. The resulting bending moment distribution is given in Table 4.2.

TABLE 4.2

Bending moment distribution in two-storey, single-bay rectangular frame at collapse

Section . . .	1	2	3	4	5	6	7	8
Bending moment (tons ft.) . .	1·825	4·775	−4·775	−2·025	14·325	−14·325	3·60	3·375

Section . . .	9	10	11	12	13	14	15	16
Bending moment (tons ft.) . .	1·350	−2·85	−9·55	−1	4·775	−9·55	−1·25	9·55

It will be seen that the bending moments in this table are safe in the sense that the fully plastic moment is not exceeded at any of the cross-sections considered. This confirms the kinematical analysis, in which the effect of varying the positions of the plastic hinges from mid-span in the beams AB and CD was not discussed. However, with the above bending moment distribution the maximum bending moment occurring anywhere within the spans of these beams, and indeed in the other members subjected to uniformly distributed loads, may exceed the fully plastic moment. The values of the maximum bending moments in these members must therefore be calculated.

Maximum bending moments in members

It will be convenient to establish general formulae from which the maximum bending moment in a member subjected to a uniformly distributed load, together with the position where this maximum occurs, can be calculated. Fig. 4.10 shows a typical member of length l which is subjected to a uniformly distributed

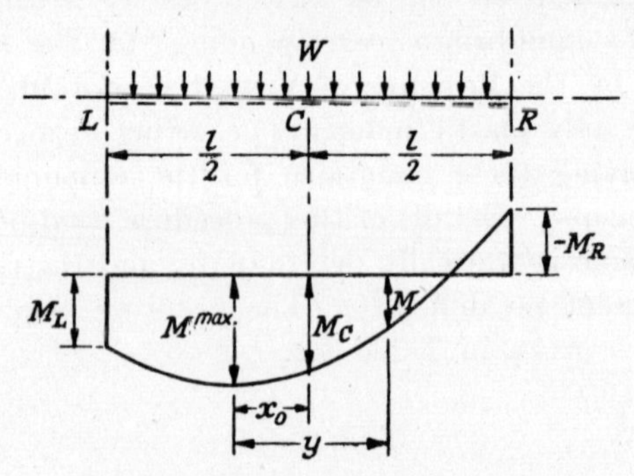

Fig. 4.10. *Bending moment diagram for member carrying a uniformly distributed load.*

load W. The end bending moments are denoted by M_R and M_L at the right-hand and left-hand ends of the beam, respectively, and the central bending moment by M_C, the bending moment diagram being parabolic. The maximum bending moment within the span is $M^{\max}$ occurring at a distance x_0 to the left of the centre of the beam. The bending moment M at a distance y from the section at which this maximum moment occurs is

$$M = M^{\max} - \frac{Wy^2}{2l} \quad . \quad . \quad . \quad 4.26$$

since the shear force must be zero at the position of maximum bending moment. Now when $y = x_0 - \frac{l}{2}$, $M = M_L$, and when $y = x_0 + \frac{l}{2}$, $M = M_R$. Hence:

$$M_L = M^{\max} - \frac{W}{2l}(x_0{}^2 - x_0 l + \tfrac{1}{4}l^2)$$

$$M_R = M^{\max} - \frac{W}{2l}(x_0{}^2 + x_0 l + \tfrac{1}{4}l^2)$$

Solving these equations for x_0 it is found that

$$x_0 = \frac{M_L - M_R}{W} \quad . \quad . \quad . \quad . \quad 4.27$$

Also, putting $M = M_C$ when $y = x_0$ in equation 4.26, it is found that

$$M^{\max} = M_C + \frac{W x_0^2}{2l} \quad . \quad . \quad . \quad 4.28$$

Thus if M_L, M_C and M_R have been calculated, the value of x_0 can be found from equation 4.27 and then the value of $M^{\max}$ follows from equation 4.28. It will be realized that the above analysis is meaningless if the value of x_0 found from equation 4.27 exceeds $\frac{l}{2}$ in magnitude; in this case there will be no position of maximum bending moment within the span, and the bending moment will increase or decrease continuously from one end of the beam to the other.

These results can now be applied to the particular example under consideration, and the calculations are summarized in Table 4.3.

TABLE 4.3

Calculation of maximum bending moments in members

Member	M_L (tons ft.)	M_C (tons ft.)	M_R (tons ft.)	W (tons)	l (ft.)	x_0 (ft.)	$M^{\max}$ (tons ft.)	Fully plastic moment (tons ft.)
AB	1·825	4·775	−4·775	2·5	20	2·64	5·21	4·775
CD	−2·025	14·325	−14·325	9	20	1·37	14·75	14·325
AC	3·375	3·60	1·825	0·8	10	1·94	3·75	4·775
BD	−4·775	−1	4·775	−0·8	10	>5	—	4·775
CE	−9·55	−2·85	1·35	0·8	12·5	$<-6·25$	—	9·55
DF	−9·55	−1·25	9·55	−0·8	12·5	$>6·25$	—	9·55

The last column in Table 4.3 gives the values of the fully plastic moments of the members when $M_p = 4{\cdot}775$ tons ft. It will be seen that $M^{\max}$ only exceeds the fully plastic moment in the beams *AB* and *CD*. The most serious discrepancy occurs in the beam *AB*, where the greatest bending moment is 5·21 tons ft. The fully plastic moment of this beam is defined as M_p, and so it follows that if the value of M_p was 5·21 tons ft. the fully plastic moment would not be exceeded in this member or in the beam

CD, whose fully plastic moment would then be $3M_p = 15{\cdot}63$ tons ft. The distribution of bending moment which has been obtained would therefore be both statically admissible and safe throughout the frame. This value of M_p is thus an upper bound on the required value, by the static theorem, and combining this with the lower bound of 4·775 tons ft. it follows that

$$4{\cdot}775 \text{ tons ft.} < M_p < 5{\cdot}21 \text{ tons ft.}$$

Exact value of M_p

For many purposes, a knowledge of the required value of M_p to within such limits would suffice. To obtain the exact value, a work equation could be written down for mechanism (iii) with the hinges in the beams *AB* and *CD* located at arbitrary distances from the centres of these beams, and the corresponding value of M_p maximized by differentiation. This was the procedure adopted in Section 3.5 for the rectangular portal frame shown in Fig. 3.6(*a*). Such an analysis would be rather tedious if carried out from first principles. However, Horne [8] has given a general analysis of the problem of determining the correct positions of the plastic hinges within the members of multi-storey, multi-bay rectangular frames, together with the corresponding value of M_p. This analysis does not lead to explicit formulae, and for most practical purposes results of sufficient accuracy can be obtained by an approximate method, which is best explained with reference to the particular example under consideration.

It is assumed that the correct positions for the plastic hinges in the spans of the beams *AB* and *CD* are those positions where the maximum bending moment occurs in the bending moment distribution of Table 4.2, these positions being given in Table 4.3. The value of M_p corresponding to this assumption can be found most conveniently by a kinematical analysis. Thus if the work equation is written down for mechanism (iii) but with the central hinges in the beams *AB* and *CD* replaced by hinges at distances 2·64 ft. and 1·37 ft. to the left of mid-span, respectively, the corresponding value of M_p is found to be 4·856 tons ft., which is a lower bound on the required value of M_p. A simple statical check reveals that the maximum bending moment in the beam *AB* in the corresponding bending moment distribution now occurs at a distance 2·47 ft. to the left of mid-span, and the value of this maximum moment is 4·858 tons ft., this being an upper bound on the required value of M_p. The closeness of these bounds is typical, and indicates that in practice the final statical computation is unnecessary.

Thus in general, the following procedure will be found to give results of considerable accuracy:

(i) Determine the "correct" collapse mechanism, assuming that any plastic hinges in those members which carry uniformly distributed loads occur at mid-span.

(ii) Perform a statical check in which the corresponding bending moments at the ends of the members, and at mid-span in the loaded members, are determined.

(iii) Determine the maximum bending moments in the members which carry uniformly distributed loads, and the positions where these maximum moments occur.

(iv) For those members carrying uniformly distributed loads in which plastic hinges occurred at mid-span in the "correct" mechanism, move these hinges to the positions of maximum moment found under (iii), and analyse the resulting mechanism kinematically. The corresponding value of M_p will then be the required value.

Simple pitched roof portal frame

Certain of the independent mechanisms which must be considered in the case of pitched roof portal frames are rather more difficult to analyse kinematically than those encountered when dealing with rectangular frames. To illustrate the technique for dealing with such frames the simple pinned-base portal frame whose dimensions are shown in Fig. 4.11(*a*) will be analysed by the combining mechanisms method. Each member of this frame is subjected to uniformly distributed loads, and the magnitudes of these loads are shown against the dotted arrows in the figure, which indicate their directions. All the members have the same fully plastic moment M_p, and it is required to find the value of M_p such that collapse would just occur under the given loads. This frame, subjected to the same loads but with the feet fixed, was analysed in Section 4.2 by the trial and error method (see Fig. 4.1(*a*)).

For this frame and loading the complete bending moment distribution is specified by the values of the seven bending

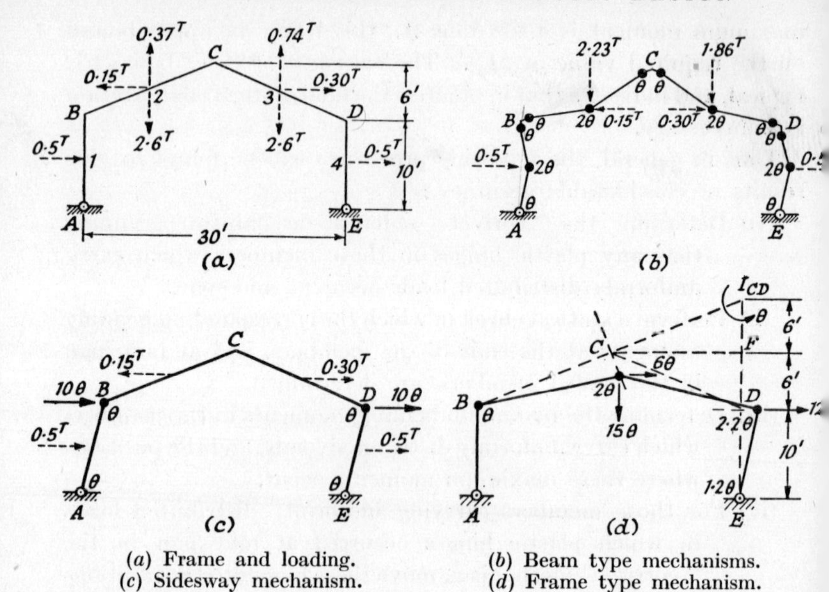

(*a*) Frame and loading. (*b*) Beam type mechanisms.
(*c*) Sidesway mechanism. (*d*) Frame type mechanism.

Fig. 4.11. *Pitched roof portal frame.*

moments at the cross-sections lettered *B*, *C* and *D* and numbered from 1 to 4 in Fig. 4.11(*a*). Since this frame has one redundancy, it follows that there must be six independent equations of equilibrium connecting these seven bending moments, and therefore six independent collapse mechanisms.

Four of these independent mechanisms are readily identified as the beam type mechanisms shown for convenience on the same diagram in Fig. 4.11(*b*). For both of the stanchions *AB* and *DE* the work equation is:

$$1{\cdot}25\theta = 3M_p\theta\,; \quad M_p = 0{\cdot}42 \text{ tons ft.} \qquad 4.29$$

For the left-hand rafter mechanism it is only necessary to note that the horizontal and vertical components of the movement of the centre of the rafter are 3θ ft. to the right and $7{\cdot}5\theta$ ft. downwards. The work equation for this mechanism is thus

$$-\tfrac{1}{2} \times 3\theta \times 0{\cdot}15 + \tfrac{1}{2} \times 7{\cdot}5\theta \times 2{\cdot}23 = 4M_p\theta$$

$$8{\cdot}14\theta = 4M_p\theta\,; \quad M_p = 2{\cdot}04 \text{ tons ft.} \qquad 4.30$$

The work equation for the right-hand rafter mechanism can be shown in a similar manner to be

$$6{\cdot}53\theta = 4M_p\theta\,; \quad M_p = 1{\cdot}63 \text{ tons ft.} \qquad 4.31$$

A fifth independent mechanism is the sidesway mechanism shown in Fig. 4.11(*c*). In this mechanism the whole roof undergoes a horizontal translation of 10θ ft. as shown, the vertical movement being zero. Thus only the horizontal loads shown in the figure do any work, and the work equation is

$$2 \times 0{\cdot}5 \times 5\theta + (0{\cdot}30 - 0{\cdot}15) \times 10\theta = 2M_p\theta$$

$$6{\cdot}5\theta = 2M_p\theta \,; \quad M_p = 3{\cdot}25 \text{ tons ft.} \quad . \quad 4.32$$

The sixth independent mechanism is most conveniently chosen as the frame type mechanism shown in Fig. 4.11(*d*), in which the loads are omitted for the sake of clarity. In this mechanism the rafter member BC rotates about B, so that the apex C is constrained to move in a direction perpendicular to BC, as indicated by the dotted line through C. The stanchion DE rotates about E, so that D is constrained to move horizontally. It follows that the *instantaneous centre of rotation* of the rafter member CD is at I_{CD}, as shown in the figure.

Now suppose that the rafter CD undergoes a small rotation θ about its instantaneous centre. The movement of C in the direction perpendicular to BC is $CI_{CD} \times \theta$, and the rotation of the member BC is thus $\dfrac{CI_{CD} \times \theta}{BC}$, or simply θ, since $BC = CI_{CD}$. The hinge rotation at B is thus θ, and the hinge rotation at C is 2θ, since BC rotates clockwise through an angle θ and CD rotates counterclockwise through the same angle.

The horizontal movement of D is $DI_{CD} \times \theta$, or 12θ ft., so that the clockwise rotation of DE is $\dfrac{12\theta}{DE} = 1{\cdot}2\theta$. Since CD rotates counterclockwise through an angle θ, the hinge rotation at D is $2{\cdot}2\theta$. It will also be seen from the figure that the horizontal and vertical components of the movement of C are $FI_{CD} \times \theta$ and $FC \times \theta$, or 6θ ft. and 15θ ft., respectively. The horizontal and vertical components of the movements of intermediate points on the rafters can be found by linear interpolation. Thus these components are 3θ ft. and $7{\cdot}5\theta$ ft. respectively for the centre of the rafter BC, and 9θ ft. and $7{\cdot}5\theta$ ft. respectively for the centre of the rafter CD.

The work equation for this mechanism can now be written down as follows:

$$2{\cdot}23 \times 7{\cdot}5\theta - 0{\cdot}15 \times 3\theta + 1{\cdot}86 \times 7{\cdot}5\theta + 0{\cdot}30 \times 9\theta + 0{\cdot}5 \times 6\theta = 5{\cdot}2M_p\theta$$

$$35{\cdot}93\theta = 5{\cdot}2M_p\theta \,; \quad M_p = 6{\cdot}91 \text{ tons ft.} \quad . \quad 4{\cdot}33$$

This is the highest value of M_p found to correspond to any of the independent mechanisms. The next highest value was 3·17 tons ft. for the sidesway mechanism, and it is natural to investigate first the possible combination of these two mechanisms to see if a higher value of M_p results. It will be seen from Figs. 4.11(c) and (d) that a direct addition of the displacements and hinge rotations of these two mechanisms results in cancellation

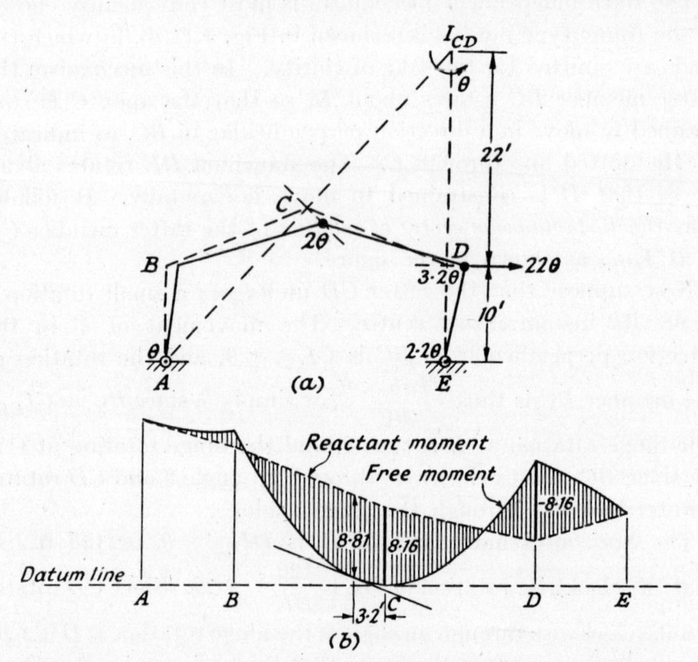

(a) Small motion of mechanism. (b) Bending moment diagram.

Fig. 4.12. *Combined mechanism.*

of the hinge at B, thus reducing the work absorbed in the combined mechanism by $2M_p\theta$. The resulting mechanism is shown in Fig. 4.12(a), and its work equation is obtained as follows:

Sidesway mechanism: $6{\cdot}5\theta = 2M_p\theta$ 4.32

Frame mechanism: $35{\cdot}93\theta = 5{\cdot}2M_p\theta$ . . . 4.33

$$42{\cdot}43\theta = 7{\cdot}2M_p\theta - 2M_p\theta = 5{\cdot}2M_p\theta$$

$$M_p = 8{\cdot}16 \text{ tons ft.}$$

This value of M_p is higher than either of the values corresponding to the mechanisms which were combined.

It will be appreciated that the direct determination of this work equation from a consideration of the kinematics of the combined mechanism of Fig. 4.12(*a*), making use of the fact that the instantaneous centre of rotation of *CD* is now 32 ft. vertically above *E*, would be far more lengthy than the above calculations, which made use of the work equations for the two combining mechanisms. The advantage of using the work equations for the combining mechanisms to obtain the work equation for their combination has not hitherto been so apparent in dealing with cases in which the kinematics of each combining mechanism has been simple.

The highest value of M_p which was found to correspond to any of the beam type mechanisms was 2·04 tons ft. for the left-hand rafter, equation 4.30. The disparity between this value of M_p and the value of 8·16 tons ft. just obtained is such that the investigation of any combinations involving the beam type mechanisms is scarcely necessary. It is therefore concluded that the combined mechanism of Fig. 4.12(*a*) is the actual collapse mechanism, subject to any small adjustments in the positions of the plastic hinges which may be revealed by constructing the bending moment diagram. Details of this construction will not be given here; the result is shown in Fig. 4.12(*b*), where it is seen that the bending moment reaches a maximum value of 8·81 tons ft. at a section 3·2 ft. to the left of the apex *C*. It is a simple matter to calculate the value of M_p corresponding to the mechanism in which the hinge at the apex *C* is replaced by a hinge 3·2 ft. to the left of the apex, and this calculation gives the required value of M_p as 8·42 tons ft.

Two-bay pitched roof portal frame

As a final example of the method of combining mechanisms the two-bay pitched roof portal frame with fixed feet whose dimensions and loading are shown in Fig. 4.13(*a*) will be considered. All the loads which are applied to this frame may be regarded as uniformly distributed along the members, which are all supposed to have the same fully plastic moment of magnitude M_p.

For this frame the number of bending moments n required to specify the complete distribution of bending moment is sixteen, namely, at the ten cross-sections numbered from 1 to 10 in Fig. 4.13(a) and at mid-span in each of the six members subjected to uniformly distributed loads. Since the number of redundancies r

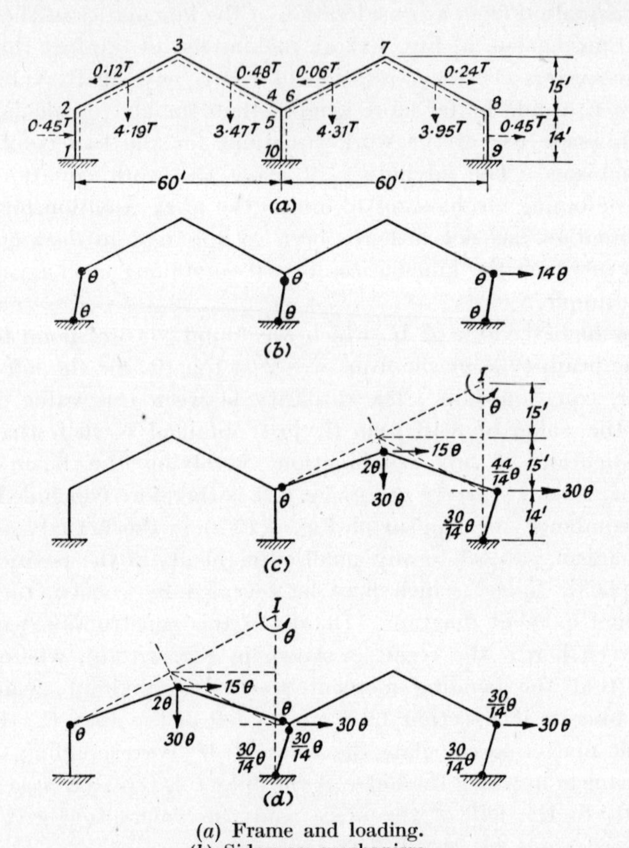

(a) Frame and loading.
(b) Sidesway mechanism.
(c) Collapse of right-hand bay.
(d) Collapse of left-hand bay with sidesway of right-hand bay.

Fig. 4.13. *Two-bay pitched roof portal frame.*

is six, the number of independent mechanisms is $(n - r) = 16 - 6 = 10$. However, six of these ten independent mechanisms may be selected as the beam type mechanisms for the six members subjected to uniformly distributed load, and as seen in the previous example these mechanisms are unlikely to feature in combination in the actual collapse mechanism. There are thus only four independent collapse mechanisms which are likely to be of any importance.

Three of these mechanisms are shown in Figs. 4.13(*b*), (*c*) and (*d*), and the fourth consists simply of a rotation of the central joint 456. The mechanism of Fig. 4.13(*b*) is of the simple sidesway type, and the work equation is

$$2 \times 0{\cdot}45 \times 7\theta + (0{\cdot}48 - 0{\cdot}12 + 0{\cdot}24 - 0{\cdot}06) \times 14\theta = 6M_p\theta$$

$$13{\cdot}86\theta = 6M_p\theta\,;\quad M_p = 2{\cdot}31 \text{ tons ft.} \qquad 4.34$$

The frame type mechanism of Fig. 4.13(*c*) is of a similar type to the mechanism of Fig. 4.11(*d*), and so requires no explanation, the kinematics being as shown in the figure.

The work equation is

$$4{\cdot}31 \times 15\theta - 0{\cdot}06 \times 7{\cdot}5\theta + 3{\cdot}95 \times 15\theta + 0{\cdot}24 \times 22{\cdot}5\theta$$

$$+ 0{\cdot}45 \times 15\theta = \frac{116}{14} M_p\theta$$

$$135{\cdot}6\theta = 8{\cdot}29M_p\theta\,;\quad M_p = 16{\cdot}37 \text{ tons ft.} \qquad 4.35$$

The frame type mechanism of Fig. 4.13(*d*) involves a collapse of the left-hand bay of the frame similar to the collapse of the right-hand bay shown in Fig. 4.13(*c*), with the right-hand bay undergoing a simple sidesway to accommodate the horizontal movement of the central joint. The work equation for this mechanism is

$$4{\cdot}19 \times 15\theta - 0{\cdot}12 \times 7{\cdot}5\theta + 3{\cdot}47 \times 15\theta + 0{\cdot}48 \times 22{\cdot}5\theta$$

$$+ (0{\cdot}24 - 0{\cdot}06) \times 30\theta + 0{\cdot}45 \times 15\theta = \frac{176}{14} M_p\theta$$

$$136{\cdot}95\theta = 12{\cdot}57M_p\theta\,;\quad M_p = 10{\cdot}89 \text{ tons ft.} \qquad 4.36$$

For reference, it may be noted that the highest value of M_p for any of the six beam type mechanisms is that obtained from the collapse of the rafter member 67. The work equation for this mechanism is:

$$32{\cdot}1\theta = 4M_p\theta\,;\quad M_p = 8{\cdot}03 \text{ tons ft.}$$

It will be realized that there are many other ways in which the independent mechanisms could be chosen. The above scheme is advantageous simply because it can be extended without difficulty to the case of a multi-bay pitched roof portal frame. For such a frame it is convenient to select a number of independent mechanisms equal to the number of bays of the frame, each of these mechanisms consisting of a collapse of one of the bays with corresponding sidesways for the rest of the bays, as in Fig. 4.13(*d*). These mechanisms, together with a single sidesway mechanism of the type shown in Fig. 4.13(*b*), and the joint and beam mechanisms, will be found to furnish the required number of independent mechanisms.

The highest value of M_p obtained from the independent mechanisms is 16·37 tons ft. for the frame type mechanism of Fig. 4.13(*c*). Since the sidesway mechanism has a corresponding value of M_p of only 2·31 tons ft., no combinations involving this mechanism need be considered. Thus the only combination which requires investigation is that which can be obtained from the mechanisms of Figs. 4.13(*c*) and (*d*). A direct addition of the displacements and hinge rotations of these two mechanisms does not lead to the cancellation of any hinges. However, such a direct combination would give rise to three hinges at the central joint at sections 4, 5 and 6. It will be seen that a clockwise rotation of this joint would tend to reduce the rotations at the hinges 5 and 6, while increasing the rotation of the hinge at 4, thus leading to a nett reduction in the total hinge rotation at this joint and therefore in the work absorbed. To take full advantage of this possible reduction in the work absorbed at the joint it is necessary to multiply the rotations and displacements involved in the mechanism of Fig. 4.13(*c*) by the factor $\frac{30}{14}$. The hinge rotations at sections 5 and 6 can then both be cancelled by rotating the joint clockwise through an angle $\frac{30}{14}\theta$. The rotation of the hinge at section 4 is thereby increased by $\frac{30}{14}\theta$, leading to a nett reduction in the work absorbed at this joint of $\frac{30}{14}M_p\theta$.

The work equation for the combination is obtained as follows:

Frame mechanism of Fig. 4.13(*c*): $\frac{30}{14} \times 135{\cdot}6\theta = \frac{30}{14} \times 8{\cdot}29M_p\theta$. 4.35

Frame mechanism of Fig. 4.13(*d*): $136{\cdot}95\theta = 12{\cdot}57M_p\theta$ . . 4.36

Combined mechanism: $427{\cdot}52\theta = \left(30{\cdot}33 - \frac{30}{14}\right)M_p\theta$

$$= 28{\cdot}19M_p\theta;$$

$$M_p = 15{\cdot}17 \text{ tons ft.}$$

Since the corresponding value of M_p is less than 16·37 tons ft., it is concluded that the mechanism of Fig. 4.13(*c*) is the actual collapse mechanism, subject to any minor variations which may be revealed in a statical check.

An application of the method of combining mechanisms to a two-bay, three-storey rectangular frame has been described by Neal and Symonds.[2] The same authors have also discussed the design of a three-bay pitched roof portal frame subjected to dead and snow loading only, as well as dead, snow and wind loading.[3] Further applications to other frames, including a saw-tooth portal and a Vierendeel girder, have also been given.[9]

4.4. Plastic moment distribution method

An account will now be given of the plastic moment distribution method developed by Horne,[4] which enables a structure to be *designed* so that a given set of loads will just cause collapse. A similar method was proposed simultaneously by English,[10] who termed his process " relaxation of yield hinges ". The previous methods which have been discussed in this chapter are primarily *analytical*, in that the ratios of the fully plastic moments must be assigned before any calculations are undertaken. Of course, once an analysis has been completed by the trial and error method or the combining mechanisms method, it is a simple matter to carry out adjustments to the sections of some of the members, either to make the required fully plastic moments equal to those of available practical sections or to see whether the total weight of steelwork involved in the frame can be reduced. However, it will be seen that it is not necessary to assign the ratios of the fully plastic moments of the members at the outset when using the plastic moment distribution method; instead, the fully plastic moments are selected in turn towards the end of the calculation. The process is therefore truly one of design.

The plastic moment distribution method involves an essentially statical approach, the object of which is to establish various distributions of bending moment throughout the frame which are in equilibrium with the given loads. If for any such distribution each member is assigned a fully plastic moment equal to the magnitude of the greatest bending moment in the member, the distribution of bending moment would be safe as well as statically admissible. It follows from the static theorem that the frame thus designed would not fail under the given loads. However, unless with this bending moment distribution enough plastic hinges were formed to transform the frame into a mechanism, this design would be capable of carrying higher loads and would therefore be inefficient. The plastic moment distribution method enables adjustments to be made to the bending moment distribution so that all the equilibrium conditions are still preserved. By carrying out such adjustments various designs are investigated, and it is easy to arrive at a design which is efficient in the sense that the bending moment distribution corresponds to the formation of sufficient plastic hinges to transform the structure into a mechanism. If the required fully plastic moments do not correspond to those of available sections, it is also possible to make further adjustments so that at least some of the fully plastic moments can be obtained by the use of such sections.

The computations are carried out in the following three stages :

(i) A set of bending moments is written down which satisfies all the equilibrium requirements except those at the joints.

(ii) The bending moments are adjusted so that all the joints are brought into rotational equilibrium, without violating the other requirements of equilibrium.

(iii) Further adjustments are made to the bending moments in such a way that all the equilibrium requirements are still fulfilled.

Two-bay rectangular portal frame

The method will be explained with reference to the two-bay rectangular portal frame whose dimensions and loading are shown in Fig. 4.14. This frame is to be designed so that it will just collapse when subjected to the loads specified in the figure. For convenience the beams *BD* and *DF* are to be of the same cross-section, and the stanchions *AB*, *HD*, and *GF* are also to be of identical cross-section, although this latter cross-section need not be the same as for the beams.

Since the basis of the whole procedure is the simultaneous satisfaction of the equations of equilibrium, the first step will be to derive these equations. In doing this a sign convention for the bending moments will be adopted which is the same as the convention used in the elastic moment distribution method, as far as the joints are concerned. Thus a clockwise bending moment acting on the end of any member is regarded as positive. In addition to this, a sagging bending moment acting at the centre of a beam is regarded as positive. The effect of employing this convention is that bending moments are regarded as positive if they cause tension in the fibres of the members adjacent to the dotted line segments in Fig. 4.14. This convention represents a change from the convention adopted previously, as can be seen by comparing this figure with Fig. 4.4(*a*), relating to a similar problem. The reason for the change is, of course, that the new convention facilitates the balancing of joints, for with this convention the

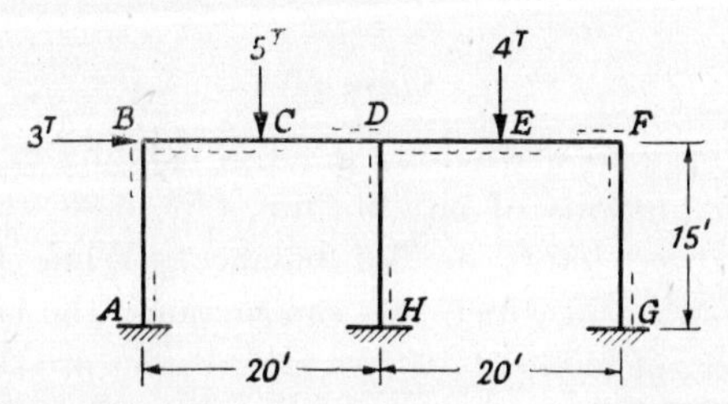

Fig. 4.14. Two-bay rectangular portal frame.

requirement for rotational equilibrium at any joint is simply that the algebraic sum of all the moments acting at the joint should be zero.

The notation for bending moments will be similar to the notation which is often employed for elastic moment distribution. Thus M_{BC} denotes the bending moment in the beam BCD at the joint B, the first suffix indicating the joint at which the bending moment acts and the suffixes taken together indicating the member on which the moment acts. M_{BA} therefore denotes the moment at joint B acting on the stanchion BA. Finally, the moments at the centres of the beams BCD and DEF are denoted by M_C and M_E, respectively.

Using the above notation and sign convention, the equations of equilibrium may be derived by ordinary statical methods or more readily by applying the Principle of Virtual Work to the various independent mechanisms. These are four in number, as was shown in Section 4.3 for the frame of Fig. 4.4(*a*), namely, two beam type mechanisms, one for each beam, one sidesway mechanism, and a rotation of the joint D. Three of the equilibrium equations derived in this way are as follows:

$$50 = -M_{BC} + 2M_C + M_{DC} \quad . \quad . \quad . \quad . \quad . \quad 4.37$$

$$40 = -M_{DE} + 2M_E + M_{FE} \quad . \quad . \quad . \quad . \quad . \quad 4.38$$

$$45 = -M_{AB} - M_{BA} - M_{HD} - M_{DH} - M_{GF} - M_{FG} \,. \quad 4.39$$

The equation of equilibrium for the joint D need not be written down, since it follows from the sign convention which has been adopted that the algebraic sum of the three moments acting at this joint must be zero. A similar condition must be fulfilled at the joints B and D, where in the early stages of the analysis the moments at these joints in the beams and the stanchions may not be equal.

Stage (i)

The first step is to write down a set of bending moments which satisfies the equations of equilibrium 4.37, 4.38 and 4.39 while leaving the three joints out of balance. While there are, of course, many ways in which this can be done, the best procedure is to use the set of moments whose magnitudes are those obtained in analysing the independent collapse mechanisms corresponding to these equations. Thus considering for example equation 4.37, which corresponds to the independent collapse of the beam BCD, it is seen that if $M_{BC} = 12{\cdot}5$, $M_C = 12{\cdot}5$ and $M_{DC} = 12{\cdot}5$ tons ft., the equation is satisfied and all the moments occurring in the beam BCD are numerically equal. 12·5 tons ft. is, in fact, the value of the fully plastic moment which would be obtained from the work equation for the independent collapse of this beam, provided of course that the fully plastic moment of the stanchion AB was at least equal to this value.

By a similar procedure equations 4.38 and 4.39 yield the following sets of moments:

$$-M_{DE} = M_E = M_{FE} = 10 \text{ tons ft.}$$

$$M_{AB} = M_{BA} = M_{HD} = M_{DH} = M_{GF} = M_{FG} = -7{\cdot}5 \text{ tons ft.}$$

Stage (ii)

The set of bending moments derived in stage (i) is such that the joints are not in rotational equilibrium. For instance, at joint B the moments are $M_{BA} = -7{\cdot}5$ tons ft. and $M_{BC} = -12{\cdot}5$ tons ft., so that $M_{BA} + M_{BC} = -20$ tons ft., whereas for rotational equilibrium the sum of these moments should be zero. In order to obtain a statically admissible distribution of moments it is therefore necessary to make adjustments in the moments such that the joints are brought into rotational equilibrium while the three equations of equilibrium 4.37, 4.38 and 4.39 are still satisfied.

The conditions which changes in the bending moments must fulfil in order that these equations should remain satisfied can be derived immediately. Thus in equation 4.37 it is evident that if changes of ΔM_{BC}, ΔM_C and ΔM_{DC} are made in the values of a set of bending moments which satisfy this equation, it will continue to be satisfied provided that

$$-\Delta M_{BC} + 2\Delta M_C + \Delta M_{DC} = 0.$$

A similar condition governs the permissible changes in the end and central bending moments for any beam subjected to any type of loading. The information contained in this equation, as applied to any beam and loading, is represented for convenience in Table 4.4, which shows the corresponding changes which can be made in any pair of moments simultaneously.

TABLE 4.4

Permissible changes in beam moments

Left-hand moment	Central moment	Right-hand moment
+1	0	+1
+1	$+\frac{1}{2}$	0
0	$-\frac{1}{2}$	+1

It remains to establish the condition which changes in the six moments at the tops and the feet of the stanchions must fulfil in order that equation 4.39 shall continue to be satisfied. This condition can be seen by inspection to be:

$$\Delta M_{AB} + \Delta M_{BA} + \Delta M_{HD} + \Delta M_{DH} + \Delta M_{GF} + \Delta M_{FG} = 0,$$

so that the sum of the changes in these six moments must be zero.

The second stage of the calculations can now be performed. The numerical work can be tabulated in a manner similar to that often employed for elastic moment distribution, as shown in Table 4.5. In this table there is a column for each moment, the moments being denoted by BA instead of M_{BA}, C instead of M_C, and so on. The columns are arranged so that all the moments at a joint are grouped together, the moments in each beam are arranged in horizontal sequence, and the moment at the foot of a stanchion appears beneath the moment at its top.

TABLE 4.5

Plastic moment distribution for minimum beam section

	BA	*BC*	*C*	*DC*	*DH*	*DE*	*E*	*FE*	*FG*	*Sway*
a	−7·5	−12·5	+12·5	+12·5	−7·5	−10	+10	+10	−7·5	
b	+20	—	—	—	+5	—	—	—	−2·5	+22·5
c	+12·5	−12·5	+12·5	+12·5	−2·5	−10	+10	+10	−10	
d	—	—	—	—	−5	+5	+2·5 −1·25	+2·5	−2·5	−7·5
e	+12·5	−12·5	+12·5	+12·5	−7·5	−5	+11·25	+12·5	−12·5	

	AB	*HD*	*GF*	*Sway*
f	−7·5	−7·5	−7·5	
g	−7·5	−7·5	−7·5	−22·5
h	−15	−15	−15	
i	+2·5	+2·5	+2·5	+7·5
j	−12·5	−12·5	−12·5	

The first rows *a* and *f* of Table 4.5 show the set of moments which satisfy the two equations of beam equilibrium and the single sidesway equilibrium equation, which were derived in stage (i). It will be seen that the out of balance moments at the joints *B*, *D* and *F* at this stage are −20, −5 and 2·5 tons ft., respectively. The joints can be balanced by applying equal and opposite moments of 20, 5 and −2·5 tons ft. respectively, but the way in which these balancing moments are divided between the various members meeting at each joint is quite arbitrary. This is in sharp contrast with elastic moment distribution, in which joint balancing moments are divided between members in proportion to their stiffness. It is helpful to have some guiding principle at this stage, and it is advantageous to investigate two extreme possibilities, the balancing moments being in the one case applied solely to the stanchions and in the other case solely to the beams.[4] The purpose of carrying out these two sets of calculations is as follows. It will be appreciated that the fully plastic moment required for the beams cannot be reduced below the value corresponding to the independent collapse mechanism for either beam, so that the minimum fully plastic moment for the beams is 12·5 tons ft., the larger of the two values recorded in row *a* of the table. By distributing the balancing moments to the stanchions, and thus leaving the moments in the beam *BCD* unchanged, the fully plastic moment required for the stanchions can be found, and in this design the fully plastic moments for the beams and stanchions will therefore have their smallest and largest possible values, respectively. If the procedure is reversed the stanchions can be assigned the least possible fully plastic moment, namely 7·5 tons ft., and in the corresponding design the beams will then have the greatest possible fully plastic moment. Having bracketed the required fully plastic moments in this way, the fully plastic moment of an available section which lies between the known limits can be assigned to either the beams or the stanchions and the required fully plastic moment of the other members deduced.

In the first place the balancing moments will be applied solely to the stanchions. Thus in row *b* of the table the joints *B*, *D* and *F* are balanced by moments of 20, 5 and $-2{\cdot}5$ tons ft., respectively, which in each case are applied wholly to the stanchions, namely *BA*, *DH* and *FG*. Since none of the beam moments are altered by this balancing, only the sidesway equation of equilibrium has been disturbed, and it will be recalled that if this equation is to remain satisfied the sum of the changes in the moments at the top and bottom of each of the three stanchions must be zero. The sum of the changes at the tops of these three stanchions is $20 + 5 - 2{\cdot}5 = +22{\cdot}5$, as recorded in row *b* of the last column of the table, which is headed *sway*. It is therefore necessary to make changes in the moments at the feet of these three members which total $-22{\cdot}5$, and since the three stanchions are to be of equal section, equal changes of $-7{\cdot}5$ tons ft. are made at these feet, as shown in row *g*. The sum of these changes, $-22{\cdot}5$ tons ft., is recorded in the sway column, in which the two entries $+22{\cdot}5$ and $-22{\cdot}5$ tons ft. add up to zero, as required. Finally, in rows *c* and *h* the changes of moment are added to the original moments to give the resultant moments. Since the joint, beam, and sidesway equilibrium conditions are all fulfilled by these resultant moments, a statically admissible solution has been found. Examination of this statically admissible solution shows that if the beams were assigned a section with a fully plastic moment of 12·5 tons ft., and the stanchions each had fully plastic moments of 15 tons ft., the fully plastic moment would not be exceeded anywhere in the frame, and so the solution would be safe as well as statically admissible. However, the bending moment distribution corresponds to a collapse condition in the beam *BCD*, for the bending moments in this beam have not been altered from those originally assigned in accordance with the independent beam collapse mechanism. Thus a possible design has been derived with fully plastic moments for the beams and stanchions of 12·5 and 15 tons ft. respectively, the collapse mechanism being partial collapse of the beam *BCD* only.

Stage (iii)

While this design is safe, it is not efficient, for it is found that the fully plastic moment of the stanchions can be reduced without requiring an increase in the fully plastic moment of the beams. Clearly the fully plastic moment of the stanchions cannot be reduced below 12·5 tons ft., for the bending moment at the joint *B* is required by the beam collapse mechanism of the beam *BCD* to be equal to this value. If the stanchion *AB* had a fully plastic moment of less than 12·5 tons ft., the joint at *B* would only be able to transmit a moment equal to the fully plastic moment of *AB*. The next step in the calculations is therefore to see whether the fully plastic moment of the stanchions can be reduced to 12·5 tons ft. Accordingly, in row *i* of the table moments of +2·5 tons ft. are added to the moments −15 tons ft. at the feet of the three stanchions, to reduce the maximum moments at these feet to −12·5 tons ft. The sway column records a change in this row of +7·5 tons ft., and so a total change of −7·5 tons ft. must be made at the tops of these stanchions, as shown in the sway column, row *d*. Of this change, −2·5 tons ft. is applied to *FG*, so that this moment is brought up to −12·5 tons ft., and the remaining −5 tons ft. is applied at *DH*, so as to leave the moment at the joint *B* unaltered. The balancing moments of +5 tons ft. and +2·5 tons ft. are applied to *DE* and

FE, respectively, and to preserve equilibrium of the beam *DEF* moments of $+\frac{1}{2} \times 5 = +2{\cdot}5$ tons ft. and $-\frac{1}{2} \times 2{\cdot}5 = -1{\cdot}25$ tons ft. are recorded at *E*, in accordance with Table 4.4. The changes in the moments have now been made to fulfil all the equilibrium requirements, and rows *e* and *j* show the resulting statically admissible set of moments. It will be seen that the moments in the beams still do not exceed the value 12·5 tons ft., and that the moments in the stanchions also do not exceed this value. Thus the best possible design, subject to the condition that the beams have the minimum possible fully plastic moment of 12·5 tons ft., is with the stanchions having this same fully plastic moment.

The other possible extreme design will now be investigated, in which the fully plastic moment of the stanchions is kept at the minimum possible value of 7·5 tons ft., and the corresponding required fully plastic moment for the beams is determined. The computations are set out in Table 4.6. In this table the first rows *a* and *f* are identical with those given in Table 4.5. In the second row *b* the joints are balanced by moments applied to the

TABLE 4.6

Plastic moment distribution for minimum stanchion section

	BA	*BC*	*C*	*DC*	*DH*	*DE*	*E*	*FE*	*FG*	*Sway*
a	−7·5	−12·5	+12·5	+12·5	−7·5	−10	+10	+10	−7·5	
b	—	+20	+10 −2·5	+5	—	—	+1·25	−2·5	—	—
c	−7·5	+7·5	+20	+17·5	−7·5	−10	+11·25	+7·5	−7·5	
d	—	—	−0·83	+1·67	—	−1·67	−0·83	—	—	—
e	−7·5	+7·5	+19·17	+19·17	−7·5	−11·67	+10·42	+7·5	−7·5	

	AB	*HD*	*GF*	*Sway*
f	−7·5	−7·5	−7·5	
g	—	—	—	—
h	−7·5	−7·5	−7·5	
i	—	—	—	—
	−7·5	−7·5	−7·5	

beams at *BC*, *DC* and *FE*, and to preserve equilibrium of the beams the appropriate moments are entered at *C* and *E* in accordance with the scheme of Table 4.4. There is of course no sway change to be carried down to the feet of the stanchions. The resultant statically admissible set of moments is shown in rows *c* and *h*. It will be seen that the largest beam moments occur in the beam *BCD*, namely +20 tons ft. at *C* and +17·5 tons ft. at *DC*, the moment at *BC* remaining at +7·5 tons ft. as demanded by the fully plastic moment of the stanchion *AB*. To determine the smallest required fully plastic moment for the beams the moments at *C* and *DC* are now made numerically equal to 19·17 tons ft. in row *d* by applying moments of −0·83 tons ft. and +1·67 tons ft. at *C* and *DC* respectively, these moment changes being made in such a way as to preserve the equilibrium of the beam *BCD* in accordance with Table 4.4. The joint *D* is then balanced by a moment of −1·67 tons ft. at *DE*, and equilibrium of the beam *DEF* is preserved by a moment of −0·83 tons ft. at *E*. The resulting moments are shown in rows *e* and *j*. It follows that if the stanchions have their minimum possible fully plastic moment of 7·5 tons ft., the beams must have a fully plastic moment of 19·17 tons ft.

The fully plastic moments of the members in tons ft. in the two extreme possible designs are thus as follows:

Beams	12·5	19·17
Stanchions	12·5	7·5

Suppose now that a convenient practical section exists with a fully plastic moment of 16·4 tons ft., lying within the range of required fully plastic moments for the beams. The beams are then given this fully plastic moment, and the corresponding required fully plastic moment for the stanchions is to be determined. The calculations are given in Table 4.7. The first rows *a* and *f* in this table are again identical with those of Table 4.5. From the two previous solutions it is evident that the fully plastic moment will be attained at *C* and *DC*, and these moments are accordingly adjusted to the value 16·4 tons ft., as shown in row *b*. The appropriate moments are carried over to *BC* to preserve equilibrium of the beam *BCD*, and all the joints are then balanced. The sway moment change of +8·3 tons ft. shown in row *b* is carried down as −8·3 tons ft. to row *g*, to preserve side-

TABLE 4.7

Plastic moment distribution for fully plastic moment of beams of 16·4 tons ft.

	BA	*BC*	*C*	*DC*	*DH*	*DE*	*E*	*FE*	*FG*	*Sway*
a	−7·5	−12·5	+12·5	+12·5	−7·5	−10	+10	+10	−7·5	
		+3·9	+3·9	+3·9		+1·1	+1·25	−2·5		
b	+8·3	+7·8			—		+0·55		—	+8·3
c	+0·8	−0·8	+16·4	+16·4	−7·5	−8·9	+11·8	+7·5	−7·5	
	—	—	—	—	−1·66	+1·66	+0·83	+1·66	−1·66	−3·32
d							−0·83			
e	+0·8	−0·8	+16·4	+16·4	−9·16	−7·24	+11·8	+9·16	−9·16	

	AB	*HD*	*GF*	*Sway*
f	−7·5	−7·5	−7·5	
g	−2·77	−2·77	−2·77	−8·3
h	−10·27	−10·27	−10·27	
i	+1·11	+1·11	+1·11	+3·33
j	−9·16	−9·16	−9·16	

sway equilibrium, and three equal moment changes of −2·77 tons ft. are made at the feet of the stanchions. The resulting statically admissible moments are shown in rows *c* and *h*.

The greatest moments in the stanchions are of magnitude 10·27 tons ft. at the feet, but the moments at *DH* and *FG* are only −7·5 tons ft., and the moment at *BA* is 0·8 tons ft. It is impossible to alter the moment at *BA* without altering the moments in the beam *BCD*, but for economy of design the moments at *DH* and *FG*, and at the feet of the three stanchions, should be equalized if possible. This can in fact be done, and the appropriate moment changes are given in rows *d* and *i*, leading to the statically admissible set of moments in rows *e* and *j*, in which the moments at *DH*, *FG*, *AB*, *HD* and *GF* are all −9·16 tons ft., while the moments in the beam *DEF* are all of magnitude less than 16·4 tons ft. Thus the optimum design if the beams are to have a fully plastic moment 16·4 tons ft. is obtained when the stanchions have a fully plastic moment of 9·16 tons ft. The corresponding collapse mechanism is shown in Fig. 4.15.

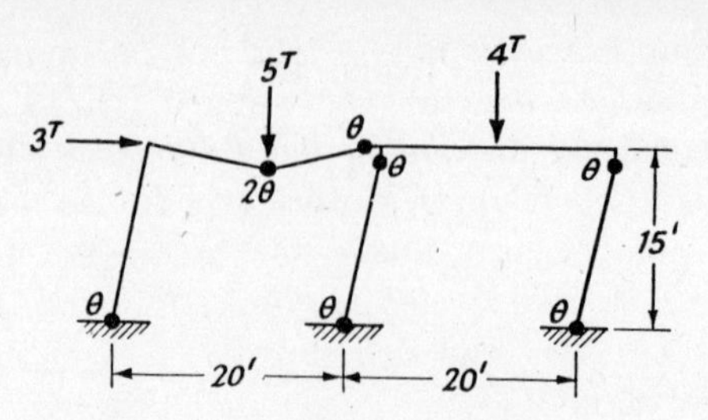

Fig. 4.15. *Collapse mechanism for two-bay rectangular frame.*

The plastic moment distribution method is particularly suited to rectangular frames. Several applications to such frames have been given by Horne,[4] and also by Baker and Horne,[11] who compared the basis of the method with that of the method of combining mechanisms. English [10] has also described the treatment of rectangular frames by his method of relaxation of yield hinges. However, when non-rectangular frames such as pitched roof portal frames are considered, the conditions which changes in the bending moments must fulfil in order that equilibrium should be preserved become more complicated. This point was considered by Horne [4] in replying to the discussion on his paper.

4.5. Other methods for determining plastic collapse loads

Several methods for the determination of plastic collapse loads other than the three methods which have already been described in this chapter have been developed. Details of these other methods will not be given, but a brief account of the basis of each method and of its limitations may be of interest.

The first attempt to deal with the problems associated with the design of large frames appears to have been made by Girkmann.[12, 13] The basic idea of his method was to construct a statically admissible bending moment distribution for the frame and loading under consideration, and then to assign to each member a fully plastic moment equal to the magnitude of the greatest bending moment in the member. In essence, therefore, his method was similar to the plastic moment distribution method, although the technique was based on the construction of bending moment diagrams rather than a tabular presentation. However, Girkmann only considered rectangular frames, and was content merely to adjust his distribution of bending moment until the maximum sagging and hogging bending moments in each beam were equal. In general, this implied that collapse would not occur under the given loads, so that the designs were inefficient in the sense that they would be capable of carrying loads in excess of the given values, by the static and uniqueness theorems.

Another method, involving an essentially statical approach, was developed by Neal and Symonds.[14] This method is based on the solution of systems of linear inequalities by the method of Dines.[15] The condition that the bending moment at any cross-section must lie within the limits of the positive and negative fully plastic moments can be expressed as a pair of inequalities for each cross-section considered. These inequalities are set up for the bending moment at each cross-section at which a plastic hinge could form. If there are n such cross-sections there will be $2n$ inequalities of this kind involving the values of the n bending moments. Since there are $(n - r)$ equations of equilibrium, r being the number of redundancies, it is possible to express $(n - r)$ of the bending moments in terms of the other r moments and the values of the external loads. This results in the formation of $2n$ inequalities involving only r bending moments as variables. Since the equations of equilibrium are linear in the moments and the applied loads, these inequalities are all linear.

The r bending moments can be eliminated in turn from the inequalities, and when these eliminations have been performed a number of inequalities on the value of W remain, it being supposed that each load is expressed as a multiple of a parameter W. Each of these inequalities sets an upper bound on the value of W, and the smallest of these upper bounds is the collapse load W_c. Although this method is completely systematic, the computations involved are extremely lengthy, and its only merit is that it might lend itself to machine computation.

The approach by means of upper and lower bounds on the collapse load which has been developed by Greenberg and Prager [16] was mentioned in Section 3.3. The procedure suggested is to assume a mechanism of collapse, and from a work equation deduce the corresponding value of the load W. From the kinematic theorem, this establishes an upper bound on the value of W_c. If the assumed collapse mechanism is of the *complete* type, the corresponding bending moment distribution can be determined throughout the structure from purely statical considerations, and the highest load at which this statically admissible bending moment distribution is also safe constitutes a lower bound on the collapse load, from the static theorem. Unless the upper and lower bounds coincide, the procedure is repeated successively with different assumed collapse mechanisms until coincidence is obtained. The disadvantage of this method is that collapse mechanisms of the *partial* type, in which the entire frame is not statically determinate at collapse, cannot be investigated very easily.

Heyman and Nachbar [17] attempted to overcome this defect in developing another upper and lower bound method, which would lend itself to systematic calculations for large frames. In this method, the lower bound is obtained in the following way. The frame is imagined to be cut at several cross-sections so that the resulting component parts of the frame are either statically determinate or redundant. Those component parts which are redundant are made sufficiently simple to enable their collapse loads under the given loading to be determined very easily. The hypothetical "collapse loads" W_c^* for each of the component parts are then determined. The actual value of W_c for collapse of the whole frame cannot then be less than the smallest collapse load thus obtained, which will be denoted by W_L. This follows at once from the static principle, for if the set of loads corresponding to W_L were applied to the whole frame, it would be known that at least one distribution of bending moments could be found which would not exceed the fully plastic moment anywhere in the frame. Such a distribution would be the one in which the distribution for each of the component parts at its collapse load W_c^* was scaled down in the ratio of W_L to W_c^*.

A lower bound W_L on W_c may therefore be obtained in this way. In order to improve the lower bound, the redundant reactions and bending moments are introduced at the cuts. These redundancies are then adjusted systematically so as to increase W_L, the lowest value of the collapse loads W_c^* for each of the component parts of the frame.

Simultaneously, an upper bound on W_c is determined, and the value of this upper bound is obtained in the first instance by inserting plastic hinges at every possible position in the frame, thus producing a mechanism with several degrees of freedom. The rotation at every plastic hinge in the mechanism will be determined in terms of the rotations of a number of plastic hinges which is equal to the number of degrees of freedom of the mechanism. When the work equation is written down for this mechanism, an expression for W will thus be obtained which contains several independent parameters, which are the rotations at those plastic hinges which have been chosen as specifying the mechanism motion. Any arbitrary choice of these plastic hinge rotations which is made in the first instance will lead to a value of W which is an upper bound on W_c. Heyman and Nachbar gave a systematic technique for reducing the value of this upper bound. As this technique is applied, the number of degrees of freedom of the mechanism is successively reduced as the rotations at the various plastic hinges which were originally taken as independent become known in relation to one another.

The calculations for the improvement of the upper and lower bounds are carried out simultaneously, and the calculations cease when these bounds are close enough together for the collapse load to be quoted to the desired degree of accuracy.

References

1. J. F. Baker. The design of steel frames. *Struct. Engr.*, **27**, 397 (1949).
2. B. G. Neal and P. S. Symonds. The rapid calculation of the plastic collapse load for a framed structure. *Proc. Instn. Civ. Engrs.*, **1**, (Part 3), 58 (1952).
3. B. G. Neal and P. S. Symonds. The calculation of plastic collapse loads for plane frames. *Prelim. Pubn. 4th Congr. Intern. Assn. Bridge and Struct. Engng.*, 75, Cambridge (1952). Reprinted in *Engineer*, **194**, 315, 363 (1952).
4. M. R. Horne. A moment distribution method for the analysis and design of structures by the plastic theory. *Proc. Instn. Civ. Engrs.*, **3**, (Part 3), 51 (1954).
5. R. A. Foulkes. A comparison between elastic and plastic designs of pitched roof portal frames. *Brit. Weld. Res. Assn. Rep.* FE.1/27A (1951).
6. *The Collapse Method of Design.* British Constructional Steelwork Association Publication No. 5, 1952.
7. A. W. Hendry. Plastic analysis and design of mild steel Vierendeel girders. *Struct. Engr.*, **33**, 213 (1955).
8. M. R. Horne. Collapse load factor of a rigid frame structure. *Engineering*, **177**, 210 (1954).

9. P. S. SYMONDS and B. G. NEAL. Recent progress in the plastic methods of structural analysis. *J. Franklin Inst.*, **252**, 383, 469 (1951).
10. J. M. ENGLISH. Design of frames by relaxation of yield hinges. *Trans. Amer. Soc. Civ. Engrs.*, **119**, 1143 (1954).
11. J. F. BAKER and M. R. HORNE. New methods in the analysis and design of structures in the plastic range. *Brit. Weld. J.*, **1**, 307 (1954).
12. K. GIRKMANN. Bemessung von Rahmentragwerken unter Zugrundelegung eines ideal-plastischen Stahles. *S.B. Akad. Wiss. Wien* (*Abt. IIa*), **140,** 679 (1931).
13. K. GIRKMANN. Über die Auswirkung der " Selbsthilfe " des Baustahls in rahmenartigen Stabwerken. *Stahlbau,* **5,** 121 (1932).
14. B. G. NEAL and P. S. SYMONDS. The calculation of collapse loads for framed structures. *J. Instn. Civ. Engrs.*, **35,** 21 (1950–51).
15. L. L. DINES. Systems of linear inequalities. *Ann. Math. Princeton* (*Series* 2), **20,** 191 (1918–19).
16. H. J. GREENBERG and W. PRAGER. On limit design of beams and frames. *Trans. Amer. Soc. Civ. Engrs.*, **117,** 447 (1952).
17. J. HEYMAN and W. NACHBAR. Approximate methods in the limit design of structures. *Proc. 1st U.S. Natl. Congr. Appl. Mech.*, 551 (1952).

Examples

Note : *It may be assumed in the following examples that unless otherwise stated all the joints in the frames are capable of developing the fully plastic moment.*

1. For the pitched roof portal frame of Fig. 4.1(a) it was shown in Section 4.2 that in the actual mechanism of collapse there are plastic hinges at the sections 2, 4 and 5, and also a plastic hinge in the left-hand rafter at a distance of about 2·5 ft. from the apex. By means of a statical analysis, write down an expression for the required value of M_p in terms of the distance x ft. of this hinge from the apex, and hence deduce the correct values for x and M_p .

2. Find the required value of M_p for the fixed-base pitched roof portal frame of Fig. 4.1(a) for the case in which only the vertical dead and snow loads of 2·6 tons uniformly distributed on each rafter are acting. Find also the required value of M_p for this frame and loading if both feet are pinned.

3. The fixed-base pitched roof portal frame shown in Fig. 4.1(a) is subjected to the same vertical dead and snow loading on the rafters, but each of the wind loads is increased by 30%. Find the value of M_p such that collapse would just occur.

4. A fixed-base pitched roof portal frame *ABCDE* is composed of two stanchions *AB* and *ED*, each of length 15 ft. and with the feet *A* and *E* 40 ft. apart, together with two rafter members *BC* and *DC* of equal length which are inclined at 15° to the horizontal. All the members of the frame are to be of the same fully plastic moment M_p. If the roof is subjected to a uniformly distributed vertical load 7 tons, find the required value of M_p such that collapse would just occur. Find also the required value of M_p if in addition there is a uniformly distributed horizontal wind load 1·5 tons acting on the stanchion *AB* in the direction *AE*.

5. A pinned-base saw-tooth portal frame *ABCDE* has two stanchions *AB* and *ED*, each of length 12 ft. and with the feet *A* and *E* 26 ft. apart. The rafter members *BC* and *CD* are of length 24 ft. and 10 ft., respectively. All the members of the frame are to be of the same fully plastic moment M_p. If the rafter *BC* is subjected to a uniformly distributed vertical load 4 tons, find the required value of M_p such that collapse would just occur.

6. A lean-to fixed-base frame *ABCD* consists of two stanchions *AB* and *DC*, whose lengths are 10 ft. and 13 ft. respectively, the feet *A* and *D* being 16 ft. apart, together with a rafter *BC*. All the members of the frame are to be of the same fully plastic moment M_p. If the rafter *BC* carries a uniformly distributed vertical load 5 tons, find the required value of M_p such that collapse would just occur. If in addition the stanchion *AB* is subjected to a uniformly distributed horizontal wind load 1·5 tons in the direction *AD*, and the rafter *BC* is subjected to a uniformly distributed wind suction 0·25 tons in the direction perpendicular to *BC*, find the required value of M_p.

7. A fixed-base pitched roof portal frame *ABCDE* has two stanchions *AB* and *ED* each of height l, the feet *A* and *E* being a distance $2l$ apart. The rafter members *BC* and *CD* are each of the same length and are inclined at 30° to the horizontal. All the members of the frame are to have the same fully plastic moment M_p. The stanchions *AB* and *ED* are each subjected to a horizontal load W in the direction *AE*, and the rafters *BC* and *CD* are subjected to vertical loads $3W$ and $4W$, respectively, each of the four loads being concentrated at the centre of the member on which it acts. Find the value of W which would cause collapse.

8. In a symmetrical pinned-base portal frame *ABCDEFGH* the stanchions *AB* and *GH* are each of height 18 ft., their bases *A* and *H* being 40 ft. apart. The rafters *BC* and *FG* are each inclined at 15° to the horizontal. *CDEF* is a rectangular monitor frame. The vertical members CD and FE are of length 6 ft., the length of the horizontal member DE being 12 ft. The rafters BC and FG are each subjected to a uniformly distributed vertical load 4 tons, while the monitor beam DE is subjected to a uniformly distributed vertical load 5 tons. If all the members of the frame have the same fully plastic moment M_p, find the value of M_p such that collapse would just occur. Show that within certain limits the required value of M_p is unaltered if the length of the vertical members CD and FE is changed, and find these limits.

9. In a two-bay fixed-base rectangular frame $ABCDEF$ the three stanchions AB, FC, and ED are each of length 15 ft. and fully plastic moment M_p, while the two beams BC and CD are each of length 15 ft. and fully plastic moment $2M_p$. A horizontal load H acts at B in the direction BC, and there are vertical loads P and Q at the centres of the beams BC and CD, respectively. Find the values of M_p such that collapse would just occur under the following load combinations

(i) $H = 1$ ton $\quad P = 2$ tons $\quad Q = 2$ tons
(ii) $H = 1$ ton $\quad P = 3$ tons $\quad Q = 4$ tons.

For case (i) show that it the loads P and Q were each 4 tons uniformly distributed over the beams instead of 2 tons concentrated at their centres, the required value of M_p would be unaltered if the hinge in the beam BC was assumed to occur at mid-span. Estimate the change in the value of M_p caused by positioning this hinge correctly.

10. In a two-storey, single-bay, fixed-base rectangular frame $ABCDEF$ the continuous stanchions ABC and FED are each of total length 20 ft., and $AB = BC = DE = EF = 10$ ft. The feet A and F are 18 ft. apart and the upper and lower beams CD and BE thus each span 18 ft. There are concentrated vertical loads V_1 and V_2 at the centres of the beams CD and BE, respectively, and concentrated horizontal loads H_1 and H_2 at D and E, respectively, acting in the directions CD and BE. If all the members of the frame have the same fully plastic moment M_p, find the value of M_p such that collapse would just occur under the following load combinations

(i) $V_1 = 2$ tons $\quad V_2 = 2$ tons $\quad H_1 = 1$ ton $\quad H_2 = 1$ ton
(ii) $V_1 = 2$ tons $\quad V_2 = 2$ tons $\quad H_1 = 0$ $\quad H_2 = 1$ ton

11. A pinned-base pitched roof portal frame $ABCDE$ has stanchions AB and ED each of height 10 ft., the feet A and E being 40 ft. apart. The rafters BC and DC are of equal length and inclined at $22\frac{1}{2}°$ to the horizontal. The knees B and D are connected by a tie-rod which cannot sustain any appreciable bending moment but is of sufficient strength to prevent any relative horizontal movement. The rafters BC and CD each carry a uniformly distributed vertical load 5 tons. If all the members of the frame have the same fully plastic moment

M_p, find the value of M_p such that collapse would just occur, and determine the tension in the tie-rod at collapse.

12. In a three-storey, single-bay, fixed-base rectangular frame *ABCDEFGH* each storey is of height 8 ft. and the span of each beam is also 8 ft. The fully plastic moments of the lowermost stanchions *AB* and *HG* are $3M_p$, those of the middle storey stanchions *BC* and *GF* are $2M_p$, while those of the uppermost stanchions *CD* and *FE* are M_p. The fully plastic moments of the beams *DE, CF* and *BG* are M_p, $2M_p$ and $2M_p$, respectively. Horizontal concentrated loads of 1 ton, 2 tons and 3 tons are applied at *E, F* and *G*, respectively, all these loads acting in the same direction. Find the value of M_p such that collapse would just occur.

13. For the two-bay, fixed-base, pitched roof portal frame of Fig. 4.13(a) the dead and snow loading consists of a uniformly distributed vertical load 4·43 tons on each rafter. For this loading find the value of M_p such that collapse would just occur, M_p being the fully plastic moment of each member of the frame.

Find also the required values of M_p for this loading, and for the dead, snow and wind loading shown in Fig. 4.13(a), if the feet of the stanchions are all pinned instead of fixed.

14. In a multi-bay, fixed-base, pitched roof portal frame the span l of each bay is the same, and the stanchion heights h and rafter slopes θ are all the same, so that all the bays are identical. Each member of the frame has the same fully plastic moment M_p. The frame is subjected to dead and snow loading so that each rafter carries the same uniformly distributed vertical load W. Show that collapse must be confined to the outermost bay at each end of the frame, regardless of the number of bays. Show also that the required value of M_p such that collapse would just occur is $\dfrac{Wlh}{4(2h + l \tan \theta)}$, neglecting the small correction due to the fact that plastic hinges do not form at the apices of the outermost frames but instead at a small distance away from the apices.

CHAPTER

5

Estimates of Deflections

5.1 Introduction

THE PLASTIC METHODS of structural analysis which have been described in Chapters 3 and 4 are concerned solely with the determination of the strength of structures, as indicated by the value of the collapse load. There are many practical instances of structures for which the primary concern of the designer is that the structure should possess adequate strength, and in such cases the design can appropriately be based on the plastic theory. However, in some types of structure it may be necessary to ensure that in addition to possessing adequate strength, certain deflections do not exceed permissible limits. Thus for instance in a crane-bearing portal frame the allowable spread of the crane rails would be limited. For such structures it may be difficult to foresee whether the frame designed by the plastic theory would possess sufficient rigidity to keep the deflections below their permissible limits. The limits of deflection are usually specified at the working loads, and in many cases a structure designed in accordance with the plastic theory would be wholly elastic under the working loads. Under these circumstances an elastic analysis would suffice for the determination of the relevant deflections. However, this kind of specification is illogical, for it is possible that rapid increases of deflection would occur if the loads were raised above their working values, thus rendering the structure useless from the point of view of deformation before the collapse load was attained, even though the deflections under the working loads appeared to be satisfactory. There is thus a clear need for methods which enable the deflections of a structure right up to the point of collapse to be determined, and the main purpose of this chapter is to discuss the development of such methods.

A further reason for discussing this question is that the plastic theory assumes that the deflections developed in a frame just prior to collapse have a negligible effect on the geometry of the frame, in the sense that the equations of equilibrium remain sensibly those for the undistorted frame. These deflections will be larger than those normally encountered in applications of the elastic methods, for collapse rather than working loads must be considered, and in addition there is a progressive loss of rigidity as the plastic hinges form and undergo rotation in succession. Thus it may be thought advisable in some cases to estimate the deflections of a frame at the point of collapse to see whether they are of such a magnitude as to invalidate the assumption of unaltered geometry.

In much of this chapter it will be assumed that the structure is initially free from stress and is subjected to proportional loading to collapse. This does not, of course, correspond to the kind of loading to which a structure is usually subjected in practice. In many practical instances a structure will experience a more or less random fluctuation of loading, as in the case of a building frame which will be subjected to wind and snow loads varying in an unpredictable manner. As will be discussed in Chapter 8, this type of loading can cause the progressive building up of deflections as the loads vary, so that even when the peak values of the loads are less than those which cause plastic collapse much larger deflections can be developed than those which occur just prior to collapse under proportional loading. This should be borne in mind when assessing the value of the deflection estimates obtained under the assumption of proportional loading.

The first step towards developing methods for determining deflections is to establish appropriate bending moment-curvature relations. This question is discussed in Section 5.2, where it is shown that these relations can be derived from the stress-strain relations for tension and compression when certain simplifying assumptions are made. Once the bending moment-curvature relation is known it is comparatively simple to derive load deflection curves for statically determinate structures ; some derivations of this kind are given in Section 5.3 for simply supported beams. Although it is simple enough in principle to extend this process to the determination of load-deflection curves for redundant frames, the computations involved are exceedingly lengthy.

Accordingly, this more general type of treatment is merely summarized in Section 5.4, where the results of some typical calculations are given. The complexity of these calculations indicates the need for an approximate method for determining the deflections of a structure at the point of collapse, and such a method is described in Section 5.5. In applying this method it is necessary to make further simplifying assumptions, and the value of the method must be assessed by comparing the deflection estimates obtained with experimental results. Comparisons of this kind have been made, and have shown that the estimates are in reasonable agreement with measured deflections. However, there are certain cases in which the estimates may be unreliable; these are also discussed briefly.

5.2 Bending moment-curvature relations

In deriving bending moment-curvature relations from the stress-strain relations in tension and compression, the most important assumptions which are made are as follows:

(i) Originally plane cross-sections remain plane, so that the longitudinal strain varies linearly across the section with distance from some neutral axis.

(ii) The only stresses acting are longitudinal normal stresses.

(iii) The relation between longitudinal strain and stress is the same in flexure as in simple tension or compression.

(iv) The material is homogeneous.

(v) The effects of shear force and axial thrust are negligible.

(vi) Residual stresses are absent.

The validity of these assumptions in respect of beams which behave elastically is discussed in the standard texts on the Strength of Materials. For beams bent into the elastic-plastic range, further points arise which will be discussed later. These assumptions were first applied to the case of bending beyond the elastic limit by Meyer,[1] who derived bending moment-curvature relations for steel beams of rectangular section based on the actual stress-strain diagrams obtained in tensile tests on the same material. The results were used to predict load-deflection curves for simply supported beams subjected to central concentrated loads, and good agreement with experimental results was found, thus confirming that the assumptions were reasonably accurate.

In the first place, the idealized type of stress-strain relation of Fig. 1.4(a), in which the upper yield stress is retained, will be assumed, and the corresponding bending moment-curvature relation for a member of rectangular cross-section will be derived. This relation is important in that it can be used to derive corresponding load-deflection curves for simply supported beams subjected to various loads. As will be seen in Section 5.3, some important qualitative conclusions can be drawn from the results obtained. Moreover, these curves are found to correspond closely to the results of tests carried out on annealed mild steel beams, provided that the strain-hardening range is not entered. Secondly, a stress-strain relation in which there is no upper yield stress, but strain-hardening occurs after a period of straining at a constant lower yield stress, will be considered. This relation is closely representative of the behaviour of the material of rolled steel joists, and the corresponding bending moment-curvature relation for an I-section, as derived by Hrennikoff,[2] is given. As will be seen in Section 5.3, it is essential to consider the effect of strain-hardening if the results of tests on simply supported beams of I-section are to be properly interpreted.

Rectangular cross-section

Consider a uniform, initially straight beam of rectangular cross-section, breadth b and depth h, which is bent by pure terminal couples M about an axis parallel to the sides of breadth b. The shear force and axial thrust will both be zero, and it can be argued from considerations of symmetry that the centre line of the beam will be bent into an arc of a circle of radius R, say. Moreover, the neutral axis will be a line bisecting the cross-section, as shown in Fig. 5.1(a). From purely geometrical considerations it can be shown that if homogeneity is assumed the longitudinal strain ε at a distance y from the neutral axis is given by

$$\varepsilon = \frac{y}{R} = \kappa y \qquad . \qquad . \qquad . \qquad . \qquad 5.1$$

where κ is the curvature of the centre line, provided that y is small in comparison with R. If the beam is initially curved, equation 5.1 is still true if κ denotes the *change* of curvature produced by the bending moment M.

It will be assumed that the stress-strain relation for each

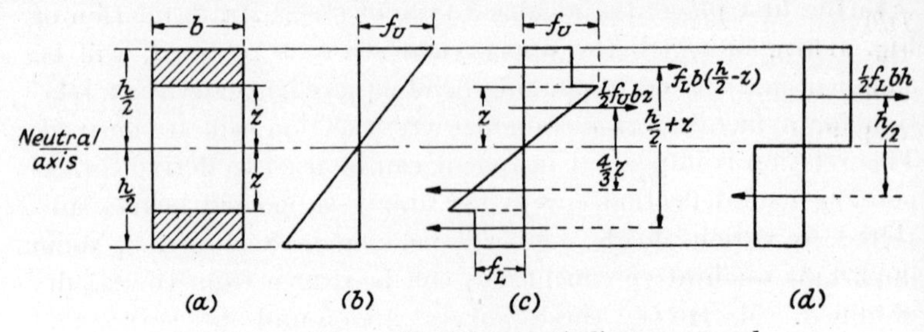

(*a*) Cross-section. Shaded areas indicate assumed plastic zones for elastic-plastic flexure.
(*b*) Elastic stress distribution at yield.
(*c*) Elastic-plastic stress distribution.
(*d*) Fully plastic stress distribution.

Fig. 5.1. *Elastic-plastic flexure of beam of rectangular cross-section.*

longitudinal fibre is the idealized relation shown in Fig. 1.4(*a*), in which the upper yield stress is retained. The linear variation of strain across the section which is implied by equation 5.1 indicates that the stress will also vary linearly across the section while the beam is behaving elastically. Fig. 5.1(*b*) shows the limiting case of this distribution in which the upper yield stress f_U is just attained in the outermost fibres. The bending moment corresponding to this distribution is defined as the yield moment M_y. If the bending moment is increased above this value, yield occurs in the outermost fibres and the stress drops to the lower yield stress f_L. The stress distribution after M has been increased to some value greater than M_y is thus as shown in Fig. 5.1(*c*), where yield has spread inwards to within a distance z of the neutral axis. It is assumed that the boundaries between the elastic and plastic zones in the beam are parallel to the neutral axis, as shown in Fig. 5.1(*a*). The bending moment M corresponding to this distribution of stress is readily evaluated by adding the moment due to the linear stress distribution in the central elastic core to the moment due to the constant lower yield stress in the outer fibres. In the upper half of the elastic core the average stress is $\frac{1}{2}f_U$ acting over an area bz, corresponding to a nett axial force $\frac{1}{2}f_U bz$ whose line of action is at a distance $\frac{2}{3}z$ from the neutral axis. The total moment due to the elastic core is thus $\frac{2}{3}f_U bz^2$. In each of the outer plastic zones the constant stress f_L acts on an area $b(\frac{1}{2}h - z)$, corresponding to a nett force

$f_L b(\frac{1}{2}h - z)$ whose line of action is at a distance $\frac{1}{2}(\frac{1}{2}h + z)$ from the neutral axis. The total moment due to these outer plastic zones is thus $f_L b(\frac{1}{4}h^2 - z^2)$. Thus the bending moment M corresponding to this stress distribution is given by

$$M = \tfrac{2}{3} f_U b z^2 + f_L b(\tfrac{1}{4}h^2 - z^2). \quad . \quad . \quad 5.2$$

The corresponding curvature κ is obtained by noting that the longitudinal strain at the elastic-plastic boundary is the yield strain ε_y, as defined in Fig. 1.4(a). This strain occurs at a distance z from the neutral axis, and it follows from equation 5.1 that the curvature κ is given by

$$\kappa = \frac{\varepsilon_y}{z}$$

The curvature κ_y which is developed as the yield moment is just attained is therefore given by

$$\kappa_y = \frac{2\varepsilon_y}{h},$$

since the bending moment distribution of Fig. 5.1(b) is a limiting case of that in Fig. 5.1(c) with $z = \frac{h}{2}$. Eliminating ε_y between these two expressions it follows that

$$z = \left(\frac{\kappa_y}{\kappa}\right)\frac{h}{2} \quad . \quad . \quad . \quad 5.3$$

When this value of z is substituted in equation 5.2, this equation becomes, after rearrangement,

$$M = \frac{1}{6} b h^2 f_U \left[\frac{3f_L}{2f_U} - \left(\frac{3f_L}{2f_U} - 1\right)\left(\frac{\kappa_y}{\kappa}\right)^2\right]. \quad . \quad 5.4$$

The value of the yield moment can be found from this expression as the value of M for which κ is equal to κ_y, or by noting that the elastic section modulus for a rectangular section is $\frac{1}{6}bh^2$. It follows that

$$M_y = \tfrac{1}{6} b h^2 f_U. \quad . \quad . \quad . \quad 5.5$$

Dividing equation 5.4 by equation 5.5 it is found that

$$\frac{M}{M_y} = \frac{3f_L}{2f_U} - \left(\frac{3f_L}{2f_U} - 1\right)\left(\frac{\kappa_y}{\kappa}\right)^2 \quad . \quad . \quad 5.6$$

This equation is the bending moment-curvature relation for the elastic-plastic condition of Fig. 5.1(c), expressed in non-dimensional form, and was first derived by Robertson and Cook.[3] An equivalent result for the case in which the upper and lower yield stresses are equal was given earlier by Saint-Venant.[4]

When the curvature becomes infinitely large, z is seen from equation 5.3 to be zero, and the stress distribution is the fully plastic distribution of Fig. 5.1(d). From equation 5.4 it is seen that the corresponding value of the bending moment is $\frac{1}{4}bh^2f_L$, agreeing with the expression for the fully plastic moment given

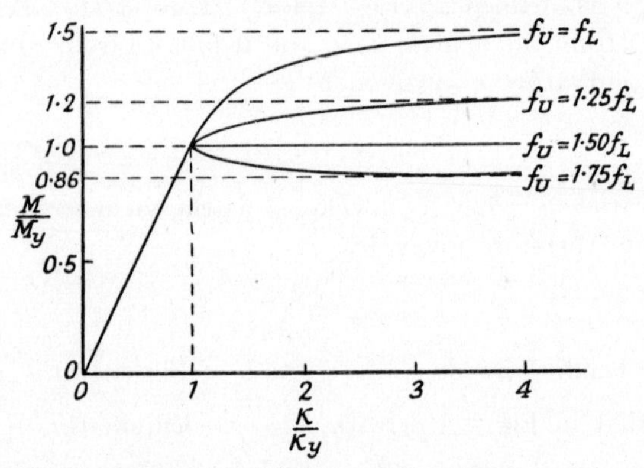

Fig. 5.2. Bending moment-curvature relations for beams of rectangular cross-section.

in equation 1.9. The ratio of M_p to M_y is thus seen with the aid of equation 5.5 to be

$$\frac{M_p}{M_y} = \frac{3f_L}{2f_U} \quad . \quad . \quad . \quad . \quad 5.7$$

If f_U is equal to f_L the value of $\dfrac{M_p}{M_y}$ is defined as the shape factor α, which is seen from equation 5.7 to have the value 1·5, agreeing with the result obtained in Section 1.4.

The bending moment-curvature relations for various values of the ratio of f_U to f_L are shown in Fig. 5.2. It will be seen from this figure that if $f_U = 1{\cdot}5f_L$, the fully plastic moment M_p is equal to the yield moment M_y, and the curvature increases indefinitely while the bending moment remains constant at the value M_y during elastic-plastic flexure. When f_U exceeds $1{\cdot}5f_L$, M_p is less than M_y, as can be seen from equation 5.7, and so an increase of curvature after the yield moment is attained is accompanied by a reduction of bending moment, as exemplified by the curve for $f_U = 1{\cdot}75 f_L$.

Since the fully plastic moment is only attained when the curvature is infinite, it will be realized that several of the assumptions which were made in developing this simple theory of flexure are violated long before this condition is attained. In the first place, the linear variation of strain across the section implied by equation 5.1 only occurs when y is small in comparison with the radius of curvature R, so that at large curvatures the theory would require modification. Some further objections to this simple theory were referred to in Section 1.4. Thus at large curvatures additional radial stresses are called into play to preserve radial equilibrium, which would affect the yield condition. It is also known that the boundaries between the elastic and plastic zones cannot be straight lines parallel to the neutral axis.

Despite these objections the bending moment-curvature relation of equation 5.6 was found by Robertson and Cook [3] to conform quite well with experimental results for annealed mild steel beams with a ratio of f_U to f_L of about 1·35. In some later tests by Cook [5] this ratio was just greater than 1·5, a reduction of moment after yield being observed, and the agreement with the simple theory was again adequate. The simple theory may therefore be accepted as explaining adequately the results of tests on annealed mild steel beams, provided that strain-hardening does not occur. Moreover, the effects of shear and axial thrust on the bending moment-curvature relation are usually assumed to be negligible when deriving load-deflection relations for beams and frames, and this assumption may well cause more serious errors than the effects just enumerated.

I-section

In order to develop a bending-moment curvature relation for beams of I-section, Hrennikoff [2] assumed that the material possessed the stress-strain relation which is shown in Fig. 5.3(a) for both tension and compression. This is in general accordance with the properties of mild steel for railway bridges as laid down in the 1940 Specifications of the American Railway Engineering Association. It will be noticed that there is no upper yield point in this relation. This neglect of the upper yield phenomenon seems reasonable in view of the fact that several tests, which are discussed in Section 6.2, have indicated that the material of rolled steel joists exhibits either a small or zero drop of stress at yield. However, the particular feature of the assumed stress-strain relation which is of especial interest is that it includes the strain-hardening range. The strain ε_s at which strain-hardening commences is 0·018 whereas the strain ε_y at the yield point is 0·0011, so that the ratio of ε_s to ε_y is $\frac{0{\cdot}018}{0{\cdot}0011}$, or 16·4. A further implicit assumption in the analysis is that of homogeneity. This is rather difficult to justify in view of the fact

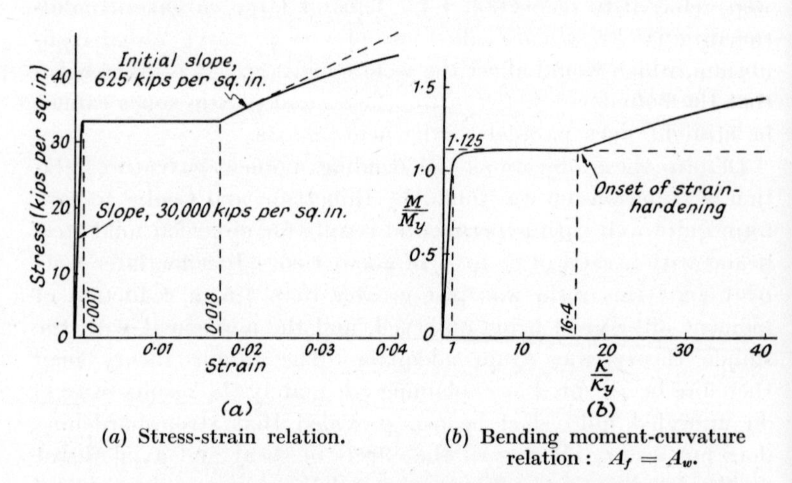

(*a*) Stress-strain relation. (*b*) Bending moment-curvature relation : $A_f = A_w$.

Fig. 5.3. *Bending moment-curvature relation for I-section with strain-hardening* (after Hrennikoff).

that the stress-strain relations obtained from tensile specimens cut from various positions in rolled steel joists are known to vary widely, as pointed out in Section 6.2.

To simplify the analysis it was assumed that the thickness of the flanges was negligible in comparison with the depth of the beam, so that each flange area could be regarded as concentrated at a constant distance from the neutral axis. With this assumption the form of the bending moment-curvature relation depends only on a single parameter, namely, the ratio of the total flange area A_f to the web area A_w. Up to the curvature at which strain-hardening develops in the outermost fibres, the analysis was a simple extension of the theory which has just been given for a beam of rectangular cross-section. To determine the bending moment for a given curvature, and thus a given linear distribution of strain across the section, it was only necessary to add the bending moment due to the rectangular web to the bending moment due to the flanges, which is equal to the product of the stress in a flange, its area, and the depth of the beam. At higher curvatures, a process of step-by-step integration became necessary.

Four values of the ratio of total flange area to web area were considered, namely, 0, 0·5, 1·0 and 1·5. The value zero corresponds to the case of a beam of rectangular cross-section, and the other values cover the range of standard I-sections. The results were presented in the form of curves, and for the purpose of accurate calculation were tabulated for the case in which this ratio is 1·0, so that $A_f = A_w$. The bending moment-curvature relation for a beam of I-section of this type, as derived from these tabulated results, is shown in Fig. 5.3(*b*). The results are plotted non-dimensionally, the ordinates being the ratio of the bending moment to the yield moment and the abscissae being the ratio of the curvature to the curvature at yield. It is readily verified that for such a cross-section the shape factor α is 1·125, so that $M_p = 1{\cdot}125\ M_y$. It will be seen from the figure that strain-hardening commences when $\dfrac{\kappa}{\kappa_y} = 16{\cdot}4$, this being the ratio of ε_s to ε_y.

Further results were also tabulated which enable the load-deflection curves of statically determinate beams and simple statically indeterminate beams and frames to be derived. Some applications of this work will be given in Sections 5.3 and 5.4.

For a more general treatment of the problem of determining bending moment-curvature relations from any assumed form of stress-strain relation, the work of Nadai [6] may be consulted. Several comparisons with experimental results for light alloy beams have been made, for instance by Rappleyea and Eastman [7] and by Dwight,[8] and the case of a light alloy beam of rectangular section which is subjected to bending moments about axes other than the principal axes has been discussed by Barrett.[9]

5.3 Load-deflection relations for simply supported beams

For a beam resting on two simple supports the bending moment distribution in the beam for a given loading is known from considerations of statics alone. Once the bending moment-curvature relation is specified the curvature at any section is known, and the deflected form of the beam can then be found by integration. In the first place beams of rectangular cross-section will be considered, with the bending moment-curvature relation of equation 5.6; subsequently, some of the results obtained by Hrennikoff [2] for joists with the bending moment-curvature relation of Fig. 5.3(b) will be given.

Beam of rectangular cross-section with central concentrated load

Consider a uniform beam of rectangular cross-section, breadth b and depth h, which is simply supported over a span l, as shown in

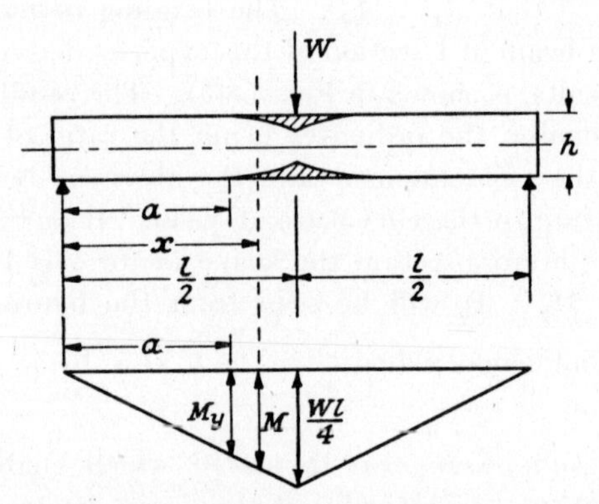

Fig. 5.4. Simply supported rectangular section beam with central concentrated load.

Fig. 5.4. It will be assumed that the relation between bending moment and curvature in the yielded regions is the relation given by equation 5.6, and for simplicity it will be assumed in the first place that $f_U = f_L$, so that

$$\frac{M}{M_y} = \frac{1}{2}\left[3 - \left(\frac{\kappa_y}{\kappa}\right)^2\right] \qquad . \quad . \quad . \quad 5.8$$

This relation corresponds to the assumption of the ideal-plastic stress-strain relation of Fig. 1.4(b).

The bending moment diagram for the beam is as shown in Fig. 5.4, the central bending moment being $\frac{1}{4}Wl$. Yield first occurs at the centre of the beam when this bending moment reaches the value M_y. The corresponding value of the load, W_y, is therefore given by the equation

$$M_y = \tfrac{1}{4} W_y l \quad . \quad . \quad . \quad . \qquad 5.9$$

If the load is increased to a value W greater than W_y, the yield moment M_y will be attained at some distance a from the supports, as shown in the figure. In the central portion of the beam where the bending moment exceeds M_y, yield occurs, and plasticity spreads inwards towards the neutral axis. The general form of the plastic zones thus created is shown in the figure ; a derivation of the shape of the elastic-plastic boundary will be given later. Eventually, collapse occurs when the central bending moment reaches the value M_p, so that plasticity has spread right down to the neutral axis at the centre of the beam. The corresponding collapse load W_c is given by the equation

$$M_p = \tfrac{1}{4} W_c l$$

Using equation 5.9 it follows that

$$\frac{W_c}{W_y} = \frac{M_p}{M_y} = 1{\cdot}5,$$

since the shape factor for a rectangular beam has the value 1·5.

From statical considerations it follows that

$$M_y = \tfrac{1}{2} W a$$

Combining this equation with equation 5.9, it is found that

$$a = \frac{l}{2}\left(\frac{W_y}{W}\right) \quad . \quad . \quad . \quad . \qquad 5.10$$

Since the slope at the centre of the beam is zero, by symmetry, the central deflection δ is seen to be given by the integral

$$\delta = \int_0^{\frac{l}{2}} x\kappa \, dx \quad . \quad . \quad . \quad . \qquad 5.11$$

where x is measured from the left-hand support. For $0 \leqslant x \leqslant a$, the beam is elastic, so that the curvature κ is equal to $\frac{x}{a}\kappa_y$. For $a \leqslant x \leqslant \frac{l}{2}$, the beam has partly yielded, so that the relation between bending moment and curvature is given by equation 5.8. Solving this equation for κ, it is found that

$$\kappa = \frac{\kappa_y}{\sqrt{3 - 2\frac{x}{a}}}, \text{ for } a \leqslant x \leqslant \frac{l}{2},$$

since $\dfrac{M}{M_y}$ is equal to $\dfrac{x}{a}$. Substituting these expressions for the curvature in equation 5.11, it is found that

$$\delta = \int_0^a \frac{\kappa_y}{a} x^2 dx + \int_a^{\frac{l}{2}} \frac{\kappa_y x}{\sqrt{3 - 2\frac{x}{a}}} dx$$

Evaluating these integrals, and eliminating a by using equation 5.10, the following result is obtained

$$\delta = \frac{l^2\kappa_y}{12}\left(\frac{W_y}{W}\right)^2\left[5 - \left(3 + \frac{W}{W_y}\right)\sqrt{3 - 2\frac{W}{W_y}}\right]$$

The deflection δ_y when yield first occurs at the centre of the beam is obtained by putting $W = W_y$ in the above expression for δ, giving $\delta_y = \dfrac{l^2\kappa_y}{12}$. Dividing δ by δ_y, the following non-dimensional result is found

$$\frac{\delta}{\delta_y} = \left(\frac{W_y}{W}\right)^2\left[5 - \left(3 + \frac{W}{W_y}\right)\sqrt{3 - 2\frac{W}{W_y}}\right].$$

This result was first obtained by Fritsche.[10]

When the collapse load $W_c = 1{\cdot}5W_y$ is just attained, but before any rotation has occurred at the central plastic hinge, the central deflection δ is seen to have the value $2{\cdot}22\delta_y$, so that this deflection is finite and of the order of elastic deflections at the point of collapse.

If the effect of the upper yield stress had been taken into account by using the bending moment-curvature relation of equation 5.6, a similar analysis would have led to the result

$$\frac{\delta}{\delta_y} = \left(\frac{W_y}{W}\right)^2\left[4\gamma^2 - 2\gamma - 1 - 2\left(2\gamma + \frac{W}{W_y}\right)\sqrt{(\gamma - 1)\left(\gamma - \frac{W}{W_y}\right)}\right] \quad 5.12$$

where
$$\gamma = \frac{3f_L}{2f_U}$$

Equation 5.12 is the relation between the central load W, expressed non-dimensionally as $\dfrac{W}{W_y}$, and the central deflection δ, expressed non-dimensionally as $\dfrac{\delta}{\delta_y}$. This relation is plotted in Fig. 5.5(a) for three values of the ratio of upper to lower yield stress, namely 1·0, 1·2 and 1·5.

The shape of the plastic zones which have developed at any

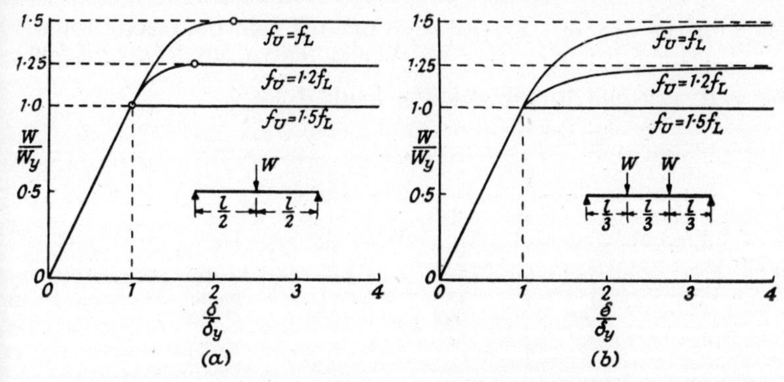

(*a*) Central concentrated load. (*b*) Symmetrical two-point loading.

Fig. 5.5. Load-deflection relations for simply supported beams of rectangular cross-section.

stage of the loading can be derived very simply. If z is the semi-depth of the elastic core at any section, as defined in Fig. 5.1(c), the relation between z and the curvature κ is as given by equation 5.3, namely,

$$z = \left(\frac{\kappa_y}{\kappa}\right)\frac{h}{2}$$

Eliminating the ratio $\frac{\kappa_y}{\kappa}$ between this equation and the bending moment-curvature relation of equation 5.8, which applies to the case in which $f_U = f_L$, it is found that

$$\left(\frac{2z}{h}\right)^2 = 3 - 2\frac{M}{M_y} \quad . \quad . \quad . \quad 5.13$$

This relation between z and the bending moment M is independent of the loading on the beam. To determine the relation between z and x, the distance measured along the beam from the chosen origin, it is only necessary to express M in terms of x. For the case of a central concentrated load this relation is seen from Fig. 5.4 and equation 5.10 to be

$$\frac{M}{M_y} = \frac{x}{a} = \frac{2x}{l}\left(\frac{W}{W_y}\right)$$

Substituting in equation 5.13, it is found that

$$\left(\frac{2z}{h}\right)^2 = 3 - \frac{4x}{l}\left(\frac{W}{W_y}\right) \quad . \quad . \quad . \quad 5.14$$

Thus for the case of a central concentrated load the plastic zones are parabolic in shape. Fig. 5.6(a) shows the shape of these zones when W is equal to the collapse load W_c.

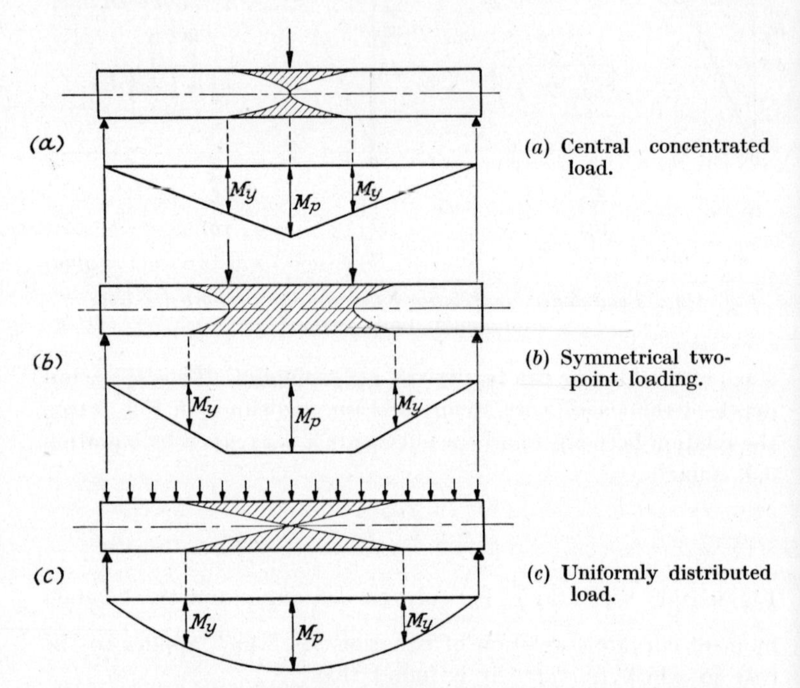

Fig. 5.6. *Shape of plastic zones for simply supported rectangular section beams :* $f_U = f_L$

Beam of rectangular cross-section with two-point loading

The analysis can readily be extended to cover cases in which the loading on the beam consists of a pair of equal loads applied at points equidistant from the centre of the span. In this case the central portion of the beam between the loads will be subjected to a constant bending moment, the bending moment diagram being as shown in Fig. 5.6(b), and will therefore bend into an arc of a circle. Load versus central deflection relations for the particular case in which the distance between the loads is equal to one-third of the length of the beam are shown in Fig. 5.5(b). In this case it is seen that before the collapse load is attained the deflection becomes infinite, in contrast with the case of a beam with a central concentrated load, for which the deflection at the point of collapse is finite. This is due to the fact that the central portion of the beam is in pure bending. As the fully plastic moment is attained in the central length, the curvature becomes infinite over this finite length, and this inevitably results in an infinite deflection. Indeed, it is clear that the deflection at the point of collapse in a statically determinate beam will be infinite whenever the maximum bending moment occurs in a finite length of the beam.

Beam of rectangular cross-section with uniformly distributed load

The final case which will be considered is a simply supported beam of rectangular cross-section which is subjected to a uniformly distributed load. The analysis is similar in principle to that which was given for the case of a central concentrated load, and will not be reproduced here. The significant feature of the results is that the load-deflection relations are similar to those shown in Fig. 5.5(*b*) for the two-point loading case, in which the deflection becomes infinite as the collapse load is attained, rather than those of Fig. 5.5(*a*) for the central concentrated load, with a finite deflection at the point of collapse. This is due to the form of the bending moment diagram, which is parabolic as shown in Fig. 5.6(*c*). The maximum bending moment occurs at the centre of the beam, where the shear force and thus the rate of change of bending moment is zero. This condition approximates more closely to the condition of pure bending over a finite central portion of the beam than the condition which obtains when there is a central concentrated load and a finite shear force at the centre of the beam, as in Fig. 5.6(*a*). For this case the elastic-plastic boundaries are linear, as shown in Fig. 5.6(*c*).

Calculations of this kind, based as they are on the bending moment-curvature relation of equation 5.8, cannot be justified strictly owing to the presence of shear forces in the beams. In the elastic range the deflections due to shear are known to be small in comparison with those due to bending unless the beam is very short. In the elastic-plastic range shear deflections might assume greater importance, for as will be shown in Section 6.4 of Chapter 6 the shear stress must be zero in the plastic zones of a rectangular beam. Thus the shear force must be carried entirely by shear stresses in the central elastic core of the beam. These shear stresses are clearly greater than the shear stresses which would occur if the beam were behaving elastically, and so would produce greater deflections. So far, no attempt has been made to evaluate the increase of deflection due to this cause, but it does not seem likely that the effect would be very marked.

Despite this and the incorrectness of some of the basic assumptions already referred to, these deflection calculations serve to illustrate an important general point. If the fully plastic moment is developed in a region of constant bending moment, as in Fig. 5.6(*b*), the deflection becomes infinite as the collapse load is attained. This also occurs if the shear force is zero at the plastic hinge, so that the bending moment is roughly constant in the neighbourhood of the hinge, as in Fig. 5.6(*c*). In contrast, if there is a finite shear force at the plastic hinge, so that the bending moment falls away at a finite rate on either side of the hinge, as in Fig. 5.6(*a*), the deflection at the point of collapse is finite. While these results were obtained for statically determinate beams, it is not difficult to see that they are qualitatively true for indeterminate frames; if there is a region of constant or nearly constant bending moment in the neighbourhood of a plastic hinge in a frame, the deflections at the point of collapse will in general be infinite.

A comprehensive investigation which showed that this simple theory of elastic-plastic flexure leads to close agreement with the results of tests on annealed mild steel beams of rectangular cross-section was reported by Roderick and Phillips.[11] The theory has been extended by Roderick and Heyman [12] to include the effect of strain-hardening; in this paper the results of experiments on beams of medium carbon steel are also described.

Beam of I-section with central concentrated load

The theoretical work of Hrennikoff [2] on beams of I-section has already been referred to in Section 5.2. This work was based on the assumption of a typical strain-hardening type of stress-strain relation for mild steel of the form indicated in Fig. 5.3(*a*), and the corresponding bending moment-curvature relation for a beam of I-section with the total flange area equal to the web area was given in Fig. 5.3(*b*). Assuming this relation, the load-deflection curve for a simply supported beam subjected to a central concentrated load is readily derived. The central deflection δ of any uniform simply supported beam of length l which is loaded symmetrically is given by equation 5.11, namely,

$$\delta = \int_0^{\frac{l}{2}} x\kappa \, dx,$$

where x is the distance measured along the beam from one of the supports. To evaluate this integral the curvature κ must be known as a function of x. Since the bending moment varies linearly with x, as in Fig. 5.4, and the bending moment-curvature relation has been obtained, the curvature κ is known as a function of x for any value of the load, thus enabling δ to be determined.

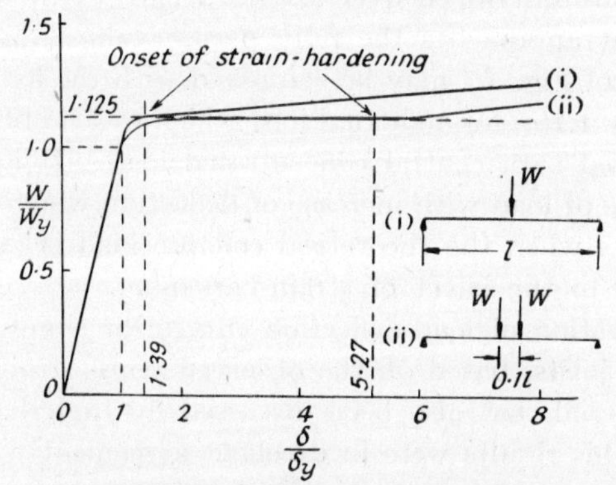

Fig. 5.7. *Load-deflection relations for simply supported beams allowing for strain-hardening.*

A procedure of this kind was adopted by Meyer [1] in his early studies of the bending of steel beams of rectangular section.

Hrennikoff's tabulated data included values of the deflection derived by evaluating this integral, and his results are plotted non-dimensionally as curve (i) in Fig. 5.7. In this figure the ordinates represent the ratio of the central load W to the load W_y at which yield first occurs, and the abscissae represent the ratio of the central deflection δ to the deflection δ_y at first yield. As already pointed out, the shape factor α for a beam of this cross-section is 1·125. Thus the fully plastic moment M_p is $1{\cdot}125M_y$, and the collapse load W_c according to the simple plastic theory would be $1{\cdot}125W_y$.

It will be seen from the figure that strain-hardening begins when the central deflection is only $1{\cdot}39\delta_y$, and that thereafter the load-deflection relation is almost linear, its slope being about 2% of the slope in the elastic region. This behaviour may be contrasted with the case of symmetrical two-point loading. Curve (ii) in the figure shows the load versus central deflection relation for this type of loading in which the distance between the loads is $0{\cdot}1l$, where l is the span of the beam, and it will be seen that in this case strain-hardening does not commence until $\delta = 5{\cdot}27\delta_y$. There is thus a considerable range of deflection in this latter case over which the load remains roughly constant at the theoretical collapse value. This is due to the fact that the central portion of the beam is subjected to a constant bending moment, so that large curvatures are developed over a finite length of the beam when this moment approaches the fully plastic value.

Curve (i) of Fig. 5.7 may be compared with the load-deflection curve of Fig. 1.1(*a*) for an actual joist which was simply supported and subjected to a central concentrated load. It appears that the slow rise of load with increase of deflection which is observed in such tests when the theoretical collapse load is exceeded can be ascribed to the effect of strain-hardening.

The derivation of load-deflection curves for simply supported rolled steel joists, based on the observed stress-strain properties of the material, has also been discussed by Roderick,[13, 14] who found that his results were in excellent agreement with tests on 8 in. × 4 in. and 10 in. × $4\frac{1}{2}$ in. British Standard beams. However, in spite of these results and the tests on beams of rectangular section already referred to,[11, 12] it must be noted that the assumption of homogeneous behaviour in mild steel beams, even assuming initially homogeneous material, is not strictly justifiable. This is due to the piecemeal nature of the yield process referred to in Section 1.3. Thus although on this assumption equation 5.1 still holds for beams of I-section, Yang, Beedle and Johnston[15] have observed local non-linear distributions of strain across WF sections bent beyond the yield point. In this paper the effect of residual stresses on the bending moment-curvature relation is also discussed.

5.4 Deflections of simple pin-based portal frame

Hrennikoff's results[2] have been applied by Horne[16] to the determination of load-deflection relations for a pin-based rectangular portal frame of height l and span $2l$, loaded as shown in the inset to Fig. 5.8. Each member of this frame was of the same I-section of depth h, with $A_f = A_w = A$, A_f being the total flange area and A_w the web area. It will be appreciated that this frame has one redundancy, so that for any given loading the redundancy had to be evaluated before the deflections of the frame could be determined. This involved a trial and error approach in which the value of the redundancy was first assumed and the spread of the feet of the frame evaluated; the correct value of the redundancy was the value which made the spread zero.

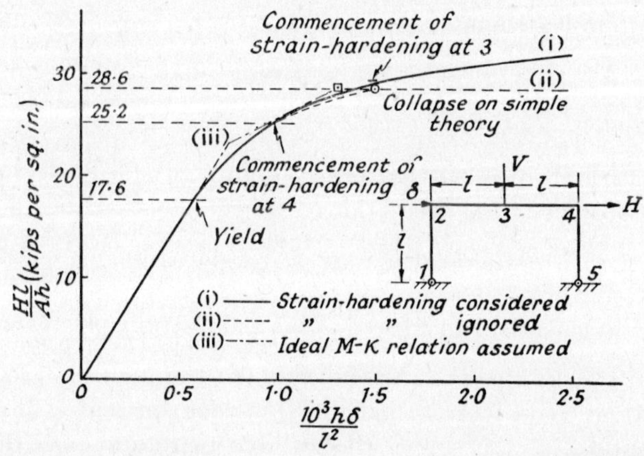

Fig. 5.8. Load-deflection relations for pin-based portal frame.

The loading programme was to apply first a vertical load V of such a value that the yield moment was just attained at the most highly stressed cross-section beneath the load V, this value of the vertical load being $70{\cdot}4\frac{Ah}{l}$ kips. The vertical load was then held constant at this value while the horizontal load was increased steadily. The computed relation between the horizontal load H and the horizontal deflection δ of the beam while V was held constant is shown as curve (i) in Fig. 5.8. The ordinates in this figure represent the load H multiplied by the factor $\frac{l}{Ah}$, the units being kips per sq. in., and the abscissae represent the horizontal deflection δ expressed non-dimensionally as $\left(\frac{10^3 h}{l^2}\right)\delta$. It will be seen that the behaviour is elastic until $H = 17{\cdot}6\frac{Ah}{l}$ kips. At this value of H the yield moment is attained at section 4. Since the bending moment at section 3 beneath the vertical load does not change when the frame is responding elastically to the horizontal load, this bending moment is also equal to the yield moment at this value of H. As H increases further, yield spreads into the webs of the members at sections 3 and 4 and also along the members for some distance from these sections, and the slope of the load-deflection relation decreases steadily. The bending moment at section 4 increases more rapidly than at section 3, and strain-hardening begins at this section when $H = 25{\cdot}2\dfrac{Ah}{l}$ kips. At section 3 strain-hardening does not occur until $H = 29{\cdot}3\dfrac{Ah}{l}$ kips.

It is readily verified that on the simple plastic theory, plastic collapse occurs with plastic hinges at the cross-sections 3 and 4 when the horizontal load H is $28{\cdot}6\dfrac{Ah}{l}$ kips, assuming the vertical load V to have the value $70{\cdot}4\dfrac{Ah}{l}$ kips and the fully plastic moment to be $24{\cdot}75Ah$ kips in. At this value of the horizontal load the deflections could increase indefinitely under constant load on the assumptions of the simple theory, due to rotations at the two plastic hinges. It will be seen from the figure that this behaviour is prevented by the influence of strain-hardening, and that there is no sharply defined collapse load. Nevertheless, it can be remarked that below the calculated plastic collapse load the deflections are not very much greater than they would have been if the frame had behaved elastically up to this load, and that above this load the deflections increase very rapidly with small increases of load.

If strain-hardening is neglected, so that an ideal-plastic stress-strain relation of the type shown in Fig. 1.4(b) is assumed, the load-deflection relation would be as shown in curve (ii) of Fig. 5.8.

It will be seen that the discrepancy between this relation and the relation that includes the effect of strain-hardening is small until the calculated plastic collapse load is reached. These two load-deflection relations may be compared with a third relation computed on the assumption that the bending moment-curvature relation for each member is of the ideal type shown in Fig. 2.1, in which the member remains elastic until the fully plastic moment is reached. This amounts to neglecting the influence on the deflections of the spread of the plastic zones along the members from the sections 3 and 4, and considering only the reduction in stiffness due to plastic hinge formation. The corresponding load-deflection relation would be as shown in curve (iii) of Fig. 5.8.

Some important qualitative conclusions may be drawn from these three load-deflection curves. In the first place, it is clear that if the effect of strain-hardening is neglected, as exemplified by curve (ii), the deflections at any given load will be increased when the loads are sufficiently large to bring the largest strains into the strain-hardening range. Thus the horizontal deflection at the plastic collapse load is found to be $1{\cdot}38 \times 10^{-3}\left(\frac{l^2}{h}\right)$ in. when strain-hardening is taken into account, but $1{\cdot}52 \times 10^{-3}\left(\frac{l^2}{h}\right)$ in. when strain-hardening is neglected, an increase of about 10%. Secondly, if the effect of the spread of plastic zones along the members is neglected, as exemplified by curve (iii), the deflections at any given load will be reduced. In this particular example the horizontal deflection at the point of collapse in curve (iii) is $1{\cdot}32 \times 10^{-3}\left(\frac{l^2}{h}\right)$ in., a reduction of about 14% as compared with curve (ii). Thus if curve (i) is considered to represent closely the way in which an actual structure would behave, the much simpler behaviour represented by curve (iii) follows from the neglect of the effects of strain-hardening and the spread of the plastic zones along the members. Of these effects, the former tends to increase, and the latter to reduce the deflections. Both the effects are small in this example, giving changes of +10% and −14% respectively in the deflection at the theoretical collapse load, and so causing a nett reduction of only 4%.

It may be conjectured that in general the magnitude of the changes in the calculated deflections at collapse caused by neglecting these two effects will usually be small, and since the effects tend to cancel one another a reasonable estimate of the deflection at the point of collapse can be made by neglecting both these effects. This conclusion forms the basis of the approximate method for estimating deflections at the point of collapse which is now to be described. The only case in which fairly serious errors might occur is when one or more of the plastic hinges forms at a section in the neighbourhood of which the bending moment only changes very slowly, for instance within the span of a beam subjected to a uniformly distributed load. As shown in Section 5.3, the deflections in such cases become infinite at collapse when strain-hardening is neglected, so that the effect of the spread of the plastic zones along the members is not small in these cases. It was for this reason that Yang, Beedle and Johnston [17] suggested that deflections should not be estimated at the point of collapse, but at the load at which the yield moment, rather than the fully plastic moment, is reached at the position of the last hinge to form. At this load the deflections must be finite.

Further comparisons between load-deflection curves taking into account the effect of strain-hardening and the corresponding curves when strain-hardening is neglected have been made by Horne,[18] who considered partially and fully built-in beams of rectangular and I-section composed of steels with various stress-strain diagrams. A survey of the various methods for estimating beam deflections has also been made by Knudsen and others.[19]

5.5 Estimates of deflections at point of collapse

For many purposes it would be unnecessary to know the full load-deflection relations of a frame loaded to collapse, but a knowledge of the deflections at the point of collapse would be of value, as discussed in Section 5.1. A method for estimating these deflections has been proposed by Symonds and Neal.[20, 21] In this method it is assumed that the effects of strain-hardening and of the spread of plastic zones along the members can be neglected, so that in effect the bending moment-curvature relation is assumed to be of the ideal type of Fig. 2.1. As just pointed out in Section 5.4, these effects are usually small, and they tend to cancel one another, so that their neglect is unlikely to cause serious errors unless one of the plastic hinges forms in a region of roughly constant bending moment.

A further basic assumption which is made is that as the loads are increased to their collapse values, the rotation at a plastic hinge never ceases once it has formed. It will be shown later that this assumption is not necessarily valid, even when attention is confined to proportional loading, and for more arbitrary loading programmes it can obviously be incorrect. Nevertheless, the method will in many cases furnish useful estimates of the deflections at the point of collapse. Acceptance of this assumption implies that all the various hinges which are involved in the collapse mechanism will form and undergo rotation in turn, while no other plastic hinges are formed at any stage. Thus just before the collapse load is attained, all except one of the plastic hinges in the collapse mechanism will have formed and undergone rotation, except in the special cases in which two or more hinges form simultaneously at collapse. Excluding such cases, it can be seen that as the collapse load is attained, the bending moment at the position of the last hinge to form must have reached its fully plastic value, but before motion of the collapse mechanism ensues the rotation at this hinge will be zero. The rotations at all the other hinges will, of course, be unknown. In this condition the structure is at the point of collapse, and it is found that the deflections can be computed quite simply by using any of the usual techniques of elastic structural analysis, the slope-deflection equations being most convenient for this purpose. This follows from the neglect of the effect of the spread of plastic zones along the members, which implies that the members of the frame remain elastic everywhere except at the plastic hinges. The principal difference between the analysis and that which would be carried out for a wholly elastic frame is that at each cross-section where a plastic hinge has undergone rotation, the bending moment is known to have its fully plastic value but the plastic hinge rotation is unknown, as compared with the condition in the wholly elastic frame where at the same cross-section the bending moment would be unknown but the hinge rotation would be zero. Thus, provided that the position where the last hinge forms can be found, the deflections can be calculated as rapidly as those for a wholly elastic frame. As will be seen, the position of this hinge can be determined quite easily. Details of the technique are best explained with reference to particular examples, but before these are discussed it is necessary to state the slope-deflection equations which will be employed.

Slope-deflection equations

Fig. 5.9 shows a uniform member AB of length L and elastic flexural rigidity EI which is subjected to an arbitrary loading and also to moments M_{AB} and M_{BA} at its ends, the sign convention for the moments being that a positive moment acts clockwise on the member. The deflection of B relative to A is δ, as shown, and the rotations at the ends of the member are ϕ_{AB} and ϕ_{BA}, each rotation being regarded as positive when clockwise. These rotations can be shown to be given by the slope-deflection equations

$$\phi_{AB} = \frac{\delta}{L} + \frac{L}{6EI}\Big[2(M_{AB} - M_{AB}^F) - (M_{BA} - M_{BA}^F)\Big] \quad . \quad 5.15$$

$$\phi_{BA} = \frac{\delta}{L} + \frac{L}{6EI}\Big[2(M_{BA} - M_{BA}^F) - (M_{AB} - M_{AB}^F)\Big] \quad . \quad 5.16$$

where M_{AB}^F and M_{BA}^F are the fixed-end bending moments which

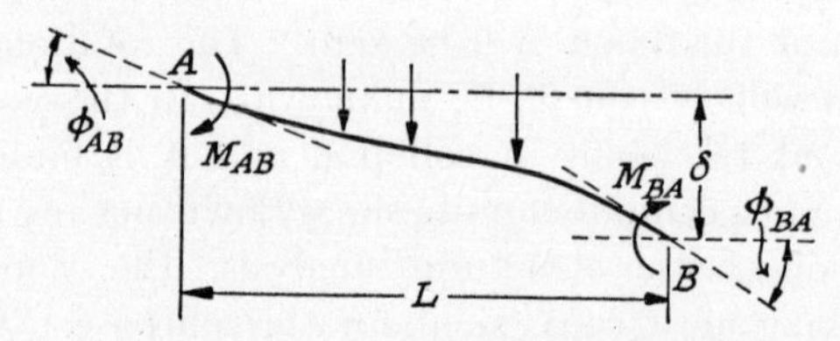

Fig. 5.9. *Definition of terms in slope-deflection equations.*

would be produced at the ends of the member if it were subjected to the same loading but both ends were held clamped in position and direction.

Fixed-ended beam with off-centre load

The first example to be considered is the fixed-ended beam of span $3l$, which is subjected to a load W at a distance $2l$ from one end, as shown in Fig. 5.10(a). The beam is supposed to be of uniform section, with a fully plastic moment M_p and elastic flexural rigidity EI. It is readily verified that the collapse load W_c has the value $\frac{3M_p}{l}$, the collapse mechanism being as shown in Fig. 5.10(b). In this figure the plastic hinges in the collapse mechanism are given in magnitude and sign, a positive hinge being defined as one which causes extension in the fibres of the member adjacent to the dotted line in Fig. 5.10(a).

During collapse the shapes of the two portions 12 and 23 of the beam between the plastic hinges remain unchanged, the deflection increases being due solely to the rotations at the three plastic hinges. Fig. 5.10(c) shows the deflected form of the beam at some stage during collapse, with rotations at each of the three plastic hinges. The slope-deflection equations 5.15 and 5.16 may be applied to each of the two portions of the beam in turn. Since these lengths are not subjected to any loading the fixed-end moments M^F are zero. The actual moments M are

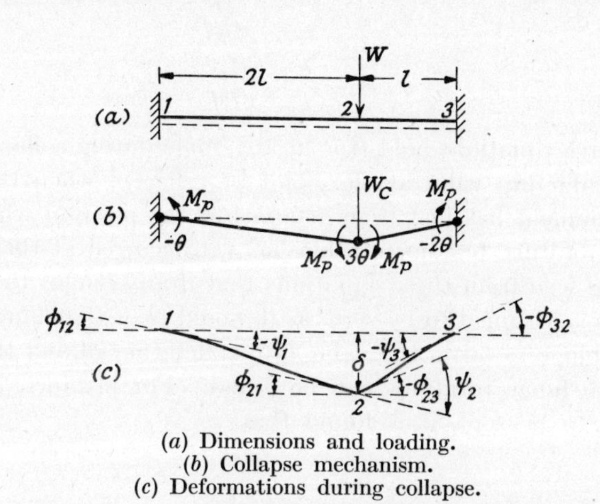

(*a*) Dimensions and loading.
(*b*) Collapse mechanism.
(*c*) Deformations during collapse.

Fig. 5.10. *Fixed-ended beam with off-centre load.*

as shown in Fig. 5.10(*b*). The four equations thus obtained are as follows:

$$\phi_{12} = \phi_{21} = \frac{\delta}{2l} - \frac{M_p l}{3EI}$$

$$\phi_{23} = \phi_{32} = -\frac{\delta}{l} + \frac{M_p l}{6EI}$$

Suppose that the rotations at the hinges are ψ_1, ψ_2 and ψ_3, as shown in Fig. 5.10(*c*). These hinge rotations should not be confused with the rotations shown in the collapse mechanism of Fig. 5.10(*b*); the latter rotations are the *increments* which occur during a small motion of the collapse mechanism, whereas the former are the *total* rotations which have occurred both prior to and during the collapse. In terms of the rotations at the joints, these hinge rotations are

$$\psi_1 = -\phi_{12}$$
$$\psi_2 = \phi_{21} - \phi_{23}$$
$$\psi_3 = \phi_{32}$$

Substituting the values of the joint rotations given by the slope-deflection equations, it is found that

$$\psi_1 = -\frac{\delta}{2l} + \frac{M_p l}{3EI} \qquad . \quad . \quad . \quad 5.17$$

$$\psi_2 = \frac{3\delta}{2l} - \frac{M_p l}{2EI} \qquad . \quad . \quad . \quad 5.18$$

$$\psi_3 = -\frac{\delta}{l} + \frac{M_p l}{6EI} \qquad . \quad . \quad . \quad 5.19$$

These three equations hold true at any stage during collapse, and in particular are valid at the point of collapse, when the fully plastic moment has just been attained at the position where the last hinge forms, but no rotation has yet occurred at this hinge. It will be seen from these equations that if any one of the hinge rotations is assumed to be zero, as it would be if that hinge were the last hinge to form, the value of δ is determined and thus the other two hinge rotations can be found. For instance, if ψ_3 is assumed to be zero, it is found that

$$\delta = \frac{M_p l^2}{6EI}$$

$$\psi_1 = \frac{M_p l}{4EI}$$

$$\psi_2 = -\frac{M_p l}{4EI}$$

$$\psi_3 = 0$$

In order to decide whether this assumption is correct, it is necessary to make use of the fact that the rotation at a hinge prior to collapse must be in the same sense as its rotation during collapse, since it is assumed that once a hinge has formed it will always continue to rotate. Reference to Fig. 5.10(b) shows that the hinge rotations must therefore have the following signs:

ψ_1	ψ_2	ψ_3
$-$	$+$	$-$

except for the last hinge to form, whose rotation at the point of collapse is zero. The assumption that ψ_3 is zero violates this condition in leading to values of ψ_1 and ψ_2 which are both of incorrect sign, and so it is concluded that the last hinge to form cannot be at section 3.

Other solutions to equations 5.17, 5.18 and 5.19 can be found similarly. Alternatively, use can be made of the fact that changes in the deflection δ during collapse are due solely to rotations at the hinges. Thus from these three equations it can be seen that a change $\Delta\delta$ in the value of δ will cause the following changes in the hinge rotations:

$$\Delta\psi_1 = -\frac{\Delta\delta}{2l}$$

$$\Delta\psi_2 = \frac{3\Delta\delta}{2l}$$

$$\Delta\psi_3 = -\frac{\Delta\delta}{l}$$

Denoting $\dfrac{\Delta\delta}{2l}$ by θ, these changes in the hinge rotations become

$$\begin{aligned}\Delta\psi_1 &= -\ \theta\\ \Delta\psi_2 &= \ \ \ 3\theta\\ \Delta\psi_3 &= -\ 2\theta\end{aligned}$$

These changes in the hinge rotations are of course precisely those due to a small motion of the collapse mechanism in which the deflection under the load increases by $\Delta\delta$, as can be seen from Fig. 5.10(*b*). It follows that any solutions to equations 5.17, 5.18 and 5.19 can be obtained by adding the hinge rotations and displacements due to an arbitrary motion of the collapse mechanism to those found by assuming ψ_3 to be zero. All possible solutions to these equations can therefore be expressed as follows:

$$\delta = \frac{M_p l^2}{6EI} + 2l\theta$$

$$\psi_1 = \frac{M_p l}{4EI} - \theta$$

$$\psi_2 = -\frac{M_p l}{4EI} + 3\theta$$

$$\psi_3 = -\ 2\theta$$

If it were supposed that the last hinge formed at section 2, it can be seen from the above equations that θ would be $\dfrac{M_p l}{12EI}$, so that ψ_1 would be $\dfrac{M_p l}{6EI}$. This assumption must also be incorrect, since the plastic hinge at section 1 is of negative sign. It is thus evident that the last hinge must form at section 1. The corresponding value of θ is $\dfrac{M_pl}{4EI}$; with this value of θ it is found that

$$\delta = \frac{2M_pl^2}{3EI}$$

$$\psi_1 = 0$$

$$\psi_2 = \frac{M_pl}{2EI}$$

$$\psi_3 = -\frac{M_pl}{2EI}$$

Here both ψ_2 and ψ_3 are of the correct sign, confirming that the last hinge does in fact form at section 1.

From this example it can be seen that one general method of calculation is to assume that one hinge is the last to form, and to work out the rotations at all the other hinges on this assumption. The correct solution can then be obtained by superposing the displacements and hinge rotations due to a motion of the collapse mechanism of such an amount that the rotations at all the hinges except one become of the correct sign and the rotation at this one hinge, which is the last hinge to form, becomes zero. Another method of calculation consists of assuming in turn that each of the hinges is the last hinge to form, and then calculating the deflection at some point; the largest of the deflections thus obtained is then the actual deflection. This result, which has been stated by Dutheil [22] without proof, can be seen to follow from the fact that any incorrect solution can be found from the correct solution by superposing the displacements and hinge rotations due to a *backwards* motion of the collapse mechanism. In applying this latter method it may be unnecessary to perform the calculations for the assumptions that certain hinges would be the last to form, for it can often be seen that when the structure is behaving elastically some sections would be more highly stressed than others, and would therefore be among the first to form hinges. For instance, in the example just considered it is easy to see that the first hinge to form would be at section 3.

Each of these methods will now be explained briefly with reference to a simple portal frame.

Rectangular portal frame

Consider the rectangular portal frame whose dimensions and loading are shown in Fig. 5.11(a). All the members of this frame are of the same uniform section, with fully plastic moment M_p and flexural rigidity EI. It is required to find the horizontal and vertical deflections h and v which occur in the directions of the loads at the point of collapse.

For this frame collapse occurs when $W = W_c = 3\frac{M_p}{l}$, plastic hinges then being formed at sections 1, 3, 4 and 5 while the

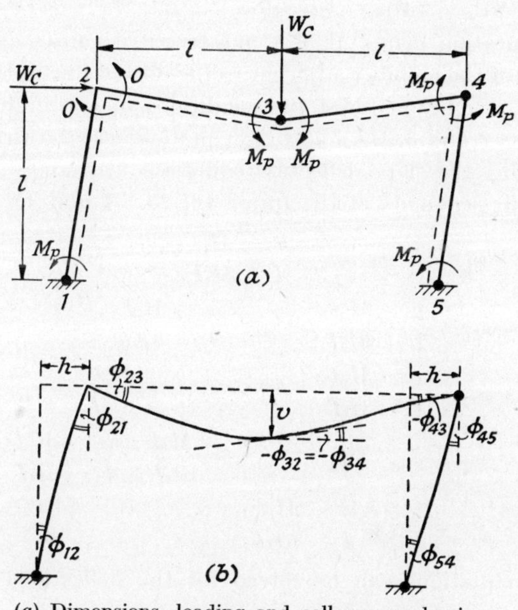

(*a*) Dimensions, loading and collapse mechanism.
(*b*) Deformations during collapse.

Fig. 5.11. *Rectangular portal frame.*

bending moment at section 2 is zero. The directions in which the fully plastic moments act upon the members are indicated. in Fig 5.11(*a*).

In the first place the frame will be analysed by assuming that the last hinge forms at a certain section, determining all the other hinge rotations, and then superposing a forwards motion of the collapse mechanism so as to give each hinge, except the one which is the last to form, a rotation of the correct sign. It is clear that when the frame is behaving elastically the largest bending moments will occur at sections 4 and 5, and so the last hinge to form will be at either section 1 or section 3. The first assumption will be that the last hinge will form at section 3. The deflected form of the frame, assuming continuity at section 3, and of course at section 2 where there is no hinge, is as shown in Fig. 5.11(*b*). The rotations of the hinges at sections 1, 4 and 5 are given in terms of the rotations at the ends of the members by the following equations :

$$\psi_1 = -\phi_{12}, \quad \psi_4 = \phi_{43} - \phi_{45}, \quad \psi_5 = \phi_{54},$$

a hinge rotation being taken as positive if it causes extension of the fibres of the member adjacent to the dotted line in Fig. 5.11(*a*). Using these relations, and the conditions of continuity of slope at sections 2 and 3, the following equations can be written down by applying the slope-deflection equations 5.15 and 5.16 to each of the four segments of the frame 12, 23, 34 and 45 in turn :

$$\psi_1 = -\phi_{12} = -\frac{h}{l} + \frac{M_p l}{3EI} \quad . \quad . \quad . \quad 5.20$$

$$\phi_{21} = \frac{h}{l} + \frac{M_p l}{6EI} = \phi_{23} = \frac{v}{l} + \frac{M_p l}{6EI} \quad . \quad . \quad . \quad 5.21$$

$$\phi_{32} = \frac{v}{l} - \frac{M_p l}{3EI} = \phi_{34} = -\frac{v}{l} + \frac{M_p l}{6EI} \quad . \quad . \quad 5.22$$

$$\psi_4 = \phi_{43} - \phi_{45} = -\frac{v}{l} + \frac{M_p l}{6EI} - \frac{h}{l} + \frac{M_p l}{6EI} \quad . \quad 5.23$$

$$\psi_5 = \phi_{54} = \frac{h}{l} - \frac{M_p l}{6EI} \quad . \quad . \quad . \quad . \quad . \quad 5.24$$

These equations can be solved for the deflections and hinge rotations, giving

$$h = v = \frac{M_p l^2}{4EI}$$

$$\psi_1 = \frac{M_p l}{12EI}$$

$$\psi_4 = -\frac{M_p l}{6EI}$$

$$\psi_5 = \frac{M_p l}{12EI}$$

Inspection of Fig. 5.11(*a*) shows that while the hinge rotations at sections 4 and 5 should be negative and positive, respectively, as in the above solution, the hinge rotation at section 1 should not be positive, but either negative or zero. Thus the last hinge must form at section 1, and the correct solution is obtained by superposing the deflections and hinge rotations due to a small motion of the collapse mechanism such that the change in the hinge rotation at section 1 is $-\frac{M_p l}{12EI}$. It will be seen from Fig. 5.11(*a*) that the changes in the deflections and hinge rotations during a small motion of the collapse mechanism which occur when the rotation of the hinge at section 1 has this value are as follows:

$$\Delta h = \Delta v = \frac{M_p l^2}{12EI}$$

$$\Delta \psi_1 = -\frac{M_p l}{12EI}$$

$$\Delta \psi_3 = \frac{M_p l}{6EI}$$

$$\Delta \psi_4 = -\frac{M_p l}{6EI}$$

$$\Delta \psi_5 = \frac{M_p l}{12EI}$$

When these increments are added to the deflections and hinge rotations obtained on the assumption that the last hinge formed at section 3, the following values are obtained:

$$h = v = \frac{M_p l^2}{3EI}$$

$$\psi_1 = 0$$

$$\psi_3 = \frac{M_p l}{6EI}$$

$$\psi_4 = -\frac{M_p l}{3EI}$$

$$\psi_5 = \frac{M_p l}{6EI}$$

These deflections and hinge rotations are those which occur when the frame is at the point of collapse.

The other method of calculation is to assume that the last hinge to form occurs at one particular cross-section and to calculate the corresponding value of one of the deflections. The process is then repeated for the assumption of continuity at other likely cross-sections where the last hinge might form, and the correct assumption is identified as the one which leads to the largest value of this deflection. In the present example it appears unlikely that the last hinge forms at either sections 4 or 5, so that it is only necessary to calculate the value of some deflection, say h, on the assumptions that the last hinge forms at either section 1 or section 3.

If the last hinge is formed at section 1, the condition of continuity at this section is that $\phi_{12} = 0$. Then from equation 5.20, the slope-deflection equation for the member 12, it is found that

$$\phi_{12} = -\frac{h}{l} + \frac{M_p l}{3EI} = 0$$

$$h = \frac{M_p l^2}{3EI}$$

If the last hinge is formed at section 3, the condition of continuity at this section is that $\phi_{32} = \phi_{34}$, so that from the slope-deflection equations 5.22

$$\phi_{32} = \frac{v}{l} - \frac{M_p l}{3EI} = \phi_{34} = -\frac{v}{l} + \frac{M_p l}{6EI}$$

$$v = \frac{M_p l^2}{4EI}$$

Then from the slope-deflection equations 5.21, expressing the condition of continuity at section 2,

$$\phi_{21} = \frac{h}{l} + \frac{M_p l}{6EI} = \phi_{23} = \frac{v}{l} + \frac{M_p l}{6EI}$$

$$h = v = \frac{M_p l^2}{4EI}$$

This value of h is smaller than the value obtained on the assumption of continuity at section 1, and so it follows that the last hinge must form at section 1. This conclusion should be checked by completing the analysis on the assumption of continuity at section 1 and confirming that the hinge rotations so obtained are of the correct sign.

Partial and over-complete collapse

In the example just considered the collapse mechanism was of the complete type, for there were $r + 1$ plastic hinges in the collapse mechanism, r being the number of redundancies, and this mechanism only possessed one degree of freedom. Thus the entire frame was statically determinate at collapse. When the collapse is partial, a statical analysis will only determine the bending moment distribution in part of the frame at collapse, but this does not give rise to any additional difficulty in determining the deflection at the point of collapse. The slope-deflection equations are applied in the usual manner, and it is found that these equations furnish sufficient information for the calculation of those bending moments which are not determined by statics, as well as the unknown deflections and hinge rotations. As compared with the deflection analysis of a case of complete collapse, it can be remarked that if the number of plastic hinges in the collapse mechanism is $r + 1 - q$, there will be q bending moments which cannot be found by statics, but there will be continuity at the corresponding q cross-sections. If there had been plastic hinges at these q cross-sections, there would have been q unknown hinge rotations to be determined, whereas in fact there are q unknown bending moments, so that the total number of unknowns is unaltered.

If the collapse mechanism for a frame is over-complete, possessing more than one degree of freedom, there will be a number of plastic hinges forming simultaneously at collapse. In such a case it would be necessary to assume continuity at a *group* of hinge positions at the point of collapse, rather than at a single position. This would lead to a considerable increase in the labour of computation, for the possible number of such groups could be quite large in all but the simplest cases. This difficulty is best overcome by making a small change in one or more of the load ratios so that the collapse is then either of the complete or partial kind. The corresponding deflections obtained would then be close to those which would occur with the actual load ratios.

Validity of assumptions

The errors introduced by assuming that the effects of strain-hardening and the spread of plastic zones along the members can both be neglected, have already been discussed. However, the method also depends on the assumption that the rotation at a plastic hinge never ceases once it has formed, with the corollary that no plastic hinges are formed which are not involved in the collapse mechanism. It might be thought that in cases of proportional loading this assumption would obviously be correct, but unfortunately this is not so. This can be demonstrated by a simple example of a beam continuous over three supports, as shown in Fig. 5.12(a). For this beam under proportional loading the first plastic hinge to form is at section 2. However, at a later stage in the loading rotation at this hinge ceases and the bending moment at this section is reduced below its fully plastic value.

The beam is supposed to be of uniform section, the flexural rigidity being EI and the fully plastic moment M_p. Fig. 5.12(b) shows the bending moment distribution at collapse, with plastic

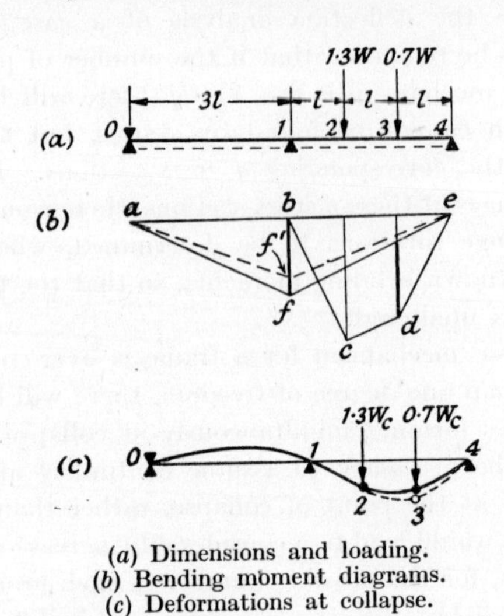

(a) Dimensions and loading.
(b) Bending moment diagrams.
(c) Deformations at collapse.

Fig. 5.12. Beam on three supports.

hinges at sections 1 and 3. In this figure the free moment diagram is shown as *abcde*, and the reactant moment diagram as *afe*. The free bending moment at section 3 is $0{\cdot}9Wl$, and from the figure it can at once be seen that the collapse load W_c can be calculated by considering conditions at this section as follows:

$$0{\cdot}9W_cl = \frac{4}{3}M_p$$

$$W_c = 1{\cdot}48\frac{M_p}{l}$$

At section 2 the free moment is $1{\cdot}1Wl$, so that at collapse the bending moment at this section is given by

$$M_2 = 1{\cdot}1W_cl - \tfrac{2}{3}M_p = 0{\cdot}96M_p$$

The deflection analysis is carried out by assuming continuity at section 2, and also at section 1, where it is fairly evident that the last hinge forms. Details of the analysis, which is very simple, need not be given. It is found that the vertical deflections v_2 and v_3 at sections 2 and 3, respectively, and the hinge rotation ψ_3 at section 3, are

$$v_2 = 1{\cdot}17\frac{M_p l^2}{EI}$$

$$v_3 = 1{\cdot}70\frac{M_p l^2}{EI}$$

$$\psi_3 = 1{\cdot}41\frac{M_p l}{EI}$$

The hinge rotation at section 3 is of the correct sign, apparently confirming the assumption of continuity at section 2.

At collapse the beam is statically determinate and the value of M_2 is less than M_p when hinges are formed at sections 1 and 3, so that the analysis gives no inkling of the fact that under proportional loading the *first* hinge forms at section 2. This can be shown by performing an elastic analysis; the corresponding reactant moment line is shown in Fig. 5.12(*b*) as the dotted line *af'e*. The slope of the line *f'e* is less than that of the portion *cd* of the free moment diagram, so that the moment at section 2 is greater than the moment at section 3. If W increases above the value which causes the formation of a plastic hinge at section 2, rotation of this hinge ensues until the bending moment at section 3 is brought up to its fully plastic value. Further increases of W cause rotation of the hinge at this section, while rotation of the hinge at section 2 ceases and the bending moment there is *reduced* below its fully plastic value. Eventually a plastic hinge also forms at section 1 and collapse occurs. This behaviour is readily investigated by means of a step-by-step analysis based on the slope-deflection equations; the results are summarized in Table 5.1.

TABLE 5.1

Proportional loading to collapse

$\frac{Wl}{M_p}$	$\frac{M_1}{M_p}$	$\frac{M_2}{M_p}$	$\frac{M_3}{M_p}$	$\frac{\psi_2 EI}{M_p l}$	$\frac{\psi_3 EI}{M_p l}$	$\frac{v_2 EI}{M_p l^2}$	$\frac{v_3 EI}{M_p l^2}$
1·32	−0·68	1	0·96	0	0	0·75	0·78
1·43	−0·86	1	1	0·36	0	0·98	0·91
1·48	−1	0·96	1	0·36	0·69	1·17	1·34

Fig. 5.12(c) shows the deflected form of the beam at collapse. In this figure the dotted line for the segment 234 of the beam shows the deflected form corresponding to the assumption of continuity at section 2. The deflected form of the segment 012 of the beam does not depend on the amount of the hinge rotation at section 2, for the *shape* of this portion of the beam is determined by its known distribution of bending moment, there being no hinge rotations within this length, and the deflections are then determined by the positions of the simple supports at 0 and 1. Thus the vertical deflection v_2 was found from the simple analysis at the point of collapse to be $1{\cdot}17\frac{M_p l^2}{EI}$, and this same value is given in Table 5.1 as a result of the step-by-step analysis. The *shapes* of the segments 23 and 34 of the beam are also determined by the known distribution of bending moment. It follows that if the deflection v_3 and the hinge rotations ψ_2 and ψ_3 calculated by the step-by-step process are subtracted from those calculated from the simple point of collapse analysis, the resulting deflection and hinge rotation differences should correspond to a small motion of a beam-type mechanism with hinges at sections 2, 3 and 4. These differences are:

$$\Delta\psi_2 = -\,0{\cdot}36\frac{M_p l}{EI}$$

$$\Delta\psi_3 = (1{\cdot}41 - 0{\cdot}69)\frac{M_p l}{EI} = 0{\cdot}72\frac{M_p l}{EI}$$

$$\Delta v_3 = (1{\cdot}70 - 1{\cdot}34)\frac{M_p l^2}{EI} = 0{\cdot}36\frac{M_p l^2}{EI},$$

and it can be seen that they do in fact correspond to a small motion of this mechanism. This result does not, of course, simplify the process of correcting the estimate of v_3 as obtained from the simple point of collapse analysis to allow for the hinge rotation at section 2, for the magnitude of this hinge rotation is not known until the step-by-step analysis has been carried through.

From this example it is seen that even if attention is confined to the case of proportional loading, it is possible for a hinge rotation to have occurred at one or more sections where the bending moment at collapse is less than the fully plastic value. No analysis which is concerned solely with conditions at collapse can possibly reveal the fact that such a hinge rotation has occurred, and so it must be concluded that the only completely safe procedure is to trace the successive formation and rotation of the plastic hinges by means of a step-by-step analysis.

It thus appears that the basic assumptions of this method for estimating deflections at the point of collapse may be incorrect in certain cases. However, it will often be possible to recognize such cases when they occur; the example just given is similar in principle to the movement of a plastic hinge along the span of a member subjected to a uniformly distributed load which often occurs (see example 2, Chapter 3). The value of deflection estimates obtained in this way can only be assessed by comparison with experimental results. Symonds [23] has compared deflection estimates of this kind with the tests of Baker and Heyman [24] on miniature rectangular portal frames; the agreement was found to be quite satisfactory in most cases. Baker and Eickhoff [25] have tested a pair of full-scale welded pitched roof portal frames

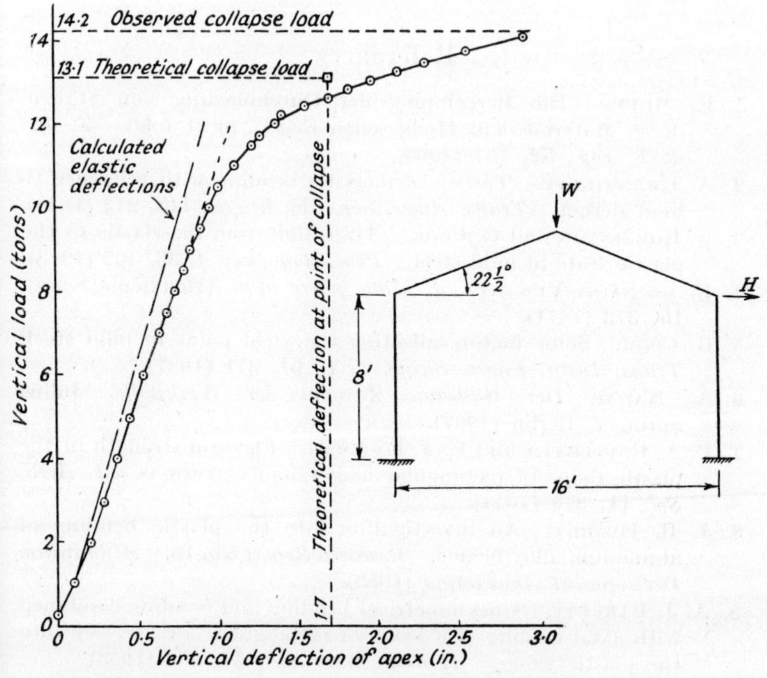

Fig. 5.13. *Test of full-scale pitched roof portal frame.*

fabricated from 7 in. × 4 in. British Standard beams, for which the centre-line dimensions were as shown in the inset to Fig. 5.13. These frames were tested together by applying first a horizontal load $H = 1{\cdot}70$ tons to each frame and then increasing the vertical load W while H was held constant. The observed relation between W and the vertical deflection of the apex for one of these frames is shown in Fig. 5.13. The theoretical value of W at collapse was 13·1 tons, but the frame continued to carry loads in excess of this value until complete failure by lateral buckling occurred at a value of W of 14·2 tons. It is evident that strain-hardening is responsible for the increase of load-carrying capacity above the theoretical value. The computed deflection of 1·70 in. at the point of collapse is shown in the figure. Although this deflection is less than the observed deflection of 2 in. at the theoretical collapse load, it should be noted that the computed deflections also underestimate the actual deflections in the elastic range quite appreciably.

References

1. E. Meyer. Die Berechnung der Durchbiegung von Stäben, deren Material dem Hookeschen Gesetz nicht folgt. *Z. ver. dtsch. Ing.*, **52**, 167 (1908).
2. A. Hrennikoff. Theory of inelastic bending with reference to limit design. *Trans. Amer. Soc. Civ. Engrs.*, **113**, 213 (1948).
3. A. Robertson and G. Cook. Transition from the elastic to the plastic state in mild steel. *Proc. Roy. Soc. A*, **88**, 462 (1913).
4. B. de Saint-Venant. *J. Math. pures appl.* (Deuxième Série). **16**, 373 (1871).
5. G. Cook. Some factors affecting the yield point in mild steel. *Trans. Instn. Engrs. Shipb. Scot.*, **81**, 371 (1937).
6. A. Nadai. *Der Bildsame Zustand der Werkstoffe.* Julius Springer, Berlin (1927).
7. F. A. Rappleyea and E. J. Eastman. Flexural strength in the plastic range of rectangular magnesium extrusions. *J. Aero. Sci.* **11**, 373 (1944).
8. J. B. Dwight. An investigation into the plastic bending of aluminium alloy beams. *Research Report* No. 16. *Aluminium Development Association* (1953).
9. A. J. Barrett. Unsymmetrical bending and bending combined with axial loading of a beam of rectangular cross-section into the plastic range. *J. Roy. Aero. Soc.*, **57**, 503 (1953).
10. J. Fritsche. Die Tragfähigkeit von Balken aus Stahl mit

Berücksichtigung des plastischen Verformungsvermögens. *Bauingenieur*, **11**, 851 (1930).

11. J. W. RODERICK and I. H. PHILLIPS. The carrying capacity of simply supported mild steel beams. *Research* (*Engng. Struct. Suppl.*), *Colston Papers*, **2**, 9 (1949).
12. J. W. RODERICK and J. HEYMAN. Extension of the simple plastic theory to take account of the strain-hardening range. *Proc. Instn. Mech. Engrs.*, **165**, 189 (1951).
13. J. W. RODERICK. The load-deflection relationship for a partially plastic rolled steel joist. *Brit. Weld. J.*, **1**, 78 (1954).
14. J. W. RODERICK and H. H. L. PRATLEY. The behaviour of rolled steel joists in the plastic range. *Brit. Weld. J.*, **1**, 261 (1954).
15. C. H. YANG, L. S. BEEDLE and B. G. JOHNSTON. Residual stress and the yield strength of steel beams. *Weld. J.*, Easton, Pa., **31**, 205-s (1952).
16. M. R. HORNE. Discussion of " Theory of inelastic bending with reference to limit design ". *Trans. Amer. Soc. Civ. Engrs.*, **113**, 250 (1948).
17. C. H. YANG, L. S. BEEDLE and B. G. JOHNSTON. Plastic design and the deformation of structures. *Weld. J.*, Easton, Pa., **30**, 348-s (1951).
18. M. R. HORNE. The effect of strain-hardening on the equalisation of moments in the simple plastic theory. *Weld. Res.*, **5**, 147 (1951).
19. K. E. KNUDSEN, C. H. YANG, B. G. JOHNSTON, L. S. BEEDLE and W. H. WEISKOPF. Plastic strength and deflections of continuous beams. *Weld. J.*, Easton, Pa., **32**, 240-s (1953).
20. P. S. SYMONDS and B. G. NEAL. Recent progress in the plastic methods of structural analysis. *J. Franklin. Inst.*, **252**, 383, 469 (1951).
21. P. S. SYMONDS and B. G. NEAL. The interpretation of failure loads in the plastic theory of continuous beams and frames. *J. Aero. Sci.*, **19**, 15 (1952).
22. J. DUTHEIL. L'exploitation des phénomène d'adaptation dans les ossatures en acier doux. *Ann. Inst. Tech. Bât. Trav. Publ.*, No. 2, Jan. 1948.
23. P. S. SYMONDS. Discussion of plastic design and the deformation of structures. *Weld. J.*, Easton, Pa., **31**, 33-s (1952).
24. J. F. BAKER and J. HEYMAN. Tests on miniature portal frames. *Struct. Engr.*, **28**, 139 (1950).
25. J. F. BAKER and K. G. EICKHOFF. A test on a pitched roof portal. *Brit. Weld. Res. Assn. Report* FE1/35 (1953).

Examples

1. A beam of uniform rectangular cross-section is simply supported over a span l and carries a uniformly distributed load W. The bending moment-curvature relation beyond the elastic limit is as given in equation 5.8, corresponding to the ideal plastic stress-strain relation with no upper yield stress. If W is increased steadily from zero to $\frac{9}{8}W_y$, show that the central deflection is

$$\left[1{\cdot}6\left(\frac{1}{\sqrt{3}} + \log\sqrt{3}\right) - \frac{2}{3}\right]\delta_y,$$

where W_y and δ_y are the values of the load and of the central deflection when yield first occurs, and sketch the form of the plastic zones.

2. A uniform fixed-ended beam is of length l and has a fully plastic moment M_p and flexural rigidity EI. It carries a load W uniformly distributed over its left-hand half and also a concentrated load W at mid-span. Estimate the central deflection at the point of collapse.

3. A fixed-base rectangular portal frame is of height l and span $3l$. All the members of the frame are of the same uniform cross-section, with fully plastic moment M_p and flexural rigidity EI. The frame is subjected to a central concentrated vertical load $3W$ and also to a concentrated horizontal load W at the top of one of the stanchions. Estimate the deflections in the directions of the two loads at the point of collapse.

4. For a frame identical in every respect with that of example 3 except that the span is $2l$ instead of $3l$, estimate the deflections in the directions of the two loads at the point of collapse.

CHAPTER

6

Factors Affecting the Fully Plastic Moment

6.1. Introduction

THE SIMPLE PLASTIC THEORY for frames is based on certain hypotheses concerning the bending moment-curvature relations for the members, which were stated in Section 1.2. The concept of the formation and indefinite rotation of a plastic hinge in a member whenever the fully plastic moment is maintained at a section is of fundamental importance in the theory, and it is assumed that for a given member the fully plastic moment is a constant. In fact, the simplicity of the plastic methods of analysis is due entirely to this concept. It must be recognized, however, that the fully plastic moment of a given member is not a definite, constant quantity. This is partly because the lower yield stress on which it is based is dependent to some extent on the manner of loading and the previous loading history. It was shown in Section 1.4 that the theoretical value of the fully plastic moment in pure bending is $Z_p f_L$, where Z_p is the plastic section modulus whose value depends solely on the shape and size of the cross-section, and f_L is the lower yield stress. Thus variations in the value of the fully plastic moment will occur whenever effects are brought into play which alter the value of the lower yield stress. As is well known, the lower yield stress of a mild steel specimen is affected by the rate of loading, and also by the temperature of testing to an extent which is small for the range of variation of temperature which will be experienced by many structures. A further factor which influences the lower yield stress is strain-ageing. A discussion of the order of magnitude of the variations in the lower yield stress which may be caused by these factors is given in Section 6.2; fortunately, it appears that the effects will usually be quite small. Although from the structural point of view the main concern is with those factors which influence the lower yield stress of a given mild steel specimen, it is also of interest to consider briefly the important metallurgical factors of chemical composition and heat treatment which govern the variations in the lower yield stress of mild steel from specimen to specimen, and a brief discussion of this topic is also included.

In calculating the fully plastic moment as $Z_p f_L$ by the methods indicated in Section 1.4, it was assumed that the member was subjected to pure bending, so that the axial thrust and shear force were both zero. As will be discussed in Sections 6.3 and 6.4, the value of the fully plastic moment is affected by the presence of axial thrust and shear force, to an extent which is calculable. Since in applications of the plastic methods, fully plastic moments and plastic hinges usually occur at positions where one or both of these influences is present, it is of considerable importance to be able to predict the changes in the values of the fully plastic moments due to these causes. It will appear that in many practical cases the effects of axial thrust and shear force will be very small, although there are certain types of structure in which it is important to make a proper allowance for these effects.

In practice, plastic hinges often occur beneath concentrated loads. Under these circumstances additional stresses other than those required to develop the fully plastic moment must be present, and the value of the fully plastic moment will be modified accordingly. These effects are discussed in Section 6.5, where it will appear that they are usually small and can in any case be allowed for on a semi-empirical basis.

6.2. Variations of lower yield stress

From the structural point of view, the most important factor which affects the lower yield stress is the rate of straining. Many investigations into this phenomenon have been carried out, for instance by Quinney,[1] Winlock and Leiter,[2] Elam [3] and Manjoine [4], and a summary of the available data on this topic was given by Cook [5] in 1937. In general, the effect of an increase in the rate of straining is to increase the value of the yield stress and to make the yield phenomenon more pronounced by increasing the length of the yield. Thus in testing a low carbon steel, Manjoine [4] found that an increase in the strain rate of nearly a thousandfold, from $9{\cdot}5 \times 10^{-7}$ per sec. to $8{\cdot}5 \times 10^{-4}$ per sec., increased the lower yield stress from 12·3 tons per sq. in. to 13·8 tons per sq. in. The higher of these strain rates is such that the specimen took just over one second to reach the yield point, and thus represents a fairly rapid rate of loading.

Tests by Manjoine [4] on the same steel showed a decrease in the lower yield stress from 13·8 tons per sq. in. to 11·8 tons per sq. in. due to an increase in the temperature of testing from room temperature to 200° C., both these tests being conducted at the same rate of straining of $8{\cdot}5 \times 10^{-4}$ per sec. Bearing in mind that the ordinary ranges of variation of rate of loading and temperature will be considerably less than the ranges which have been cited, it can be seen that these effects will not usually cause variations of more than one or two per cent in the value of the lower yield stress.

The lower yield stress of mild steel may be affected by strain-ageing. If a tensile specimen is subjected to loads which cause some degree of plastic straining, it is found that after a fairly considerable lapse of time at atmospheric temperature the elastic limit observed in a further loading in the same sense is increased. The amount of this increase has been stated by Elam [3] to be the same whether the rest takes place under load or with the load removed. The phenomenon is accelerated enormously by a small increase of temperature, and in consequence many investigations have been concerned with ageing for a short time of the order of an hour at temperatures of the order of 100–250° C. Thus in the early work of Muir [6] overstrained specimens were immersed in boiling water for 10 min.; this caused an increase in the elastic limit of the order of 4 tons per sq. in. if the second loading was in the same sense as the original loading. However, if the second loading was in the opposite sense, a *reduction* in the elastic limit of similar magnitude was observed due to the Bauschinger [7] effect. Edwards, Jones and Walters,[8] working with a 0·08% carbon killed steel, found that if the specimen was strained through the yield up to the point where strain-hardening just commenced, and was then aged at room temperature, the yield stress was raised from 12·0 tons per sq. in. to 13·1 tons per sq. in. by ageing for 3 days, and to 14·8 tons per sq. in. by ageing for one year. Similar results were obtained by Griffis, Kenyon and Burns.[9] The effects of larger amounts of cold work were discussed by Elam.[3] Summaries of the available information on ageing in steel have been given by Kenyon and Burns,[10] and by Davenport and Bain.[11]

Two main conclusions can be drawn from this evidence concerning the variation of lower yield stress with rate of straining, temperature of testing and the degree of strain-ageing. In the first place, it is clear that the variations of yield stress, and therefore of fully plastic moment, due to the variations in rate of loading and of temperature which may be experienced by structures will usually be small. Also, strain-ageing effects will only occur in a structure if it is subjected to loads of sufficient magnitude to cause the formation of one or more plastic hinges. In structures designed to load factors of the order of 2, such loadings are rather unlikely to occur, and in any event the effect would only be to raise the fully plastic moments at the hinge positions concerned by 10–20%, depending on the time which had elapsed since the overload. Thus the collapse load, depending as it does on the values of the fully plastic moments at all the hinges involved in the collapse mechanism, would not be greatly affected. It therefore seems reasonable to conclude that in all but exceptional cases the assumption of constant fully plastic moments for the members of steel frames is not likely to lead to serious error.

The second conclusion to be drawn is that any attempt to assess the value of the fully plastic moment by determining the lower yield stress from a tensile test and then computing the fully plastic moment as $Z_p f_L$ must rest on empirical evidence as to its validity. In flexure the strain at any instant, and therefore the rate of straining, must vary across the section in a roughly linear manner. Since the lower yield stress depends on the rate of straining, it follows that some variation of lower yield stress across the section is to be expected. Moreover, constant rates of strain rarely occur in either full-scale structures or in model structures subjected to dead loads. Increments of load are found to cause initially rapid rates of deflection followed by a progressively slower approach towards final equilibrium. It follows that the specification of the strain rate at which a tensile test should be conducted in order to allow the prediction of the fully plastic moment is almost purely empirical. In this connection it is also interesting to note that for a rolled steel section in the " as received " condition, the properties of the material vary widely for specimens cut from different locations in the section. Thus Robertson[12] performed tensile tests on small specimens cut from a 24 in. × $7\frac{1}{2}$ in. British Standard beam, and found yield stresses varying from 13 to 18 tons per sq. in. There was also a considerable variation in the general form of the stress-strain diagrams. Similar tests have been reported by Luxion and Johnston [13] and by Roderick,[14] the variations in yield stress being smaller in these later tests. These variations are apparently due to the differences in amounts of plastic working and in rates of cooling during the rolling process, but they seem to occur rather unpredictably as far as location in the flanges and web is concerned. This creates a further difficulty in the selection of the position in the section from which a tension specimen should be cut in order to determine the lower yield stress. It may be concluded that the only sure way of determining the fully plastic moment of a beam is by means of a bending test on a simply supported length.

Although from the point of view of the plastic theory the principal concern is with the variations of lower yield stress which may occur in a beam of a given steel, it is also of some interest in relation to laboratory experiments to examine briefly the factors which govern the variations of lower yield stress between different mild steel specimens. Of these, the most important factors are the chemical composition, in particular the carbon content, and the heat treatment, these factors being interrelated. The standard heat treatment for structural mild steel is to heat the specimen in a furnace to a uniform temperature of the order of 900° C., depending on the carbon content, and then cool to room temperature. If the cooling takes place in the furnace, the treatment is referred to as *annealing*, and if air-cooling is used the treatment is termed *normalizing*. In both these treatments a complete recrystallization takes place, so that the properties become largely independent of any previous heat treatment or cold working. Another treatment which is sometimes used is *process annealing*, in which the soaking temperature is about 600° C. Here there may be no recrystallization, but the specimen is partly relieved of stresses caused by any previous cold working or machining operations, according to the time of heating.

Large changes in the lower yield stress can be produced by heat treatment. Bullens [15] has summarized data from various sources concerning the variation of lower yield stress with carbon content for both annealed and normalized steels. From his curves it appears that typical values of the lower yield stress for a 0·15% carbon steel are 13·4 tons per sq. in. when annealed and 15·2 tons per sq. in. when normalized. Although it will be appreciated that there is a wide degree of scatter in the observations owing to the influence of other factors, these figures indicate that an annealing treatment results in a somewhat smaller yield stress than a normalizing treatment.

From the same data it is found that for annealed specimens the effect of raising the carbon content from 0·15% to 0·3% is to raise the lower yield stress from 13·4 tons per sq. in. to 17·0 tons per sq. in. However, while an increase in the carbon content raises the lower yield stress, it also makes the yield phenomenon less pronounced, the ratio of upper to lower yield stress and the length of the yield both being reduced. Thus Roderick and Heyman [16] carried out bending tests on four annealed steels of different carbon content, and from these tests derived values for the lower yield stress f_L, the ratio of upper to lower yield stress, the ratio of the strain ε_s at the onset of strain-hardening to the strain ε_y at the yield point, and the ratio of Young's Modulus to the initial slope E_s of the strain-hardening portion of the stress-strain diagram. The values of these quantities were as shown in Table 6.1.

TABLE 6.1

% C.	f_L (tons per sq. in.)	$\frac{f_U}{f_L}$	$\frac{\varepsilon_s}{\varepsilon_y}$	$\frac{E}{E_s}$
0·28%	22·0	1·33	9·2	26·9
0·49%	25·0	1·28	3·7	17·3
0·74%	29·0	1·19	1·9	14·2
0·89%	34·0	1·04	1·5	10·2

6.3. Effect of normal force

In considering the influence of a normal force, either tensile or compressive, on the value of the fully plastic moment of a beam, the assumptions which were mentioned in Section 1.4 will be retained, in particular that the neutral surface is plane and parallel to the axis of bending. It will also be assumed that the cross-section of the beam has two axes of symmetry, and that flexure is taking place about one of the axes of symmetry, as in Fig. 6.1(a).

Suppose that at a particular cross-section in a beam of this type, at which there is a tensile or compressive normal force N, a condition of full plasticity has been attained, and the fully plastic moment, which was M_p in the absence of any normal force, is M_N. The stress distribution will be as shown in Fig. 6.1(b), the neutral axis being displaced by a distance e from the relevant axis of symmetry.

For the purpose of calculating the total normal force and moment it is convenient to consider separately the effect of the stresses on the central portion of the cross-section within a distance e from the relevant axis of symmetry, these stresses being

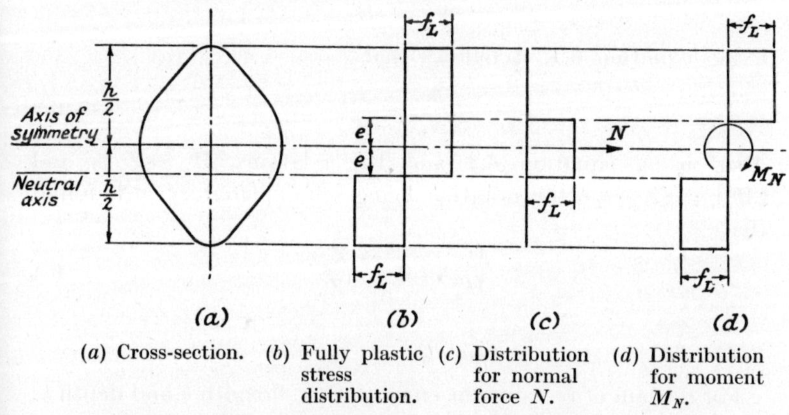

(a) Cross-section. (b) Fully plastic stress distribution. (c) Distribution for normal force N. (d) Distribution for moment M_N.

Fig. 6.1. *Effect of normal force on fully plastic moment.*

indicated in Fig. 6.1(c), and the effect of the stresses on the remainder of the cross-section, as indicated in Fig. 6.1(d). It will be appreciated that the former stresses have a resultant which is a pure normal force through the centroid of the cross-section, and that the resultant of the latter stresses is a pure couple about the axis of symmetry. Thus the central portion of the cross-section of depth $2e$ can be regarded as carrying the normal force N, leaving the remainder of the section for the development of the moment M_N. It follows that the fully plastic moment is reduced by an amount equal to the fully plastic moment of the central portion of the section of depth $2e$. If the area of this central portion of the section is denoted by A_e, it follows that

$$N = A_e f_L \quad . \quad . \quad . \quad . \quad 6.1$$

Also, if M_p is the fully plastic moment of the section in the absence of any normal force, and $(M_p)_e$ is the fully plastic moment of the central portion of the section of depth $2e$,

$$M_N = M_p - (M_p)_e \quad . \quad . \quad . \quad 6.2$$

For a given value of the normal force N, equation 6.1 defines the value of e and thus the position of the neutral axis; the fully plastic moment is then found from equation 6.2.

It is convenient to define the value of the normal force at which the cross-section would become fully plastic in the absence of any bending moment as N_p, the *fully plastic thrust*. If the total area of the cross-section is A,

$$N_p = Af_L$$

Using equation 6.1, it follows that

$$\frac{N}{N_p} = \frac{A_e}{A} \quad . \quad . \quad . \quad . \quad 6.3$$

Also, using equation 6.2 and the relations $M_p = Z_p f_L$ and $(M_p)_e = (Z_p)_e f_L$, the notation being self-explanatory, it is found that

$$\frac{M_N}{M_p} = 1 - \frac{(Z_p)_e}{Z_p} \quad . \quad . \quad . \quad 6.4$$

Rectangular cross-section

For a beam of rectangular cross-section, breadth b and depth h, which is bent about an axis parallel to the sides of breadth b, the general equations 6.3 and 6.4 give the following results:

$$\frac{N}{N_p} = \frac{2eb}{hb} = \frac{2e}{h}$$

$$\frac{M_N}{M_p} = 1 - \frac{be^2}{\frac{1}{4}bh^2} = 1 - \frac{4e^2}{h^2}$$

Eliminating e between these equations it is found that

$$\frac{M_N}{M_p} = 1 - \left(\frac{N}{N_p}\right)^2 \quad . \quad . \quad . \quad 6.5$$

This result was first obtained by Girkmann.[17]

I-section

I-sections may be regarded as consisting of three rectangles, as indicated in Fig. 6.2, without any great loss of accuracy. If under a normal force N the neutral axis lies in the web, as shown in the figure, the value of A_e is $2et_1$, so that from equation 6.3

$$\frac{N}{N_p} = \frac{2et_1}{A} \quad . \quad . \quad . \quad . \quad 6.6$$

where A is the total area of the cross-section, the other symbols being defined in Fig. 6.2. This relation holds true so long as A_e does not exceed the web area A_w, so that $\frac{N}{N_p} \leqslant \frac{A_w}{A}$. The value of $(Z_p)_e$ is that appropriate to a rectangular section of breadth t_1

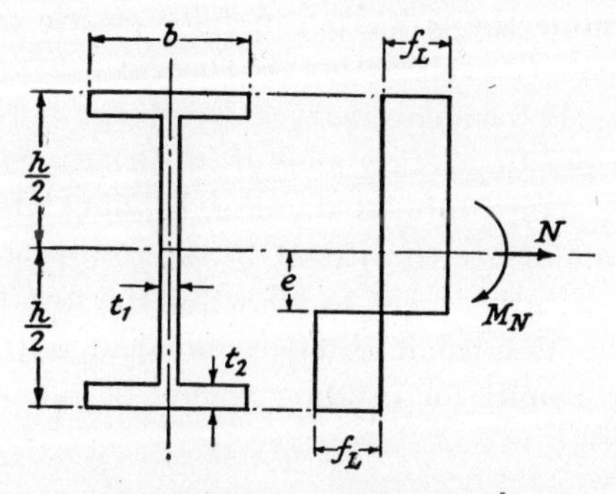

Fig. 6.2. *Effect of normal force on fully plastic moment of I-section.*

and depth $2e$, namely t_1e^2, and the value of Z_p was shown in Section 1.4, equation 1.13, to be

$$Z_p = \tfrac{1}{2}A_f(h - t_2) + \tfrac{1}{4}A_w(h - 2t_2)$$

where A_f is the total flange area and A_w is the web area. Thus from equation 6.4,

$$\frac{M_N}{M_p} = 1 - \frac{t_1e^2}{\frac{1}{2}A_f(h - t_2) + \frac{1}{4}A_w(h - 2t_2)} \quad . \quad . \quad 6.7$$

Eliminating e between equations 6.6 and 6.7, the following result is obtained after some rearrangement:

$$\frac{M_N}{M_p} = 1 - \left(\frac{N}{N_p}\right)^2 \left[\frac{1}{1 - \left(\frac{A_f}{A}\right)^2\left(1 - \frac{t_1}{b}\right)}\right] \quad . \quad 6.8$$

so long as $\frac{N}{N_p} \leqslant \frac{A_w}{A}$.

If $\frac{N}{N_p} > \frac{A_w}{A}$, so that the neutral axis lies in one flange, it can be shown that

$$\frac{M_N}{M_p} = 1 - \left[\frac{\left(\frac{N}{N_p}\right)^2 - \left(1 - \frac{t_1}{b}\right)\left(\frac{N}{N_p} - \frac{A_w}{A}\right)^2}{1 - \left(\frac{A_f}{A}\right)^2\left(1 - \frac{t_1}{b}\right)}\right] \quad . \qquad 6.9$$

so long as $\frac{A_w}{A} \leqslant \frac{N}{N_p} \leqslant 1$. Equivalent results were first obtained by Girkmann.[17] Beedle, Ready and Johnston [18] have carried out tests to determine the reduction of fully plastic moment due to thrust which agreed well with this theory.

In Appendix B formulae derived by Horne [19] for the reduced values of the plastic section moduli of British Standard beams under axial load are tabulated. For bending about the major axis these formulae are equivalent to the results in equations 6.8 and 6.9, with an adjustment to allow for the fact that the actual shape of each section is not exactly equivalent to three rectangles. Corresponding results for bending about the minor axis are also given.

For British Standard beams the ratio $\frac{A_f}{A}$ varies from about 0·55 to 0·8. The 7 in. × 4 in. joist lies roughly in the middle of this range, with a value of $\frac{A_f}{A}$ of about $\frac{2}{3}$, and for this section the ratio $\frac{t_1}{b}$ is 0·063. The variation of $\frac{M_N}{M_p}$ with $\frac{N}{N_p}$ for this section is plotted as curve (ii) in Fig. 6.3, curve (i) representing the parabolic relation of equation 6.5 for a rectangular section.

Effect of normal force in practical cases

It will be seen from Fig. 6.3 that the reduction in fully plastic moment due to normal force is more marked for a beam of I-section than for a beam of rectangular cross-section. However, even in the former case the reduction is not appreciable unless $\frac{N}{N_p}$ is greater than about 0·1, for at this value of $\frac{N}{N_p}$ the value of $\frac{M_N}{M_p}$ is 0·983, so that the reduction in fully plastic moment is only 1·7%. For

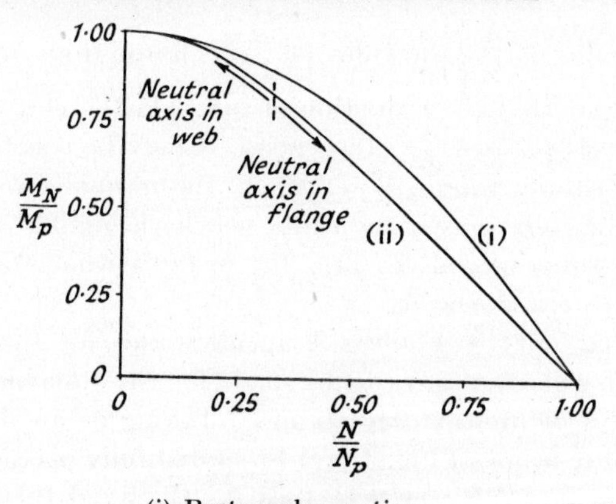

(i) Rectangular section.
(ii) Typical I-section, $A_f = \frac{2}{3}A$.

Fig. 6.3. *Reduction of fully plastic moment by normal force.*

single-storey frames the value of $\frac{N}{N_p}$ is usually rather less than 0·1. Thus taking the pitched roof portal frame illustrated in Fig. 4.1(a) as a typical example, it was shown by the trial and error method in Section 4.2 that if the feet of the vertical members were both fixed the required value of the fully plastic moment was 5·73 tons ft. Assuming each member of the frame to be a 6 in. × 3 in. British Standard beam, with a plastic section modulus of 8·084 in³., the fully plastic moment in the absence of axial thrust would have the value

$$M_p = \frac{8{\cdot}084 \times 15{\cdot}25}{12} = 10{\cdot}27 \text{ tons ft.}$$

taking the lower yield stress as 15·25 tons per sq. in. The use of this member would thus provide a load factor of $\frac{10{\cdot}27}{5{\cdot}73} = 1{\cdot}79$.

The greatest axial thrust occurs in the member 45, this thrust being 1·97 tons. The cross-sectional area of a 6 in. × 3 in. British Standard beam is 3·53 sq. in. Thus the fully plastic thrust N_p has the value

$$N_p = 3{\cdot}53 \times 15{\cdot}25 = 53{\cdot}8 \text{ tons.}$$

At collapse the thrust in the member DE would be 1·79 × 1·9ʼ = 3·53 tons, so that for this member

$$\frac{N}{N_p} = \frac{3{\cdot}53}{53{\cdot}8} = 0{\cdot}066\,.$$

With this value of $\frac{N}{N_p}$, the value of $\frac{M_N}{M_p}$ is found from Appendix B to be 0·993, so that the reduction of fully plastic moment due to axial thrust is only 0·7%. In general, it may be concluded that for a single-storey frame the effect of the normal forces on the values of the fully plastic moments will be negligible unless the frame has some special feature, for instance that the vertical members are crane-bearing.

As pointed out by Foulkes,[20] an allowance for the effect of axial thrust will often have to be made for the stanchions in the lower storeys of multi-storey frames. Taking as an illustration the two-storey frame of Fig. 4.7, the required fully plastic moment for the stanchion DF was 9·71 tons ft., the axial thrust in this member being 6·68 tons. If this member was an 8 in.×4 in. British Standard beam, with a plastic section modulus of 16.02 in³., the fully plastic moment in the absence of axial thrust would be

$$M_p = \frac{16{\cdot}02 \times 15{\cdot}25}{12} = 20{\cdot}4 \text{ tons ft.}$$

giving a load factor of $\frac{20{\cdot}4}{9{\cdot}71} = 2{\cdot}10$. For this member the cross-sectional area is 5·30 sq. in., so that the fully plastic thrust is

$$N_p = 5{\cdot}30 \times 15{\cdot}25 = 80{\cdot}8 \text{ tons.}$$

At collapse the thrust in DF would be $2{\cdot}10 \times 6{\cdot}68 = 14{\cdot}0$ tons, so that

$$\frac{N}{N_p} = \frac{14{\cdot}0}{80{\cdot}8} = 0{\cdot}173\,.$$

The corresponding value of $\frac{M_N}{M_p}$ is found from Appendix B to be 0·953, so that the reduction of fully plastic moment due to axial thrust is 4·7% for this member. It would be necessary to make due allowance for this effect when making a final selection of the cross-section of this member.

The case of beams whose cross-sections have one axis of symmetry which coincides with the plane of flexure has been discussed in detail by Eickhoff.[21] In developing this theory it is of importance to specify the axis about which the resultant couple is measured, and Eickhoff chooses for this purpose the equal area axis. When there is a second axis of symmetry, as in the cases which have just been considered, the equal area axis coincides with the centroidal axis, and the resultant force obtained from the stress distribution in Fig. 6.1(c) passes through the centroid of the cross-section.

An important type of structure for which the influence of axial thrust on the value of the fully plastic moment must often be taken into account is the arch. Swida [22] has described a statical approach for this type of structure, assuming that the arch member has two axes of symmetry, one of which coincides with the plane of the arch. This author also showed that the influence of the deflections which develop prior to collapse on the value of the collapse load could be of importance, and Johansen [23] has given the results of experiments on slender model steel arches in which this effect was appreciable. However, the tests of Hendry [24] on model two-pinned steel arches, which were less slender, gave results in conformity with the theory assuming the deflections to be negligible. The analysis of arches was studied extensively by Onat and Prager,[25] who gave both a statical and a kinematical method of analysis. The collapse of thin rings in which the effect of axial thrust was assumed to be negligible has been discussed by Hwang.[26] Eickhoff [21] has also discussed the analysis of arches for the case in which the arch member has only one axis of symmetry which coincides with the plane of the arch.

In conclusion, it should be emphasized that the additional moments due to normal force have been tacitly assumed to be negligible in the applications of the results which have been obtained. This amounts to assuming that instability effects are absent.

6.4. Effect of shear force

Even within the framework of the simple theory of flexure no precise estimate of the effect of a transverse shear force on the fully plastic moment of a beam has yet been made, but results are available which enable the reduction of fully plastic moment due to shear to be estimated to a close degree of approximation. The problem was first considered by Stüssi,[27] whose treatment was defective in that no allowance was made for the influence of shear stresses on the yield stress in tension or compression. More recently, Horne [28] considered the case of a beam of rectangular cross-section which is bent about one of the axes of symmetry of the cross-section, and derived an approximate solution which is valid only for shear forces less than a certain value. This solution was also applied to the case of a beam of I-section bent about its major axis. Leth [29] has pointed out that this latter solution is not quite accurate in certain respects, and has given the necessary modifications, but the errors involved appear to be quite small. Unfortunately, this solution was not extended to cover the full range of shear forces from zero to the fully plastic value causing complete shear failure in the web. However, the same author developed a solution which gave a lower bound on the value of the fully plastic moment in the presence of shear force which is valid over the full range of shear forces. This solution was supplemented by an upper bound on the fully plastic moment when the shear force has its fully plastic value; this upper bound agreed closely with the lower bound obtained for this value of the shear force. Heyman and Dutton [30] have proposed a semi-empirical theory which is also valid over the full range of shear forces. This theory gives results agreeing quite well with Leth's analysis, and it is suggested that it should be used in practice on account of its simplicity. Green [31] has derived an upper bound on the value of the fully plastic moment in the presence of shear force for beams of both rectangular and I-section, assuming a plastic-rigid material; this upper bound agrees closely with Leth's lower bound for large values of the shear force. Upper bound solutions were also derived independently by Green and by Onat and Shield [32] for the case of a wide beam of rectangular cross-section, treated as a problem in plane strain and assuming the material to be plastic-rigid.

Rectangular cross-section

The general character of the problem is best illustrated by considering the case of a cantilever of rectangular cross-section. Suppose that a cantilever of breadth b, depth h and length l is subjected to an end shear load F, as shown in Fig. 6.4, the value of F being such that a condition of full plasticity is attained at the clamped end under the combined action of the shear force F and the bending moment Fl. It is required to estimate this fully plastic moment Fl, which will be denoted by M_F, and in particular to compare its value with the fully plastic moment $M_p = \frac{1}{4}bh^2 f_L$ in the absence of shear. It will be assumed that the stress-strain relation is of the ideal plastic type shown in Fig. 1.4(b), the upper yield stress being neglected.

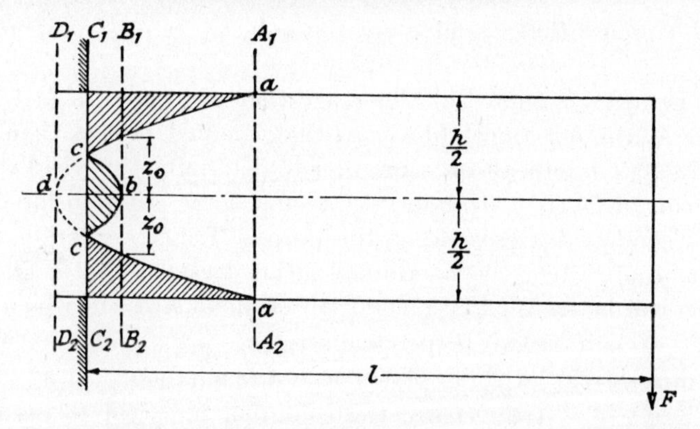

Fig. 6.4. Plastic zones in cantilever of rectangular cross-section.

In the elastic portions of the cantilever the distribution of longitudinal normal stress σ is assumed to be linear across the section, so that at section A_1A_2, where yield just occurs, this distribution is as shown in Fig. 6.5(a). On this assumption the distribution of shear stress τ can be derived from considerations of longitudinal equilibrium and the fact that the shear stress must be zero on the boundaries; this distribution is parabolic, as shown in the figure. For such a distribution the average shear stress is $\frac{2}{3}\tau_0$, where τ_0 is the maximum shear stress at the centre of the beam. Since the shear force is F and the area of the cross-section is bh, it follows that

$$\tfrac{2}{3}\tau_0 bh = F$$

$$\tau_0 = \frac{3F}{2bh} \quad . \quad . \quad . \quad . \quad 6.10$$

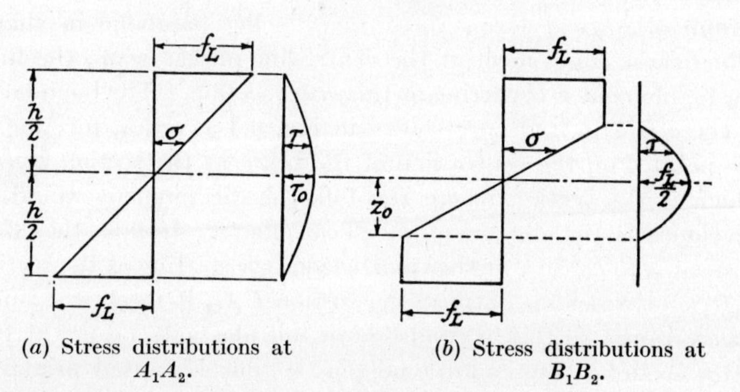

(a) Stress distributions at A_1A_2.

(b) Stress distributions at B_1B_2.

Fig. 6.5. Stress distributions in cantilever of rectangular cross-section.

It is assumed that σ and τ are the only two stress components which are not zero.

It is tacitly assumed at this stage that yield does in fact first occur at the outermost fibres, so that τ_0 must be less than the yield stress in pure shear. In order to determine this yield stress in terms of f_L, it is necessary to assume some criterion for yield under combined stress, and for this purpose Tresca's criterion, that yield occurs when the maximum shear stress reaches a critical value, will be used. For a condition of plane stress in which the direct stresses on two perpendicular planes are σ and zero, while the shear stress on these planes is τ, the maximum shear stress τ^{max} is $\frac{1}{2}\sqrt{\sigma^2 + 4\tau^2}$. Hence the condition for yield to occur is that

$$\tau^{max} = \tfrac{1}{2}\sqrt{\sigma^2 + 4\tau^2} = \tfrac{1}{2}f_L \quad . \quad . \quad . \quad 6.11$$

since in pure tension yield occurs when $\sigma = f_L$, $\tau = 0$. It follows from equation 6.11 that the yield stress in pure shear is $\frac{1}{2}f_L$.

The assumption that τ_0, as given by equation 6.10, is less than the yield stress in pure shear thus implies that

$$\frac{3F}{2bh} \leqslant \frac{f_L}{2}$$

$$F \leqslant \tfrac{1}{3}bhf_L \quad . \quad . \quad . \quad . \quad 6.12$$

It is, of course, also necessary to show that τ^{max} does not exceed $\frac{1}{2}f_L$ over the entire cross-section A_1A_2. It is easily shown that this is the case.

Now consider the conditions pertaining in those portions of the cantilever to the left of the section A_1A_2. As was shown in Section 5.3, in the absence of shear effects zones of tensile and compressive yield form, these zones being parabolic in shape. When these zones meet at the centre-line of the beam, the fully plastic moment is occurring at this cross-section. The boundaries of these plastic zones are shown in Fig. 6.4 as *acdca*, meeting at the point d on the centre line of the beam at the section D_1D_2, which is the section where the fully plastic moment would be developed in the absence of any shear effects. In fact, the effect of shear is to cause a further plastic zone *cbc* starting at the section B_1B_2 and widening until at the section C_1C_2 it meets the outer plastic zones, so that a condition of full plasticity occurs at this latter section. This additional plastic zone is caused primarily by the higher shear stresses which occur in the central elastic core of the beam when yield has occurred in the outer fibres due to the bending stresses; to understand the reason for its formation it is thus necessary to determine the distribution of shear stress across the section in this region.

As shown by Prager and Hodge,[33] and also by Horne,[28] under the simplifying assumptions which have been made the shear stress τ must be zero in the outer plastic regions, so that the longitudinal normal stress σ is still f_L in these regions. Thus the distribution of σ across the section is as shown in Fig. 6.5(b) for the section B_1B_2, where the depth of the elastic core has been reduced to $2z_0$. The shear stress τ is still distributed parabolically over the elastic core, and so its maximum value, occurring on the centre line, must be inversely proportional to the depth of the elastic core. At the section B_1B_2 it is supposed that the shear stress just attains the yield value $\frac{1}{2}f_L$ on the centre-line; the distribution of shear stress is thus as shown in Fig. 6.5(b). It follows that

$$\tau_0\left(\frac{h}{2z_0}\right) = \frac{f_L}{2}.$$

Combining this result with equation 6.10 it is found that

$$z_0 = \frac{3F}{2bf_L} \quad . \quad . \quad . \quad . \quad 6.13$$

Consider now the variation of $\tau^{\max}$ across the section B_1B_2. The value of $\tau^{\max}$ has its yield value at b and throughout the outer plastic zones. Its value at any distance y from the centre-line in the elastic core must follow from the assumed distribution of σ and τ and from equation 6.11. Thus from Fig. 6.5(b),

$$\left.\begin{aligned}\sigma &= \left(\frac{y}{z_0}\right)f_L \\ \tau &= \left(1 - \frac{y^2}{z_0{}^2}\right)\frac{f_L}{2}\end{aligned}\right\} \quad - z_0 \leqslant y \leqslant z_0,$$

and inserting these results in equation 6.11, it is found that

$$\tau^{\max} = \tfrac{1}{2}f_L\sqrt{1 - \frac{y^2}{z_0{}^2} + \frac{y^4}{z_0{}^4}}\,; \; - z_0 \leqslant y \leqslant z_0 \quad . \quad 6.14$$

It is readily shown that within the range of values of y for which equation 6.14 is valid, $\tau^{\max}$ is always less than $\frac{1}{2}f_L$, and that the least value of $\tau^{\max}$ is $\frac{\sqrt{3}}{4}f_L$, occurring when $y = \frac{z_0}{\sqrt{2}}$. The variation

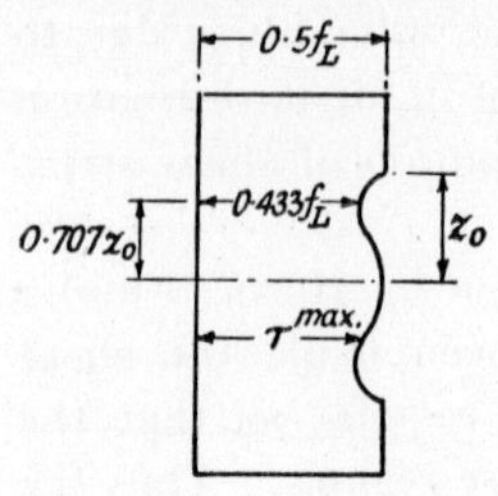

Fig. 6.6. *Variation of* τ^{max} *across section* B_1B_2.

of τ^{max} across the section B_1B_2 is thus as shown in Fig. 6.6, and it is clear that at this section the state of stress is close to one of full plasticity. A reasonably good but conservative estimate of the fully plastic moment as reduced by the effect of shear can thus be obtained by evaluating the bending moment at this section.

From equation 5.13, Section 5.3, the bending moment M'_F at B_1B_2 is given by

$$M'_F = M_p\left[1 - \frac{1}{3}\left(\frac{2z_0}{h}\right)^2\right],$$

since the stress distribution as indicated in Fig. 6.5(*b*) is identical with that used in developing equation 5.13. Substituting the value of z_0 given by equation 6.13, the following result is obtained :

$$\frac{M'_F}{M_p} = 1 - 3\left(\frac{F}{bhf_L}\right)^2.$$

For convenience let F_p denote the shear force obtained by multiplying the area of the cross-section bh by the yield stress in pure shear $\frac{1}{2}f_L$. It then follows that

$$\frac{M'_F}{M_p} = 1 - 0{\cdot}75\left(\frac{F}{F_p}\right)^2 \quad . \quad . \quad . \quad 6.15$$

where $$F_p = \tfrac{1}{2}bhf_L.$$

From equation 6.12, it is seen that equation 6.15 is only valid so long as

$$\frac{F}{F_p} \leqslant \frac{2}{3} \quad . \quad . \quad . \quad . \quad 6.16$$

Equation 6.15 within the limits set by the condition 6.16 provides an over-estimate of the reduction in the fully plastic moment due to shear. Horne [28] attempted to improve this estimate by making an approximate analysis of the conditions between the sections B_1B_2 and C_1C_2 shown in Fig. 6.4. In this region it was necessary to introduce a third stress component consisting of the transverse direct stress acting in a direction perpendicular to σ. The solution also involved certain difficulties in the neighbourhood of the points *c* where the two plastic zones meet, notably that the condition of longitudinal equilibrium could not be satisfied at these points. Despite these defects it seems likely that the analysis gives a reasonably accurate estimate of the reduction in the fully plastic moment. The results are similar in form to those which have just been obtained, differing only in the values of the numerical constants. Thus if M_F denotes the fully plastic moment as reduced by the effect of a shear force F, Horne's analysis gave the following results:

$$\frac{M_F}{M_p} = 1 - 0{\cdot}44\left(\frac{F}{F_p}\right)^2 ; \quad \frac{F}{F_p} \leqslant 0{\cdot}79 \quad . \quad . \quad 6.17$$

I-section

These results were extended by Horne [28] to cover the case of a beam of I-section bent about its major axis. To simplify the resulting expressions it will be assumed that the flange thickness t_2 is negligible in comparison with the depth h of the section (see Fig. 6.2). In the absence of shear effects the fully plastic moment is then seen from equation 1.13, Section 1.4, to be

$$\begin{aligned} M_p &= (bht_2 + \tfrac{1}{4}t_1h^2)f_L \\ &= (\tfrac{1}{2}hA_f + \tfrac{1}{4}hA_w)f_L \quad . \quad . \quad 6.18 \end{aligned}$$

where A_f is the total flange area $2bt_2$ and A_w is the web area ht_1. In this equation the first term represents the contribution of the flanges and the second term the contribution of the web. Since the web is rectangular in section its contribution to the fully plastic moment will be modified by shear in accordance with equation 6.17, while the flange contribution remains unchanged. It follows that

$$M_F = \tfrac{1}{2}hA_f f_L + \tfrac{1}{4}hA_w f_L\left[1 - 0{\cdot}44\left(\frac{F}{F_p}\right)^2\right].$$

Combining this result with equation 6.18, and rearranging, it is found that

$$\frac{M_F}{M_p} = 1 - 0{\cdot}44\left(\frac{A - A_f}{A + A_f}\right)\left(\frac{F}{F_p}\right)^2 ; \quad \frac{F}{F_p} \leqslant 0{\cdot}79 \quad . \quad 6.19$$

where

$$F_p = \tfrac{1}{2}A_w f_L \quad . \quad . \quad . \quad . \quad 6.20$$

In equation 6.19 A denotes the total area of the cross-section, $A_f + A_w$. The relation between $\dfrac{M_F}{M_p}$ and $\dfrac{F}{F_p}$ implied by equation

6.19 is shown as curve (i) in Fig. 6.7, for the case in which $A_f = \frac{2}{3}A$, corresponding roughly to a 7 in. × 4 in. British Standard beam.

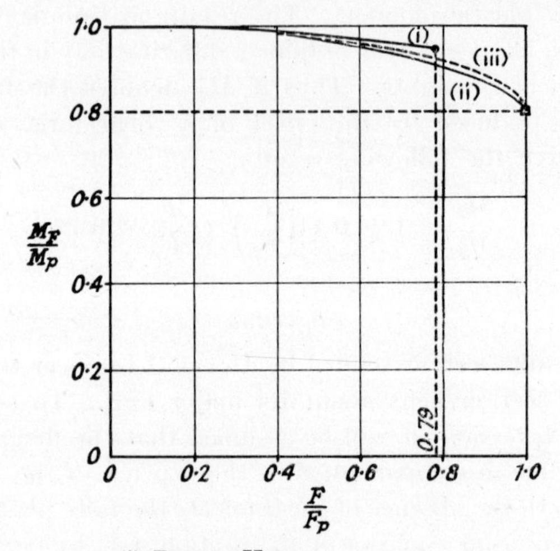

(i) Due to Horne.
(ii) Due to Leth.
(iii) Due to Heyman and Dutton.

Fig. 6.7. *Effect of shear force on fully plastic moment of I-section:* $A_f = \frac{2}{3}A$.

Leth [29] has pointed out that for many practical I-sections, additional plastic zones will form a short distance away from the clamped end of the cantilever, at the junction between the flange and the web, for values of $\frac{F}{F_p}$ less than 0·79. When the effects of these plastic zones are taken into account, the analysis becomes more complicated, but the value of $\frac{M_F}{M_p}$ for a given value of $\frac{F}{F_p}$ is not affected to any great extent.

For the case of a beam of I-section it is not unreasonable to suppose that the value of F_p given in equation 6.20, representing a condition of shear yield throughout the web, is in fact close to the maximum shear load which can be supported, since the shear stress is not required to be zero at the junctions between the web and the flanges. It is therefore desirable to be able to estimate the effect of shear forces beyond the upper limit of $0{\cdot}79F_p$ for which equation 6.19 is applicable. Leth [29] has given a lower bound solution which enables this to be done. This solution is based essentially on the lower bound theorem of Drucker, Prager and Greenberg,[34] which as pointed out in Section 3.2 is the generalization of the static theorem for frames. The procedure was to derive distributions of longitudinal stress σ and shear stress τ throughout the cantilever which satisfied the equations of equilibrium; the bending moment corresponding to the stress distribution at the clamped end was then a conservative estimate of the bending moment which could be carried in the presence of the shear force in the cantilever. The results obtained cannot be expressed in a convenient explicit form, but curve (ii) in Fig. 6.7 shows the relation between $\frac{M_F}{M_p}$ and $\frac{F}{F_p}$ obtained in this way for a 7 in. × 4 in. British Standard beam. In this analysis the von Mises yield criterion was assumed; this amounts to replacing the factor 4 by 3 in equation 6.11.

An upper bound, based on the generalization of the kinematic theorem, was also obtained by Leth [29] for the particular case in which $\frac{F}{F_p} = 1$; this upper bound on M_F has the value $0{\cdot}814M_p$ for a 7 in. × 4 in. British Standard beam, as compared with the lower bound of $0{\cdot}808M_p$. The closeness of these bounds indicates that the lower bound result is very accurate when $\frac{F}{F_p}$ is unity. It is also of interest to note that for this case Green's upper bound solution [31] for a plastic-rigid material gives the value $0{\cdot}799\ M_p$.

Heyman and Dutton [30] have proposed empirical distributions of σ and τ at the cross-section where full plasticity is occurring, consisting of a constant shear stress q and a constant longitudinal normal stress f throughout the web, together with zero shear stress and a uniform longitudinal normal stress f_L in the flanges. In addition, it was assumed that f and q satisfy the yield condition, which was assumed to be that of von Mises, so that

$$f^2 + 3q^2 = f_L^2.$$

It is readily shown that on this basis the fully plastic moment M_p as reduced by the effect of a shear force F is given by

$$\frac{M_F}{M_p} = 1 - \left(\frac{A - A_f}{A + A_f}\right)\left[1 - \sqrt{1 - \left(\frac{F}{F_p}\right)^2}\right]; \quad \frac{F}{F_p} \leqslant 1. \qquad 6.21$$

where F_p has the value $\frac{1}{\sqrt{3}}A_w f_L$. The effect of assuming the

Tresca yield criterion is to leave equation 6.21 unchanged but to alter the value of F_p to $\frac{1}{2}A_w f_L$, as in equation 6.20. The variation of $\frac{M_F}{M_p}$ with $\frac{F}{F_p}$ according to equation 6.21 is shown as curve (iii) in Fig. 6.7, for the case in which $A_f = \frac{2}{3}A$.

It will be seen from this figure that the curve given by Heyman and Dutton's semi-empirical analysis does not differ much from the lower bound solution of Leth, and in addition it gives rather more conservative values than Horne's analysis where the latter is applicable. It is therefore suggested that the effect of shear should be allowed for by means of Heyman and Dutton's result, equation 6.21, in view of the analytical simplicity of this expression. This suggestion gains support from the fact that Green's upper bound result [31] for a beam of I-section of plastic-rigid material, which is subjected to a large shear force, can be cast into a form almost identical with equation 6.21, the only difference being that the factor $\frac{2}{\sqrt{3}}$ appears before the radical in this case.

Comparison with experiments

It is not easy to compare these theoretical predictions with experiments, since until F approaches the value F_p the effect of shear is quite small for I-sections, whereas when F is in the neighbourhood of F_p the load-deflection curves for beams often show no appreciable horizontal portion corresponding to plastic hinge action. Instead, it is found that the load-deflection curves continue to rise fairly steadily for loads such that the theoretical fully plastic moment is exceeded, so that there is no sharply defined failure condition. This was observed in the tests by Baker and Roderick [35] on simply supported beams of $1\frac{1}{4}$ in. × $1\frac{1}{4}$ in. × $\frac{1}{8}$ in. web H-section. These beams were subjected to symmetrical two-point loading in which the distance between the loads was held constant while the total span was varied from test to test. Hendry [36, 37] also found the same effect in testing simply supported beams of various sections subjected to central concentrated loads. However, in testing simply supported model plate girders subjected to central concentrated loads, Heyman and Dutton [30] observed sharply defined failure loads even when F was close to or equal to F_p, the observed values of the fully plastic moments agreeing closely with the values predicted from equation

6.21. Further tests on plate girders by Longbottom and Heyman [38] provided additional confirmation of the theory. It will of course be realized that in applying the theory to the design of plate girders, due regard must be paid to the possibility of web buckling when the depth to thickness ratio of the web is large.

Effect of shear force in practical cases

The effect of shear on the fully plastic moment will usually be negligible for frames. The pitched roof portal frame of Fig. 4.1(*a*) will again be taken as a typical example. When the feet of this portal are fixed the required value of the fully plastic moment was found in Section 4.2 to be 5·73 tons ft. The corresponding value of the maximum shear force is 1·69 tons at section 2. As pointed out in Section 6.3, a 6 in.×3 in. British Standard beam gives a load factor of 1·79 in the absence of any effects of shear or axial thrust. If this section is used the shear force at collapse is thus 1·79 × 1·69 = 3·03 tons. The area of the web of this section is 1·38 sq. in., so that with a tensile yield stress of 15·25 tons per sq. in. the value of F_p is found from equation 6.20 to be

$$F_p = \tfrac{1}{2} \times 1{\cdot}38 \times 15{\cdot}25 = 10{\cdot}5 \text{ tons.}$$

Hence
$$\frac{F}{F_p} = \frac{3{\cdot}03}{10{\cdot}5} = 0{\cdot}289 .$$

From equation 6.21, with $A_f = 0{\cdot}65A$, it follows that $M_F = 0{\cdot}991 M_p$, so that the fully plastic moment is reduced by only 0·9% at the section where the greatest shear force occurs.

In certain cases, notably when designing continuous beams, the effect of shear can be quite appreciable. Consider for example the design of a beam which is to be of uniform section and continuous over several supports, the largest intermediate span being 16 ft. carrying a total working load of 27 tons. As far as this span is concerned the design is the same as for a fixed-ended beam ; the required value of the fully plastic moment, neglecting shear effects, is $\frac{W_c l}{16}$, where W_c is the total load under collapse conditions and l is the span. Thus with a load factor of 2, $W_c = 54$ tons and the required plastic section modulus Z_p is given by

$$15{\cdot}25 Z_p = \frac{54 \times 16 \times 12}{16}$$

$$Z_p = 42{\cdot}5 \text{ in.}^3,$$

assuming a yield stress of 15·25 tons per sq. in. This is precisely the value of the plastic section modulus of a 12 in. × 5 in. British Standard beam.

At the centre of the span, where a plastic hinge forms, the shear force is zero, but at each end where hinges form the shear force at collapse is $\frac{1}{2}W_c$, or 27 tons. For this section the web area is 4·20 sq. in., so that

$$F_p = \tfrac{1}{2} \times 4{\cdot}20 \times 15{\cdot}25 = 32{\cdot}0 \text{ tons}$$

$$\frac{F}{F_p} = \frac{27}{32{\cdot}0} = 0{\cdot}844$$

The ratio $\dfrac{A_f}{A}$ for this section is 0·59; it follows from equation 6.21 that

$$\frac{M_F}{M_p} = 1 - \frac{0{\cdot}41}{1{\cdot}59}[1 - \sqrt{1 - 0{\cdot}844^2}] = 0{\cdot}880$$

Thus with this section the effect of shear is to cause a reduction of 12% in the fully plastic moment at the supports, so that the 12 in. × 5 in. joist would in fact have a load factor appreciably less than 2. It is of interest to note that Bull [39] has described a test to destruction of a six-panel Vierendeel girder of 90 ft. span in which the effect of shear force on certain of the fully plastic moments was even larger than in this example.

In conclusion, it must be noted that the reduction of fully plastic moment due to the combined effects of shear force and axial thrust acting simultaneously have not been discussed in this or the preceding Section. The only analysis of this problem is due to Green,[31] who discussed the case of a wide rectangular section beam of plastic-rigid material, treated as a problem in plane strain. However, although a plastic hinge will often form at a section where there is both thrust and shear force, it will only be in rare cases that these are both of sufficient magnitude to have an appreciable effect on the fully plastic moment. In Green's paper the influence of the nature of the end clamping arrangement of a cantilever on the fully plastic moment as reduced by shear force was also discussed.

6.5. Contact stresses beneath loads

The fully plastic moment of a structural member is determined more logically by conducting beam tests rather than by determining f_L from tensile tests and multiplying by the calculated value of Z_p, as pointed out in Section 6.2. The usual arrangement in laboratory beam tests is to use a simply supported beam of rectangular cross-section subjected to either a central concentrated load or symmetrical two-point loading. It has often been observed that the fully plastic moment obtained from the former type of test is somewhat higher than that derived from the latter type of test on beams of similar section and material. The reason for this is that with the symmetrical two-point loading the central length of the beam between the loads is subjected to a constant bending moment, so that the stresses in this portion of the beam are only required to resist a pure couple, as was assumed in the simple theory of Section 1.4 which led to a value of the fully plastic moment of $Z_p f_L$. In contrast, the actual stress distribution at collapse for the case of a beam with a central concentrated load must be much more complicated than that envisaged in the simple theory, not only because the stresses must balance shear forces on either side of the mid-section, but because they must also, in the immediate vicinity of the load, balance the contact stresses due to the load.

An approximate elastic solution for the stresses in a simply supported rectangular beam subjected to a central concentrated load was developed by Stokes to explain the results of photo-elastic tests performed by Carus Wilson,[40] the analysis being published in Wilson's paper. This solution indicates that the local stresses due to the proximity of the concentrated load consist principally of two stress components. The first of these components is the compressive stress acting on planes parallel to the axis of the beam which is obviously required to support the load, and the second is a longitudinal normal stress which modifies the linear distribution of stress across the section which is obtained by the ordinary Bernoulli-Euler theory of elastic flexure. The effect of this latter stress is to multiply the ordinary bending stresses by a factor $\left(1 - k\frac{h}{l}\right)$, where h and l are the depth and length of the beam, respectively, and k is a positive factor whose value varies across the section and is small except within a distance of the order of h on either side of the load, where it is of the order of unity. This lends support to the view put forward by Roderick and Phillips [41] that the contact stresses tend to annul the ordinary bending stresses near the load even when yield has occurred.

Since the elastic theory suggests that the disturbance due to the load occurs in a total length of about h at the mid-section, Roderick and Phillips suggested that in analysing test results for beams subjected to central concentrated loads it should be assumed that collapse will not occur until the bending moment reaches the value $M_p = Z_p f_L$ at a distance $\frac{1}{2}h$ on either side of the concentrated load.

On this assumption the central bending moment at collapse is $\dfrac{M_p}{\left(1 - \dfrac{h}{l}\right)}$, or approximately $M_p\left(1 + \dfrac{h}{l}\right)$, so that the fully plastic moment M'_p beneath a concentrated load has the value

$$M'_p = M_p\left(1 + \frac{h}{l}\right) \quad . \quad . \quad . \quad 6.22$$

Roderick and Phillips [41] tested beams of rectangular section cut from the same stock, and found that the fully plastic moments derived from their central concentrated load tests were from 5% to 8% greater than those derived from two-point loading tests; these observations were satisfactorily explained on the basis of equation 6.22. In these tests the effects of shear was negligible. Further evidence in support of this semi-empirical result has been furnished by Heyman,[42] who used relaxation methods for an approximate determination of the stress distribution in the neighbourhood of a central concentrated load on a beam at the point of collapse. The photo-elastic tests of Baes [43] on beams of rectangular section and of Hendry [36] on beams of I-section also indicate that equation 6.22 may be expected to be reasonably correct in many cases.

In the case of beams of I-section, bearing stiffeners are usually provided at those sections where concentrated loads act, and a correction of the kind given in equation 6.22 is then not applicable.

References

1. H. Quinney. Further tests on the effect of time of testing. *Engineer*, **161**, 669 (1936).
2. J. Winlock and R. W. E. Leiter. Some factors affecting the plastic deformation of sheet and strip steel and their relation to the deep drawing properties. *Trans. Amer. Soc. Metals*, **25**, 163 (1937).

3. C. F. ELAM. The influence of rate of deformation on the tensile test with special reference to the yield point in iron and steel. *Proc. Roy. Soc. A.*, **165**, 568 (1938).
4. M. J. MANJOINE. Influence of rate of strain and temperature on yield stresses of mild steel. *J. Appl. Mech.*, **11**, 211 (1944).
5. G. COOK. Some factors affecting the yield point in mild steel. *Trans. Instn. Engrs. Shipb. Scot.*, **81**, 371 (1937).
6. J. MUIR. On the overstraining of iron by tension and compression. *Proc. Roy. Soc. A.*, **77**, 277 (1906).
7. J. BAUSCHINGER. Die Veränderungen der Elastizitätsgrenze. *Mitt. mech.-tech. Lab. tech. Hochschule*, München (1886).
8. C. A. EDWARDS, H. N. JONES and B. WALTERS. A study of strain-age-hardening of mild steel. *J. Iron St. Inst.*, **139**, 341 (1939).
9. R. O. GRIFFIS, R. L. KENYON and R. S. BURNS. The ageing of mild steel sheets. *Year Book, Amer. Iron and Steel Inst.* (1933), 142.
10. R. L. KENYON and R. S. BURNS. Ageing in Iron and Steel. Symposium : Age Hardening of Metals. *Amer. Soc. Metals* (1939), 262.
11. E. S. DAVENPORT and E. C. BAIN. The ageing of steel. *Trans. Amer. Soc. Metals*, **23**, 1047 (1935).
12. A. ROBERTSON. The use of small specimens in the testing of steel. 1*st Report Steel Struct. Res. Cttee.*, H.M.S.O., 194 (1931).
13. W. W. LUXION and B. G. JOHNSTON. Plastic behaviour of wide flange beams. *Weld. J.*, Easton, Pa., **27**, 538-s (1948).
14. J. W. RODERICK and H. H. L. PRATLEY. The behaviour of rolled steel joists in the plastic range. *Brit. Weld. J.*, **1**, 261 (1954).
15. D. K. BULLENS *et alia*. *Steel and its Heat Treatment*. John Wiley (N.Y.), Chapman & Hall (London), 5th Ed. (1949), 5.
16. J. W. RODERICK and J. HEYMAN. Extension of the simple plastic theory to take account of the strain-hardening range. *Proc. Instn. Mech. Engrs.*, **165**, 189 (1951).
17. K. GIRKMANN. Bemessung von Rahmentragwerken unter Zugrundelegung eines ideal-plastischen Stahles. *S. B. Akad. Wiss. Wien* (*Abt. II*a), **140**, 679 (1931).
18. L. S. BEEDLE, J. A. READY and B. G. JOHNSTON. Tests of columns under combined thrust and moment. *Proc. Soc. exp. Stress Anal.*, **8**, 109 (1950).
19. M. R. HORNE. The plastic moduli of British Standard rolled steel joists. *Brit. Weld. Res. Assn. Report* FE. 1/33 (1953).
20. R. A. FOULKES. Discussion of " The rapid calculation of the collapse load for a framed structure. *Proc. Instn. Civ. Engrs.*, (*Part III*), **1**, 79 (1952).
21. K. G. EICKHOFF. To be published.
22. W. SWIDA. Die Berechnung von stählernen Bögen unter

Berücksichtigung der Tragfähigkeits-reserve im elastisch-plastischen Zustand. *Stahlbau*, **19**, 17, 29 (1950); **20**, 25 (1951).

23. K. W. JOHANSEN. Studies on the load carrying capacity of steel structures (Danish). *Laboratoriet for Bygningsteknik, Danmarks Tekniske Høјskole.* Meddelelse Nr. 3 (1954).
24. A. W. HENDRY. The plastic design of two-pinned mild steel arch ribs. *Civil Engng.*, **47**, 38 (1952).
25. E. T. ONAT and W. PRAGER. Limit analysis of arches. *J. Mech. Phys. Solids*, **1**, 77 (1953).
26. C. HWANG. Plastic collapse of thin rings. *J. Aero. Sci.*, **20**, 819 (1953).
27. F. STÜSSI. Über den Verlauf der Schubspannungen in auf Biegung beanspruchten Balken aus Stahl. *Schweiz. Bauztg.*, **98**, 2 (1931).
28. M. R. HORNE. The plastic theory of bending of mild steel beams with particular reference to the effect of shear forces. *Proc. Roy. Soc. A.*, **207**, 216 (1951).
29. C-F. A. LETH. The effect of shear stresses on the carrying capacity of I-beams. *Tech. Rep.* A11–107, Brown Univ. (1954).
30. J. HEYMAN and V. L. DUTTON. Plastic design of plate girders with unstiffened webs. *Welding and Metal Fabrication*, **22**, 265 (1954).
31. A. P. GREEN. A theory of the plastic yielding due to bending of cantilevers and fixed-ended beams. *J. Mech. Phys. Solids*, **3**, 1, 143 (1954).
32. E. T. ONAT and R. T. SHIELD. The influence of shearing forces on the plastic bending of wide beams. *Tech. Rep.* A11–103/10, Brown Univ. (1953).
33. W. PRAGER and P. G. HODGE. *Theory of Perfectly Plastic Solids.* John Wiley (N.Y.), Chapman & Hall (London), 51 (1951).
34. D. C. DRUCKER, W. PRAGER and H. J. GREENBERG. Extended limit design theorems for continuous media. *Quart. Appl. Math.*, **9**, 381 (1952).
35. J. F. BAKER and J. W. RODERICK. Further tests on beams and portals. *Trans. Inst. Weld.*, **3**, 83 (1940).
36. A. W. HENDRY. The stress distribution in a simply supported beam of I-section carrying a central concentrated load. *Proc. Soc. exp. Stress Anal.*, **7**, 91 (1949).
37. A. W. HENDRY. An investigation of the strength of certain welded portal frames in relation to the plastic method of design. *Struct. Engr.*, **28**, 311 (1950).
38. E. LONGBOTTOM and J. HEYMAN. Tests on full-size and on model plate girders. *Struct. Paper No.* 49, *Instn. Civ. Engrs.* (1956).
39. F. B. BULL. Tests to destruction on a Vierendeel girder. Prelim. Vol., Conference on the correlation between calculated and

observed stresses and displacements in structures, *Instn. Civ. Engrs.*, 135 (1955).

40. CARUS WILSON. The influence of surface loading on the flexure of beams. *Phil. Mag.*, (Ser. 5) **32,** 481 (1891).
41. J. W. RODERICK and I. H. PHILLIPS. The carrying capacity of simply supported mild steel beams. *Research* (*Engng. Struct. Suppl.*), *Colston Papers*, **2,** 9 (1949).
42. J. HEYMAN. Elasto-plastic stresses in transversely loaded beams. *Engineering*, **173,** 359, 389 (1952).
43. L. BAES. Les palplanches plates beval P pour constructions cellulaires. *Ossat. métall.*, **13,** 75 (1948).

Examples

1. A 12 in. × 8 in. British Standard beam has mean flange and web thicknesses of 0·904 in. and 0·43 in., respectively. Assuming the section to be composed of three rectangles, find the fully plastic moment for bending about the minor axis in the absence of axial thrust and also when there is an axial thrust of 150 tons, assuming a yield stress of 15·25 tons per sq. in. Compare the results with those obtained from the Tables in Appendix B, and show that the percentage reduction in fully plastic moment due to the thrust is the same in both cases.

2. A 6 in. × 3 in. × 12·41 lb. British Standard channel has mean flange and web thicknesses 0·38 in. and 0·25 in., respectively. Assuming the section to be composed of three rectangles, find the fully plastic moment for bending about the minor axis, assuming a yield stress of 15·25 tons per sq. in. If there is an axial thrust of 9 tons, find the fully plastic moment, and show that its value does not depend on whether the tips of the flanges are in tension or compression. If there is an axial thrust of 18 tons, show that the value of the fully plastic moment depends on whether the tips of the flanges are in tension or compression, and find the two possible values of the fully plastic moment in this case.

In each case take moments about the equal area axis.

3. A fixed-ended beam is of length 12 ft. and carries a concentrated load 13 tons at a position 4 ft. from one end. The beam is a 10 in. × $4\frac{1}{2}$ in. British Standard beam. Find the load factor against failure by plastic collapse, neglecting the effect of shear on the fully plastic moment and assuming a lower yield stress of 15·25 tons per sq. in. Find also the load factor if the effect of shear is allowed for according to equation 6.21, assuming the yield stress in pure shear to be half the value in pure tension or compression.

Neglect the effect of contact stresses beneath the load.

4. In a two-storey single-bay fixed-base rectangular frame each storey is of height 10 ft. and the span is 20 ft. The lower stanchions *AB* and *FE* and the lower beam *BE* are all 12 in. × 5 in. British Standard beams, and the upper stanchions *BC* and *ED* and the upper beam *CD* are all 9 in. × 4 in. British Standard beams. The beams *CD* and *BE* carry central concentrated loads of 5 tons and 10 tons respectively, and there are concentrated horizontal loads each of 2 tons acting at *D* and *E* in the directions *CD* and *BE*, respectively. Find the load factor against plastic collapse, neglecting the effects of shear and axial thrust and assuming a lower yield stress of 15·25 tons per sq. in. Determine the shear force and axial thrust at the foot *F* at collapse, and find the percentage reductions in the fully plastic moment at this section due to their separate action.

CHAPTER

7

Minimum Weight Design

7.1. Introduction

THE PLASTIC DESIGN METHODS which were described in Chapter 4 are not really design methods in the strictest sense of the term. In applying these methods to a frame it is assumed that the working loads are prescribed, and that these loads are then multiplied by the specified load factor to give a set of loads which the frame must be capable of carrying without collapse. It is then necessary to assume the fully plastic moments of all the members in terms of the fully plastic moment M_p of any one of the members, so that at the outset the *relative* values of the fully plastic moments are chosen. The ensuing analysis then determines the value of M_p and thus the *absolute* values of all the fully plastic moments. In this way the frame is designed, in the sense that if each member is chosen so that its fully plastic moment is not less than the value determined by the analysis, the frame can safely withstand the factored loads. However, there is no guarantee that by this procedure the best design has been achieved, for the initial assignment of the ratios of the fully plastic moments of the members can be made in many ways, each way leading to a different final design. Thus for any given shape and size of frame together with a prescribed loading there will be a large number of possible designs, and it is relevant to consider which of these designs represents the optimum which can be achieved.

It will be clear that the design which involves the use of the least possible weight of material has a fair claim to be regarded as the best possible design, but to assert that minimum weight is the only important criterion in design is to disregard the numerous other economic factors which must always be considered. However, a discussion of these factors cannot be entered into here, and the present chapter will be concerned solely with the problem of designing for minimum structural weight.

Throughout the chapter, it will be assumed that the shape and size of each frame considered, together with its loading, are prescribed. Furthermore, it will be assumed that the frames are composed of uniform prismatic members. The nature of the problem of designing such frames for minimum weight has been clarified considerably by the work of Foulkes,[1, 2] who used a geometrical analogue in discussing certain simple problems. By means of this analogue Foulkes explained the meaning of several important theorems for which he gave general proofs. After a preliminary discussion of the fundamental assumptions has been given in Section 7.2, this analogue and the theorems are described in detail in Section 7.3. A few applications of the theorems are then given in Section 7.4. Finally, the principal methods of minimum weight design are discussed briefly in Section 7.5, where an example is also worked through in detail. It will appear that techniques for the solution of complicated minimum weight problems have not yet been developed fully, but nevertheless the results which have already been obtained are of considerable interest and significance.

It will be realized that greater economies of material can be achieved if varying sections are allowed, as for instance by welding additional flange plates at critical locations; this point was referred to briefly in Section 3.6. This type of problem has been discussed by Horne [3] for the case of a fixed-ended beam subjected to uniformly distributed loading and also to a rolling load.

7.2. Assumptions

The assumptions which were stated in Section 3.1 for the calculation of plastic collapse loads are naturally retained when considering minimum weight design, and in addition certain other assumptions are made. In the first place, it is assumed throughout that the fully plastic moments of the members are unaffected by shear force and axial thrust. Also, despite the fact that in practice there is only a finite number of rolled sections, it is assumed that an infinite range of sections is available. In addition, it is assumed that there is a smooth curve connecting the weight per unit length of a beam with its fully plastic moment. That this latter assumption is reasonably valid can be seen from Fig. 7.1.

In this figure the plastic section modulus Z_p of each of the British Standard beams is plotted against the weight per unit length w. Points representing those sections which are intended primarily for use as beams are indicated by circles, and points representing those sections which are intended primarily for use as stanchions are indicated by crosses. It will be seen that the latter points are fairly widely scattered, whereas the former points lie closely on a

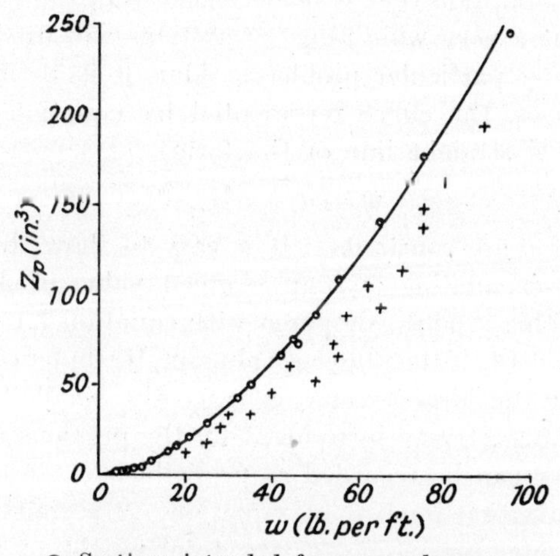

⊙ Sections intended for use as beams.
+ Sections intended for use as stanchions.

Fig. 7.1. *Weights of British Standard beams.*

smooth curve. If the dimensions of Z_p are in³., and w is measured in lb. per ft., the equation to this curve is

$$w = 3{\cdot}4 Z_p^{0{\cdot}6},$$

a form of equation suggested by Heyman.[4] If M_p is the fully plastic moment in tons ft., and the lower yield stress is taken to be 15·25 tons per sq. in., it follows that M_p has the value $15{\cdot}25Z_p$ tons in., or $1{\cdot}27Z_p$ tons ft., so that

$$w = 2{\cdot}95 M_p^{0{\cdot}6} \quad . \quad . \quad . \quad . \quad 7.1$$

Thus if attention is confined to those sections intended for use as beams, it is reasonably accurate to assume that w and M_p are related in accordance with the curve represented by equation 7.1. However, the assumption that there is an infinite range of sections available is more questionable.

It is interesting to note that if a series of geometrically similar cross-sections is considered, the cross-sectional area and thus the weight per unit length w is proportional to d^2, where d is any typical dimension such as the total depth of the section, whereas the plastic section modulus Z_p is proportional to d^3. It follows that for such a series of sections, $w \propto Z_p^{\frac{2}{3}}$. The index 0·6 in equation 7.1 is thus close to the value which would be expected for a series of geometrically similar cross-sections.

A final assumption that is made stems from the fact that it is unlikely that a very wide range of sections will need to be considered in any particular problem. Thus it is assumed that in any given case the curve represented by equation 7.1 can be replaced by a straight line of the form

$$w = a + bM_p \quad . \quad . \quad . \quad . \qquad 7.2$$

where a and b are constants. It is easy to show that the error involved in calculating w from a given value of M_p from an equation of this type as compared with equation 7.1 will only be of the order of 1% if the range of values of M_p under consideration is such that the largest value is twice the smallest value. In view of the uncertainty introduced by the previous assumptions such an error can be regarded as negligible.

The linearization implied by equation 7.2 permits the formation of a simple expression for the total weight of a structure. If l is the length of any member, the total structural weight X is given by

$$X = \Sigma wl$$

where the summation includes all the members of the structure. Using equation 7.2

$$\begin{aligned} X &= \Sigma(a + bM_p)l \\ &= a\Sigma l + b\Sigma M_p l \quad . \quad . \quad . \qquad 7.3 \end{aligned}$$

In equation 7.3 the term $a\Sigma l$ is a constant for given dimensions of the structure, and so X is minimized when the term $\Sigma M_p l$ is minimized. This term has been called the *weight function*, and it will be denoted by the symbol x, so that

$$x = \Sigma M_p l \quad . \quad . \quad . \quad . \qquad 7.4$$

The minimum weight problem is thus to design a frame so that the weight function x, as given by equation 7.4, is a minimum. It is of interest to note that similar problems arise in the theory of linear programming.[5]

7.3. Geometrical analogue and minimum weight theorems

The geometrical analogue is best explained in terms of a simple example, and for this purpose the rectangular portal frame whose dimensions and loading are as shown in Fig. 7.2 will be used. This example was used by Foulkes,[2] and his treatment will be followed with only minor variations. In this problem it is supposed that the beam is to have a fully plastic moment β_1,

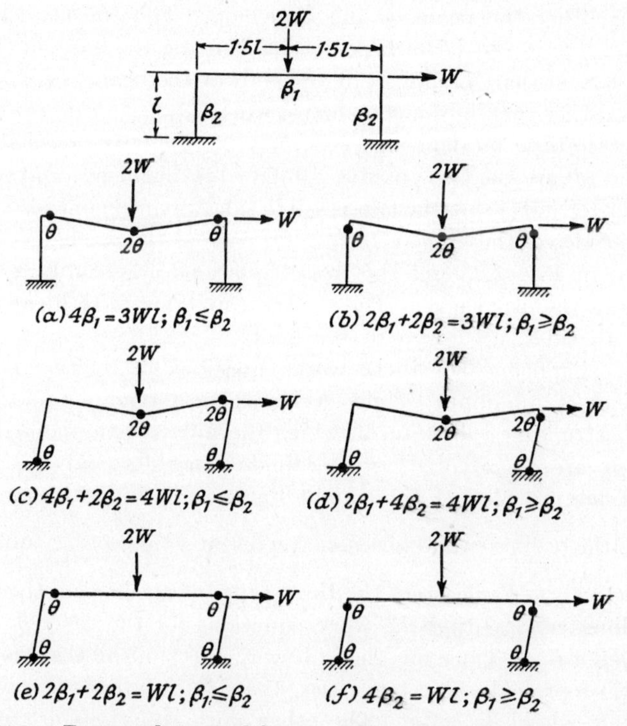

Fig. 7.2. Rectangular portal frame : Mechanisms and work equations.

and the two stanchions are each to have the same fully plastic moment β_2. The value of the weight function x is seen from equation 7.4 and Fig. 7.2 to be

$$x = 3l\beta_1 + 2l\beta_2 \quad . \quad . \quad . \quad . \qquad 7.5$$

and it is required to find the values of β_1 and β_2 which minimize this value of x while just enabling the loads to be supported without collapse occurring.

It will be recalled from Section 4.3 that for this type of frame there are two independent mechanisms, the beam and sidesway mechanisms, and that these can be combined to form a third mechanism. Thus there are three possible collapse mechanisms, but since it is not known *ab initio* whether the fully plastic moment of the beam is greater than or less than that of the stanchions, each of these three mechanisms can take two forms, with each hinge which occurs at a joint between the beam and a stanchion appearing in either the beam or the stanchion. Thus mechanism (*a*) of Fig. 7.2 is the beam-type mechanism for the case in which $\beta_1 \leqslant \beta_2$, so that the hinges at the ends of the beam occur in the beam rather than in the stanchions, whereas mechanism (*b*) shows the beam-type mechanism for $\beta_1 \geqslant \beta_2$. Similarly, mechanisms (*c*) and (*d*) are the two possible combined mechanisms, and mechanisms (*e*) and (*f*) are the two possible sidesway mechanisms. The magnitudes of the hinge rotations for each of the mechanisms are shown in Fig. 7.2, and the work equations are readily written down in the usual way. These work equations, with θ cancelled on both sides, are given in the figure.

The significance of these work equations in relation to the problem of minimum weight design is most readily understood by constructing a diagram in which the fully plastic moments β_1 and β_2 are represented as coordinate variables and the work equations are plotted as straight lines, as in Fig. 7.3. In this figure the ordinates and abscissae represent values of $\frac{\beta_1}{Wl}$ and $\frac{\beta_2}{Wl}$, respectively. For each of the three types of mechanism there are two lines representing the work equations for the cases $\beta_1 \leqslant \beta_2$ and $\beta_1 \geqslant \beta_2$. Thus for the sidesway mechanism the lines (*e*) and (*f*) represent these two cases, these lines intersecting at the point N where $\beta_1 = \beta_2$. The other work equations are shown similarly as the lines (*a*), (*b*), (*c*) and (*d*), the lettering corresponding to the lettering of the mechanisms and their work equations in Fig. 7.2.

A simple deduction can be made from this figure with the aid of the kinematic theorem. For any given ratio of β_1 to β_2 this theorem states that the actual collapse mechanism is the one for which the corresponding work equation gives the highest values for the fully plastic moments. Consider, for instance, the case for which $\beta_1 = \frac{3}{4}\beta_2$. This condition is represented by a straight line OA' through the origin of slope $\frac{3}{4}$. It will be seen that this line intersects the three work equation lines (e), (c) and (a). The intersection with (a) at the point A' is furthest from the origin and thus represents the highest values for the fully plastic moments. Thus for this ratio of β_1 to β_2 this point represents the required values of β_1 and β_2. By considering all the possible ratios of β_1 to β_2 in this way, it is seen that all possible conditions of collapse are represented by the line segments BU_2U_1MC shown shaded in Fig. 7.3. Furthermore, it is clear that any points

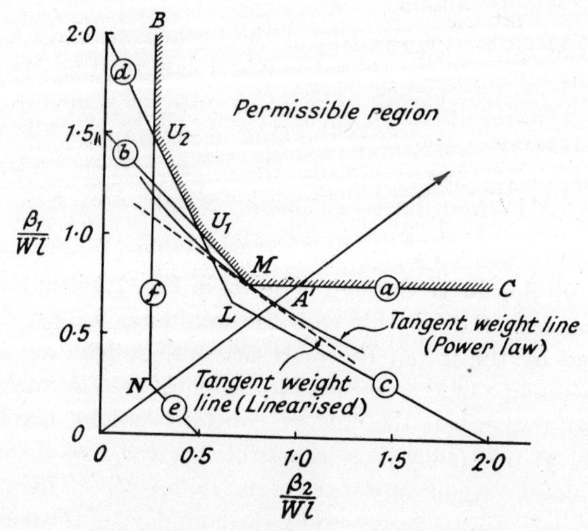

Fig. 7.3. Geometrical analogue for rectangular portal frame.

further from the origin than these line segments represent designs which would not collapse under the given loads; the region so bounded is termed the *permissible region.* Points nearer to the origin than the boundary of the permissible region represent structures which could not support the loads.

The minimum weight problem thus reduces to locating that point on the boundary of the permissible region for which the total structural weight is a minimum. To do this it is only necessary to note that from equation 7.5, any straight line of the form

$$3\beta_1 + 2\beta_2 = \text{constant}$$

represents designs of constant weight, the perpendicular distance from the origin to the line being proportional to the weight function x. Hence the minimum weight design is found by determining where such a line, of slope $-\frac{2}{3}$, just touches the boundary of the permissible region. The appropriate *tangent weight line* is shown dotted in Fig. 7.3; this line touches the boundary of the permissible region at the point M. The co-ordinates of M are

$$\beta_1 = \tfrac{3}{4}Wl, \ \beta_2 = \tfrac{3}{4}Wl,$$

and these values of the fully plastic moments represent the minimum weight design.

It is of interest to notice that if the linearization of the weight function is not carried out, but a power law relating the weight per unit length to the fully plastic moment of a member of the form of equation 7.1 is used, the lines of constant weight become curves which are convex towards the origin. If the index 0·6 in equation 7.1 is used, lines of constant weight are of the form

$$3\beta_1^{0\cdot 6} + 2\beta_2^{0\cdot 6} = \text{constant}$$

and the full line curve through point M in Fig. 7.3 is of this form. It will be seen that in this case the minimum weight design is unaffected by the use of the more accurate weight relationship.

The boundary of the permissible region is convex towards the origin, so that when the line of constant weight touches this boundary at one point it is impossible for the line to pass into the permissible region anywhere along its length. This property of convexity is always possessed by the boundary of the permissible region. A design can therefore be proved to be of minimum weight by making only *local* tests to show that the weight is increased by any small changes in the ratios of the fully plastic moments. However, Prager[6] has pointed out that this is not true if a power law weight relationship such as equation 7.1 is used.

Another general result of considerable importance and utility is exemplified in Fig. 7.3. At the minimum weight point M in this figure the lines corresponding to mechanisms (a) and (b) intersect to form a corner of the boundary of the permissible region. This implies that the minimum weight structure can fail in either of these two mechanisms. These two alternative mechanisms are seen from Fig. 7.2 to be the two possible beam-type mechanisms, the hinges at the ends of the beam forming either in the beam [mechanism (a)] or in the stanchions [mechanism (b)], depending on whether the fully plastic moment of the beam β_1 is less than or greater than the fully plastic moment of the stanchions β_2. In the minimum weight structure the fully plastic moments of the beam and the stanchions are equal, so that the hinges at the ends of the beam can form either in the beam or in the stanchions without affecting the collapse load. This implies that there exists a collapse mechanism with two degrees of freedom, as shown in Fig. 7.4. In this figure plastic hinges are shown in the stanchions, the angle of rotation being ϕ, and at each end of the beam, the angle of rotation being ψ. Each half of the beam then rotates through an angle $(\psi + \phi)$ and the hinge rotation at the centre of the beam is $2(\psi + \phi)$. The only restriction on ψ and ϕ is that both these angles should be positive in order that the bending moments at the ends of the beam should be hogging. The work equation for this mechanism is seen from Fig. 7.4 to be

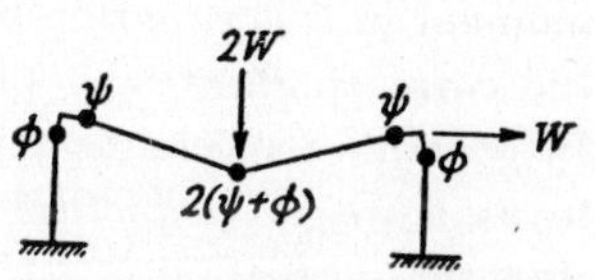

Fig. 7.4. Mechanism with two degrees of freedom.

$$3Wl(\psi + \phi) = (4\psi + 2\phi)\beta_1 + 2\phi\beta_2 \quad . \quad . \quad 7.6$$

If $\beta_1 = \beta_2$, this equation yields the result $\beta_1 = \beta_2 = \frac{3}{4}Wl$, which was obtained previously.

For any given ratio of ψ to ϕ equation 7.6 represents a straight line in Fig. 7.3. When $\phi = 0$ the line becomes line (a) corresponding to mechanism (a) of Fig. 7.2, and when $\psi = 0$ the line becomes line (b) corresponding to mechanism (b). The slope of the tangent weight line lies between the slopes of these two mechanism lines (a) and (b), and it follows that for some intermediate condition equation 7.6 must represent a line of the same slope as the tangent weight line. The equation of this line is seen from equation 7.5 to be

$$x = 3l\beta_1 + 2l\beta_2 = \text{constant}$$

and for equation 7.6 to represent a line of the same slope it follows by comparing coefficients of β_1 and β_2 that

$$\frac{4\psi + 2\phi}{3} = \frac{2\phi}{2}$$

so that $\psi = \frac{1}{4}\phi$. Substituting this value of ψ in equation 7.6, it is found that

$$3{\cdot}75Wl\phi = \phi(3\beta_1 + 2\beta_2) \quad . \quad . \quad 7.7$$

and the straight line represented by this equation is clearly parallel to the lines of constant weight.

The particular beam-type mechanism obtained by choosing $\psi = \frac{1}{4}\phi$ fulfils the condition that if its work equation is written down, as in equation 7.7, the coefficient of each fully plastic moment in this equation bears the same ratio to the coefficient of the corresponding fully plastic moment in the weight equation. In general, a mechanism fulfilling this condition will be referred to as a *weight compatible* mechanism, and it will be seen later that the derivation of such mechanisms plays an important role in minimum weight theory.

In this particular case the collapse mechanism for the minimum weight structure possessed two degrees of freedom, this number being the same as the number of different fully plastic moments β_1 and β_2 which specified the design. This result can be generalized for the case in which the design is specified by n different fully plastic moments, leading to the following theorem, which was proved by Foulkes.[2]

Theorem 1

If the design of a frame is specified by the values of n different fully plastic moments, it is always possible to find at least one minimum weight design for which the collapse mechanism has at least n degrees of freedom.

The wording of this theorem includes those cases in which there is a *range* of values of the fully plastic moments for which the weight of the structure is constant and equal to the minimum weight. A case of this kind is illustrated in Fig. 7.5. This figure refers to the design of the rectangular portal frame shown, with span $2l$ and height l, subjected to vertical and horizontal loads $3W$ and W, respectively. The fully plastic moment of the beam is β_1 and that of the stanchions is β_2. The figure shows lines corresponding to the work equations of the six possible mechanisms of the same type as shown in Fig. 7.2. The resulting boundary to the permissible region is indicated as in Fig. 7.3 by the shaded line segments. For this frame the lines of constant weight, as obtained from the linearized relation of equation 7.4, are

$$x = 2l(\beta_1 + \beta_2) = \text{constant} . \qquad 7.8$$

and it will be seen that the tangent weight line, shown dotted in Fig. 7.5, touches the boundary of the permissible region along the

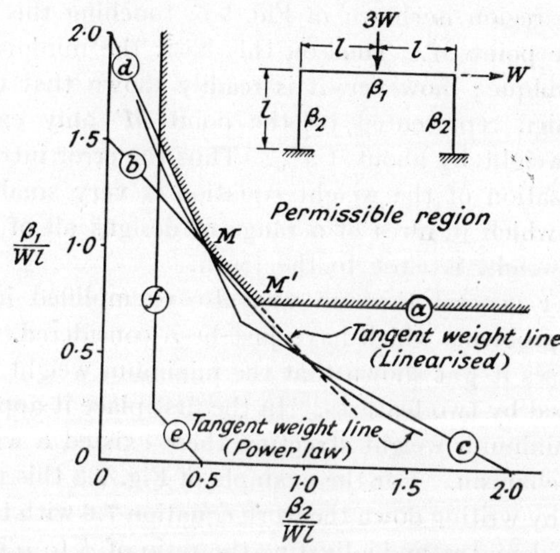

Fig. 7.5. *Range of designs of minimum weight.*

segment MM'. Thus in this case there is a range of designs, represented by MM', all of which have minimum weight. This is because the line MM' and its prolongation represents the work equation for the mechanism of type (*b*) in Fig. 7.2, and it is easily seen that for the loading and frame dimensions now under consideration the work equation for this mechanism is

$$3Wl\theta = (2\beta_1 + 2\beta_2)\theta$$

This is seen to fulfil the conditions of weight compatibility by comparison with the weight equation 7.8, thus confirming that there should be a range of designs whose weights are all equal to the minimum weight. Of these designs, the two particular designs represented by the points M and M' are such that Theorem 1 is fulfilled. For instance, the design represented by M could fail either by mechanism (*b*) or mechanism (*d*), so that a mechanism with two degrees of freedom could be formed; similarly, the design represented by M' could fail either by mechanism (*b*) or mechanism (*a*).

It is of interest to note that if the power law of equation 7.1, relating the weight per unit length of each member to its fully plastic moment, is used, lines of constant weight are of the form

$$\beta_1^{0 \cdot 6} + \beta_2^{0 \cdot 6} = \text{constant}$$

The line of this form which is tangential to the boundary of the permissible region is shown in Fig. 7.5, touching this boundary only at the point M. Thus on this basis the minimum weight design is unique; however, it is readily shown that the weight of the design represented by the point M' only exceeds the minimum weight by about 1·5%. Thus the error introduced by the linearization of the weight equation is very small, and the indication which it gives of a range of designs all of which are minimum weight is close to the truth.

A further general theorem was also exemplified in the two particular problems which have just been considered. For both of these cases it was shown that the minimum weight frame was distinguished by two features. In the first place it appeared that for each minimum weight structure there existed a weight compatible mechanism. For the example of Fig. 7.3 this mechanism was found by writing down the work equation 7.6 with two degrees of freedom ϕ and ψ, and adjusting the ratio of ϕ to ψ so that the coefficients of β_1 and β_2 were in the same ratio as in the weight equation 7.5, whereas for the example of Fig. 7.5 the weight compatible mechanism had only one degree of freedom. The other distinguishing feature of each of the minimum weight designs was the obvious one that the points representing them lay on the boundary of the permissible region; this implied that in each case a safe and statically admissible distribution of bending moment throughout the structure could be found. It can be shown that in general a design will be of minimum weight if it satisfies the two conditions that a weight compatible mechanism of collapse can be found and that a corresponding safe and statically admissible distribution of bending moment throughout the frame can be shown to exist. Before stating this result as a theorem a general definition of what is meant by a weight compatible mechanism must be given.

Consider a frame for which n different fully plastic moments $(\beta_1, \beta_2 \ldots \beta_n)$ are to be chosen, and suppose that the work equation of any mechanism is

$$\text{Work done} = [a_1\beta_1 + a_2\beta_2 + \ldots a_n\beta_n]\theta$$

where $a_r\theta$ is the total hinge rotation in all the members whose fully plastic moment is β_r.

The linearized weight equation for the frame is supposed to be

$$x = [l_1\beta_1 + l_2\beta_2 + \ldots l_n\beta_n]$$

where l_r is the total length of all the members whose fully plastic moment is β_r. The mechanism is then said to be weight compatible if the coefficients in the work equation and the linearized weight equation are related as follows:

$$\frac{a_1}{l_1} = \frac{a_2}{l_2} = \ldots = \frac{a_n}{l_n},$$

implying that the total hinge rotation in each set of members with the same fully plastic moment is proportional to their total length.

The second theorem can now be stated as follows:

Theorem 2

If for any design of a frame a weight compatible mechanism can be formulated, and a corresponding safe and statically admissible distribution of bending moment throughout the frame can be found, the design will be of minimum weight.

This theorem, which was also established by Foulkes,[2] does not imply that there is a unique minimum weight design. As in the example of Fig. 7.5, there may be a range of designs which all have the same minimum weight.

Upper and lower bounds

It is clear that the problem of designing for minimum weight is complicated, and for this reason it would be useful to be able to establish upper and lower bounds on the minimum weight. If these bounds could be made to approach one another sufficiently closely, it would scarcely be necessary to determine a completely accurate minimum weight design, particularly when it is recalled that in practice there is only a finite number of sections available, for it is unlikely that any of these sections would have the precise values of the fully plastic moments demanded by theory. In fact, it will be seen later when an actual example is considered that it is often too difficult to increase the lower bound to make this type of technique practicable. Nevertheless, the upper and lower bound theorems which have been developed are of intrinsic interest, and will be given here.

The establishment of an upper bound on the minimum weight is a comparatively trivial matter. It is only necessary to assign any arbitrary values to the ratios of the fully plastic moments to one another, so that all the fully plastic moments are known in terms of a single one. The frame is then analysed to find the value of this fully plastic moment such that the frame would then just collapse under the prescribed loads. This was, of course, precisely the procedure adopted in Chapters 3 and 4. The weight of the frame whose design is thus determined is clearly an upper bound on the minimum weight, since the frame is capable of carrying the prescribed loads. For the sake of completeness this result will be stated as a theorem:

Theorem 3

If for any design of a frame it can be shown that collapse would just occur in a certain mechanism while a corresponding safe and statically admissible distribution of bending moment throughout the frame can be found, the weight of this design is greater than or equal to the minimum weight.

It will be appreciated that the mechanism condition could be omitted from this theorem without affecting its validity, for the design would then be over-safe. In geometrical terms, the point representing the design would then be within the permissible region, whereas with the mechanism condition included the point is on the boundary of the permissible region. Thus the theorem as it stands gives the lowest possible value of the upper bound.

A lower bound on the minimum weight can be established with the aid of the following theorem, due to Foulkes.[2]

Theorem 4

If for any design of a frame a weight compatible mechanism can be formulated, but no corresponding safe and statically admissible distribution of bending moment throughout the frame can be found, the weight of this design will be less than the minimum weight.

The meaning of this theorem is best understood with reference to Fig. 7.3. At the point L in this figure there are two possible collapse mechanisms, namely, mechanisms (c) and (d) of Fig. 7.2. The slopes of the lines (c) and (d) in Fig. 7.3, representing the work equations for these two mechanisms, are respectively less than and greater than the slope of the lines of constant weight. Thus by combining these two mechanisms it must be possible to form a weight compatible mechanism. However, the point L is outside the permissible region. Thus the design represented by L fulfils the conditions of Theorem 4, and its weight is seen to be less than the minimum weight. In contrast, at the point N there are two possible mechanisms whose work equations are represented by the lines (e) and (f), and both these lines are of greater slope than the lines of constant weight. Thus no weight compatible mechanism could be formed for this design, and so it is impossible to state on this basis alone whether the weight of this design is less than or greater than the minimum weight.

7.4. Applications of theorems

Two-bay rectangular frame

The significance of the above theorems will now be explained by means of three examples, the first being the two-bay rectangular

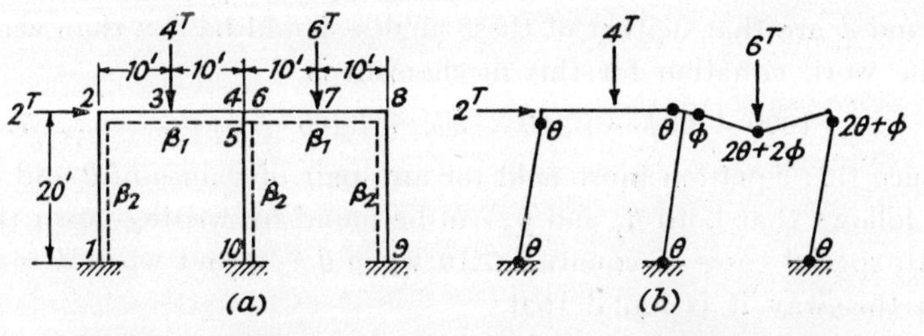

(a) Frame and loading. (b) Mechanism with two degrees of freedom.

Fig. 7.6. *Two-bay rectangular portal frame.*

portal frame whose dimensions and loading are shown in Fig. 7.6(a). In this frame the two beams each have the same fully plastic moment β_1 and the three stanchions each have the same fully plastic moment β_2.

This frame was analysed in Section 4.3 by the method of combining mechanisms for the particular case in which $\beta_1 = 2\beta_2$ (see Fig. 4.4). It was shown that the required values of β_1 and β_2 were 17·14 tons ft. and 8·57 tons ft., respectively, the collapse mechanism being simply collapse of the right-hand beam. From Theorem 3, this design gives an upper bound on the minimum weight. The linearized weight function x is found from equation 7.4 to be

$$x = 40\beta_1 + 60\beta_2 \quad . \quad . \quad . \quad . \quad 7.9$$

It follows that the minimum weight value of x, which will be denoted by x^{min}, is bounded from above as follows:

$$x^{min} \leqslant 40 \times 17{\cdot}14 + 60 \times 8{\cdot}57 = 1{,}200.$$

The combining mechanisms analysis also revealed that apart from the right-hand beam mechanism, the mechanism which gave the highest required values for the fully plastic moments was that shown in Fig. 4.5(b), in which the right-hand beam mechanism is combined with the sidesway mechanism together with a rotation of the central joint. The design which results from making this mechanism an alternative to the right-hand beam mechanism will now be considered.

The appropriate mechanism with two degrees of freedom, θ and ϕ, is shown in Fig. 7.6(b). It will be seen that if $\phi = 0$ this mechanism reduces to that of Fig. 4.5(b), whereas if $\theta = 0$ it reduces to the right-hand beam mechanism. The only restrictions on θ and ϕ are that neither of these angles should be less than zero. The work equation for this mechanism is

$$100\theta + 60\phi = \beta_1(3\theta + 3\phi) + \beta_2(6\theta + \phi) \quad . \quad . \quad 7.10$$

Since this equation must hold for any pair of values of θ and ϕ, it follows that both β_1 and β_2 can be found by writing down the two special cases of equation 7.10 when $\theta = 0$ and when $\phi = 0$. In this way it is found that

$$60 = 3\beta_1 + \beta_2$$
$$100 = 3\beta_1 + 6\beta_2,$$

so that $\beta_1 = 17{\cdot}33$ tons ft., $\beta_2 = 8$ tons ft.
By Theorem 4, this design will give a lower bound on x^{min} if it is possible to adjust the ratio of ϕ to θ in the work equation 7.10 so as to produce a weight compatible mechanism. Comparing this equation with the weight equation 7.9, the condition for the work equation to be weight compatible is seen to be

$$\frac{3\theta + 3\phi}{40} = \frac{6\theta + \phi}{60}$$

$$\phi = \frac{3}{7}\theta.$$

As a check, if this value of ϕ is substituted in equation 7.10 it is found that

$$\frac{880}{7}\theta = \theta\left(\frac{30}{7}\beta_1 + \frac{45}{7}\beta_2\right),$$

and this equation is readily seen to be weight compatible by comparison with equation 7.9. Moreover, since θ is positive, ϕ has been found to be positive, which does not violate the restriction that this angle should not be negative. Thus for this design a weight compatible mechanism has been found; from Theorem 4 it is concluded that the corresponding value of x is a lower bound on $x^{\min}$. Hence,

$$x^{\min} \geqslant 40 \times 17{\cdot}33 + 60 \times 8 = 1{,}173.$$

Combining this with the upper bound on $x^{\min}$ which has just been found, it follows that

$$1{,}173 \leqslant x^{\min} \leqslant 1{,}200.$$

The closeness of these bounds indicates that the ratio of β_1 to β_2 in either of these designs must be near the value necessary to give minimum weight. From Theorem 1 the second design may be suspected to be the actual minimum weight design, for this design satisfies the condition that there is a possible mechanism with two degrees of freedom. To test this, Theorem 2 will be used. From this Theorem it follows that the second design will be the minimum weight design if a corresponding safe and statically admissible distribution of bending moment throughout the frame can be found, for the condition of weight compatibility has already been fulfilled.

The equations of equilibrium for this frame and loading were shown in Section 4.3 to be

$$40 = 2M_3 - M_2 - M_4 \quad . \quad . \quad . \quad . \quad 4.13$$

$$60 = 2M_7 - M_6 - M_8 \quad . \quad . \quad . \quad . \quad 4.14$$

$$40 = M_2 - M_1 + M_5 - M_{10} + M_9 - M_8 \quad . \quad 4.15$$

$$0 = M_4 + M_5 - M_6, \quad . \quad . \quad . \quad . \quad 4.16$$

with the usual sign convention that a positive bending moment causes tension in the fibres of a member adjacent to the dotted line in Fig. 7.6(a). For the mechanism of Fig. 7.6(b) the fully plastic moments, in tons ft., are

$$M_1 = -8, \quad M_2 = 8, \quad M_4 = -17{\cdot}33, \quad M_6 = -17{\cdot}33,$$
$$M_7 = 17{\cdot}33, \quad M_8 = -8, \quad M_9 = 8, \quad M_{10} = -8.$$

Substituting these values in equations 4.13–4.16, it is found that the bending moments at the two sections 3 and 5 where plastic hinges do not occur have the values

$$M_3 = 15{\cdot}33, \quad M_5 = 0,$$

and that with these values the four equations of equilibrium are satisfied identically. Since these two bending moments are less than the fully plastic values, it follows that a safe and statically admissible set of bending moments corresponding to the mechanism of Fig. 7.6(*b*) has been found. Thus from Theorem 2, the design $\beta_1 = 17{\cdot}33$ tons ft., $\beta_2 = 8$ tons ft. is the minimum weight design.

The collapse mechanism for this minimum weight frame was shown in Fig. 7.6(*b*), with two degrees of freedom, so that the collapse of the frame is of the over-complete type referred to in Section 3.6. It is evident from Theorem 1 that in many cases the collapse of a minimum weight frame will be over-complete, although it frequently happens that the requisite number of degrees of freedom of the collapse mechanism is obtained by joint rotations, as in the next example.

Pitched roof portal frame

The second example to be considered is the design of the pinned-base pitched roof portal frame whose dimensions and loading are as shown in Fig. 4.11(*a*). This frame was analysed in Section 4.3 by the combining mechanisms method on the assumption that every member of the frame possessed the same fully plastic moment, and it was shown that the required value of the fully plastic moment was 8.42 tons ft. It will now be shown that this design is of minimum weight, given that the two rafter members are required to be of the same section, with fully plastic moment β_1, and that the two stanchions are also required to be of the same section, with fully plastic moment β_2.

The collapse mechanism for this frame, on the assumption that $\beta_1 = \beta_2 = M_p$, was found to be as shown in Fig. 4.12(*a*), apart from the small adjustment necessitated by the fact that the statical check of Fig. 4.12(*b*) reveals that the hinge at the apex should be moved about 3·2 ft. down the left-hand rafter. In what follows this adjustment will be disregarded, so that it will be assumed that the mechanism of Fig. 4.12(*a*) is the actual collapse mechanism. For this mechanism the work equation was shown to be:

$$42{\cdot}43\theta = 5{\cdot}2M_p\theta\,;\quad M_p = 8{\cdot}16 \text{ tons ft.}$$

Since for the design under consideration β_1 and β_2 are equal, the plastic hinge at the right-hand knee of the frame, whose rotation is shown in Fig. 4.12(a) as $3{\cdot}2\theta$, can form in either the stanchion or the rafter, providing a mechanism with two degrees of freedom. Thus if a hinge rotation ϕ is assumed to occur in the rafter at this section, the hinge rotation in the stanchion would be $(3{\cdot}2\theta - \phi)$, where ϕ is restricted to lie within the limits of zero and $3{\cdot}2\theta$, so that each hinge at this section should be closing when viewed from within the frame. This does not affect the work done by the loads; the work equation thus becomes

$$42{\cdot}43\theta = \beta_1(2\theta + \phi) + \beta_2(3{\cdot}2\theta - \phi)$$

Equating coefficients of θ and ϕ in turn in this equation merely reproduces the original work equation together with the result that $\beta_1 = \beta_2$.

Since the total length of the rafter members is 32·3 ft., and the length of each stanchion is 10 ft., the linearized weight equation is

$$x = 32{\cdot}3\beta_1 + 20\beta_2$$

The condition for weight compatibility is therefore

$$\frac{2\theta + \phi}{32{\cdot}3} = \frac{3{\cdot}2\theta - \phi}{20},$$

from which it is found that $\phi = 1{\cdot}21\theta$, lying as required between the limits of zero and $3{\cdot}2\theta$. With this value of ϕ the mechanism becomes weight compatible, and although the apex hinge was not positioned quite correctly it is clear that there would be no difficulty in fulfilling the condition of weight compatibility if this hinge were moved down the left-hand rafter to its correct position. Since it is possible to derive a weight compatible mechanism for this design, and a corresponding safe and statically admissible distribution of bending moment throughout the frame can be found, as shown in Section 4.3, it can be concluded from Theorem 2 that thc design $\beta_1 = \beta_2 = 8{\cdot}42$ tons ft. is of minimum weight.

In this case an extra degree of freedom for the collapse mechanism was furnished by an arbitrary rotation of the knee joint, made possible by the fact that the two members meeting at this joint possessed the same fully plastic moment. It is thus scarcely worth while classifying this collapse mechanism as over-complete, despite the fact that it does possess more than one degree of freedom.

Simple rectangular frame

In the preceding example it appeared that the minimum weight design was a frame of uniform section throughout, the condition of weight compatibility being achieved by means of the degree of freedom furnished by the joint rotation as a direct consequence of the uniformity of the section. The provision of a uniform frame for a single-bay portal frame, whether of rectangular or pitched roof construction, will often ensure minimum weight owing to the possibility thus afforded for using joint rotations to satisfy

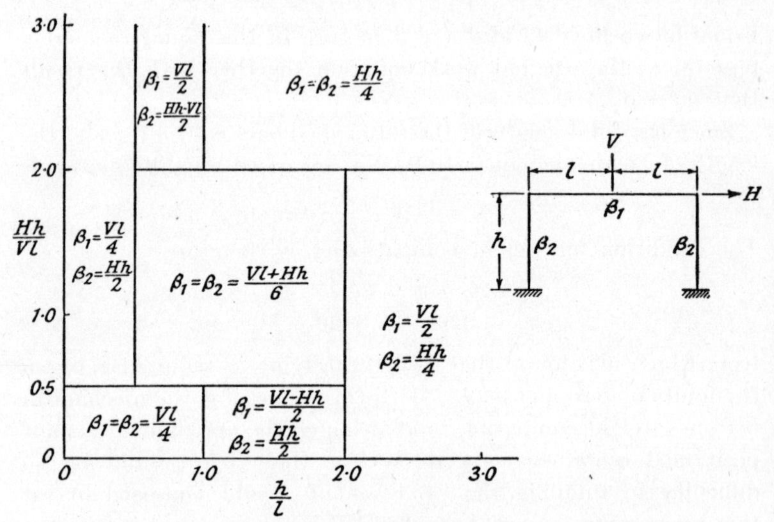

Fig. 7.7. *Design chart for rectangular portal frame.*

the conditions of weight compatibility. This observation was borne out by the calculations made by Foulkes [2] on the minimum weight design of a simple rectangular portal frame with fixed feet, which are summarised in the chart of Fig. 7.7. The frame is taken to be of height h and span $2l$, and is subjected to a central vertical load V and a horizontal load H at beam level. The beam has a fully plastic moment β_1, and each stanchion has a fully plastic moment β_2.

The interpretation of the chart of Fig. 7.7 is simply that for given values of the loads V and H and the frame dimensions l and h, the ratios $\frac{Hh}{Vl}$ and $\frac{h}{l}$ are computed and the point with these coordinates is located on the chart. This point will lie within one of the seven regions which are bounded by straight lines parallel to the axes, and the values of β_1 and β_2 appropriate to the particular region then determine the minimum weight design. It will be seen from Fig. 7.7 that a large proportion of the designs represent uniform frames; different values for β_1 and β_2 tend to occur only when the frame is of unusual proportions or is subjected to an unlikely combination of loads.

A design chart of this kind is readily constructed by assuming a particular mechanism with two degrees of freedom and determining under what conditions this would lead to a minimum weight design. Suppose for instance that a mechanism of the type illustrated in Fig. 7.4 is assumed. With the loads and dimensions as given in Fig. 7.7 the work equation is

$$Vl(\phi + \psi) = \beta_1(4\psi + 2\phi) + \beta_2(2\phi)$$

Equating coefficients of ψ and ϕ in turn, it is found that

$$\beta_1 = \beta_2 = \frac{Vl}{4}$$

The linearized weight equation for this frame is

$$x = 2l\beta_1 + 2h\beta_2,$$

so that the mechanism will be weight compatible if

$$\frac{4\psi + 2\phi}{2l} = \frac{2\phi}{2h}$$

$$\psi = \left(\frac{l-h}{2h}\right)\phi$$

From Fig. 7.4 it is seen that for the hinges at the ends of the beam to correspond to hogging bending moments, both ϕ and ψ must be positive. It follows that

$$\frac{l-h}{2h} \geqslant 0$$

so that

$$\frac{h}{l} \leqslant 1$$

Apart from the condition of weight compatibility, it must be possible to find a safe and statically admissible distribution of bending moment throughout the frame in order for the design to be of minimum weight, by Theorem 2. This can be done by writing down the equations of equilibrium, substituting those fully plastic moments which occur in the beam mechanism, and then establishing the condition that the moments at the feet of the stanchions must not exceed the fully plastic moment $\beta_2 = \frac{Vl}{4}$. An equivalent procedure is to establish the condition that the assumed mechanism gives the highest required values of the fully plastic moments. By either of these methods it is found that the following restriction applies to the loads and dimensions of the frame:

$$\frac{Hh}{Vl} \leqslant \frac{1}{2}$$

Thus it follows that the design will be of minimum weight if this condition, together with the condition $\frac{h}{l} \leqslant 1$ which is necessary for weight compatibility, is fulfilled. The corresponding region is shown in Fig. 7.7.

7.5. Methods of solution

The first method proposed for the solution of minimum weight problems was due to Heyman,[7] this method being based on the inequalities method for the collapse analysis of a given frame, which was described briefly in Section 4.5. The conditions that the bending moment at every cross-section of a frame where a plastic hinge might form should not exceed the fully plastic moment, when coupled with the equations of equilibrium, can be expressed as a set of linear inequalities involving the bending moments as variables. In analysing a given frame the bending moments are eliminated from the inequalities in turn, and when all the bending moments have been eliminated each of the resulting inequalities sets a lower limit on the value of M_p. The greatest of these lower limits is then the required value of M_p.

Heyman adapted this method to the solution of minimum weight problems by using the linearized weight equation to eliminate one of the bending moments from the set of linear inequalities before proceeding with the elimination of the remaining bending moments. When all the bending moments have been eliminated each of the resulting inequalities then sets a lower limit on the linearized weight function x, and the largest of these lower limits is the minimum possible value of x. By working back through the inequalities the corresponding fully plastic moments are then determined. This technique was also applied by Heyman to the solution of problems of minimum weight design in which the structure is to be capable of withstanding two or more different loadings.

The method is lengthy in its application to all but the simplest types of problem. This is primarily because it consists in essence of investigating every possible collapse mechanism and its associated designs. The possibility of making an intelligent guess as to the minimum weight design, testing the design to see whether it is in fact of minimum weight, and if it is not improving the design by a series of rational steps, is thereby excluded. Trial and error procedures of this kind have been suggested by both Foulkes [1] and Heyman.[8] The techniques advocated by these authors are essentially the same in principle, differing only in the details of their application. In the first place, the fully plastic moment of each member is expressed as a simple multiple of the fully plastic moment M_p of one of the members, and the frame is then analysed to find the value of M_p such that collapse would just occur under the prescribed loads. This determines an initial design which will be of minimum weight if there is a corresponding weight compatible mechanism, by Theorem 2, and otherwise will have a weight exceeding the minimum weight, by Theorem 3. In this latter case a fresh guess is made as to the ratios of the fully plastic moments to one another, and a new collapse analysis is carried out to determine the corresponding design. As will be seen, this second guess is not made in a random manner, for the results of the first analysis will suggest how the design should be modified to lower the weight. Proceeding in this manner a succession of designs is investigated, until a design is achieved to which there corresponds a mechanism which is very nearly weight compatible. It is then a fairly simple matter to formulate a weight compatible mechanism, which when analysed will almost certainly give the minimum weight design.

Two-storey single-bay rectangular frame

The technique will be illustrated by an application to the two-storey single-bay rectangular frame whose dimensions and loading are as illustrated in Fig. 7.8(*a*). In this frame the upper beam has a fully plastic moment β_1, the two upper stanchions each have the same fully plastic moment β_2, the lower beam has a fully plastic moment β_3 and the two lower stanchions each have the same fully plastic moment β_4; subject to these conditions the frame is to be designed so that the given loads would just cause failure by plastic collapse and the linearized weight function x, whose value is

$$x = 20\beta_1 + 30\beta_2 + 20\beta_3 + 30\beta_4 \quad . \quad . \quad 7.11$$

has its least possible value.

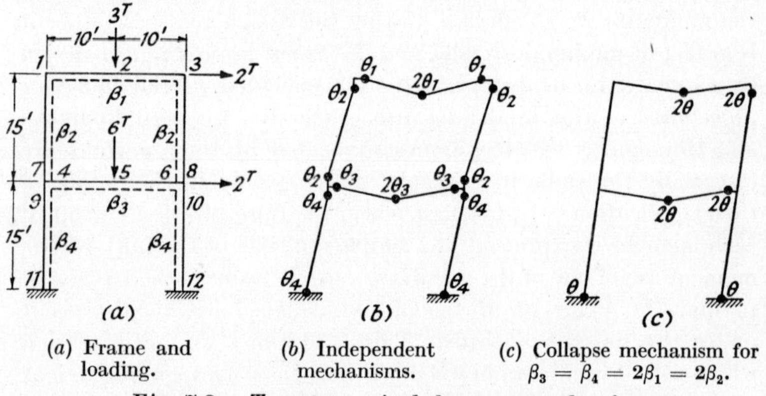

(a) Frame and loading. (b) Independent mechanisms. (c) Collapse mechanism for $\beta_3 = \beta_4 = 2\beta_1 = 2\beta_2$.

Fig. 7.8. *Two-storey single-bay rectangular frame.*

The collapse analyses for the various designs which will be investigated will be carried out by means of the combining mechanisms technique, so that it is first necessary to establish the number of independent mechanisms. For this frame and loading the number of bending moments n required to specify the distribution of bending moment throughout the frame is twelve, since the bending moment must vary linearly between those cross-sections numbered from 1 to 12 in Fig. 7.8(a). The number of redundancies r is six, since the frame becomes statically determinate if the frame is imagined to be cut at two cross-sections such as 2 and 5, one in the upper half and one in the lower half of the frame, and the values of the thrust, shear force and bending moment at each cut are specified. It follows that the number of independent equations of equilibrium is $(n - r) = 6$. There must therefore be six independent mechanisms for this frame. Four of these are indicated in Fig. 7.8(b), consisting of two beam-type mechanisms together with sidesway mechanisms for the upper and lower storeys, respectively. In this figure the hinge rotations involved in each of the four mechanisms are given in terms of a single parameter which is different for each mechanism, so that the figure may be regarded as representing a mechanism with four degrees of freedom. To derive any one of the independent mechanisms it is only necessary to isolate the mechanism by equating to zero the hinge rotations associated with the other three mechanisms. Thus if $\theta_1 = \theta_2 = \theta_3 = 0$, and θ_4 is non-zero, the sidesway mechanism for the lower storey is obtained. The remaining two independent mechanisms consist of rotations of the joints 4,7,9 and 6,8,10 [Fig. 7.8(*a*)].

The work equation for the mechanism of Fig. 7.8(*b*) is as follows :

$$30\theta_1 + 30(\theta_2 + \theta_4) + 60\theta_3 + 30\theta_4 = 4\beta_1\theta_1 + 4\beta_2\theta_2 + 4\beta_3\theta_3 + 4\beta_4\theta_4 \quad . \quad . \quad 7.12$$

Equating in turn the coefficients of θ_1, θ_2, θ_3 and θ_4, the corresponding values of the fully plastic moments are found to be

$$\beta_1 = \beta_2 = 7{\cdot}5 \text{ tons ft.}, \ \beta_3 = \beta_4 = 15 \text{ tons ft.}$$

It is easily shown that the work equation 7.12 can be made weight compatible ; if $\theta_2 = \theta_4 = 1{\cdot}5\theta_1 = 1{\cdot}5\theta_3$ in this equation, the coefficients of the fully plastic moments are seen to be in the same ratios as those in the weight equation 7.11. However, it is clear that the design which is represented by these fully plastic moments would be inadequate to withstand the applied loads, for the corresponding bending moment distribution is not statically admissible. For instance, at the top left-hand joint, both the beam and the stanchion are seen from Fig. 7.8(*b*) to be exerting clockwise moments on the joint, so that this joint is not in rotational equilibrium. It can also be seen that the joints 4,7,9 and 6,8,10 are not in equilibrium ; fortuitously, the top right-hand joint 3 is in equilibrium. It follows from Theorem 4 that the weight of this design is a lower bound on the minimum weight. Thus if x^{min} is the minimum value of the linearized weight function, it is found from equation 7.11 that

$$x^{min} \geqslant 20 \times 7{\cdot}5 + 30 \times 7{\cdot}5 + 20 \times 15 + 30 \times 15 = 1{,}125.$$

This derivation of a lower bound on the minimum weight is equivalent to isolating various portions of the frame in turn and designing each portion for minimum weight on the assumption that the rest of the frame is completely rigid. The sum of these minimum weights is then a lower bound on the minimum weight of the whole frame, as pointed out by Heyman.[8]

First trial

It may be conjectured that the fully plastic moments thus obtained are in roughly the correct ratios to give a design whose weight is not much in excess of the minimum weight. The frame is therefore analysed on the assumption that

$$\beta_1 = M_p, \quad \beta_2 = M_p, \quad \beta_3 = 2M_p, \quad \beta_4 = 2M_p,$$

these fully plastic moments being in the same ratio as those obtained from equation 7.12.

Details of this analysis will not be given. It is found that the collapse mechanism is as shown in Fig. 7.8(*c*), and the corresponding value of M_p is 11·25 tons ft., so that the fully plastic moments are

$$\beta_1 = \beta_2 = 11{\cdot}25 \text{ tons ft.}, \quad \beta_3 = \beta_4 = 22{\cdot}5 \text{ tons ft.}$$

For this design the value of the linearized weight function x is seen from equation 7.11 to be given by

$$x = 20 \times 11{\cdot}25 + 30 \times 11{\cdot}25 + 20 \times 22{\cdot}5 + 30 \times 22{\cdot}5$$
$$= 1{,}687{\cdot}5$$

This value of x is, of course, an upper bound on $x^{\min}$, as stated in Theorem 3, since it is derived from the fully plastic moments for a design which can withstand the prescribed loads. Combining this upper bound with the lower bound obtained previously, it follows that

$$1{,}125 \leqslant x^{\min} \leqslant 1{,}687{\cdot}5$$

The collapse mechanism for this design must now be tested to see whether it is weight compatible. To do this the work equation is written down, preserving the identity of each fully plastic moment β_1, β_2, β_3 and β_4. This work equation is seen from Fig. 7.8(*c*) to be

$$180\theta = (2\beta_1 + 2\beta_2 + 4\beta_3 + 2\beta_4)\theta \quad . \quad . \quad 7.13$$

Comparing this equation with the linearized weight equation 7.11, it is seen that the coefficients of the fully plastic moments in the two equations are not in the same ratios, so that the mechanism is not weight compatible. In order to achieve weight compatibility the coefficients of β_2 and β_4 in the work equation need to be increased, and the coefficient of β_3 decreased in comparison with that of β_1. When a fresh guess as to the ratios of the fully plastic moments is made, β_2 and β_4 should therefore be *decreased,* and β_3 *increased,* in comparison with β_1. The reason for this is seen by considering the effect of a change in any one of the fully plastic moments. For instance, an increase of β_3 renders it less likely that the corresponding beam-type mechanism for the lower beam will feature in combination in the collapse mechanism, and so tends to reduce the coefficient of β_3 in the work equation.

Second trial

In choosing fresh ratios of the fully plastic moments for the purpose of developing a second design of lower weight, it is borne in mind that at least one design of minimum weight will involve a mechanism with four degrees of freedom, by Theorem 1, since the number of fully plastic moments required to specify a design is four. Accordingly, the fully plastic moments are selected as follows:

$$\beta_1 = M_p\,,\quad \beta_2 = M_p\,,\quad \beta_3 = 2{\cdot}5M_p\,,\quad \beta_4 = 1{\cdot}5M_p$$

It will be seen that β_1 and β_2 have been kept equal, although the indication from the work equation 7.13 was that β_2 should be increased. This was done in order that a degree of freedom corresponding to a rotation of the top right-hand joint 3 should be available; if $\beta_1 = \beta_2$ a hinge at this joint can appear either in the beam or the stanchion, thus providing a degree of freedom. In Fig. 7.8(c) the hinge was placed in the stanchion so as to make the coefficient of β_2 in equation 7.13 as large as possible. The only other point involved in this fresh selection of the fully plastic moments is that β_3 was chosen to be equal to the sum of β_2 and β_4; as will be seen, this enables a rotation of the joint 6,8,10 to feature as an alternative mechanism.

When a fresh analysis of the frame by the method of combining mechanisms is made with these new values of the ratios of the fully plastic moments, it is found that the collapse mechanism is still the mechanism of Fig. 7.8(c), and that the corresponding value of M_p is 10·59 tons ft., so that the fully plastic moments in this second design are:

$$\beta_1 = \beta_2 = 10{\cdot}59 \text{ tons ft.},\quad \beta_3 = 26{\cdot}47 \text{ tons ft.},\quad \beta_4 = 15{\cdot}88 \text{ tons ft.}$$

For this design the value of the linearized weight function x is seen from equation 7.11 to be

$$\begin{aligned} x &= 20 \times 10{\cdot}59 + 30 \times 10{\cdot}59 + 20 \times 26{\cdot}47 + 30 \times 15{\cdot}88 \\ &= 1{,}535{\cdot}3 \end{aligned}$$

This value of x is an upper bound on $x^{\min}$; it is seen to be less than the previous upper bound of 1,687·5.

The collapse mechanism must now be tested for weight compatibility. This mechanism is shown in Fig. 7.9(a), and is identical with that of Fig. 7.8(c) except that the top right-hand joint 3 is rotated through an angle ϕ clockwise and the bottom right-hand joint 6,8,10 is rotated through an angle ψ counterclockwise. These joint rotations can be made without disturbing the joint equilibrium equations owing to the fact that $\beta_1 = \beta_2$ and $\beta_3 = \beta_2 + \beta_4$ in this design. It will be appreciated that both ϕ and ψ must lie between certain limits. Thus the hinges at the top right-hand

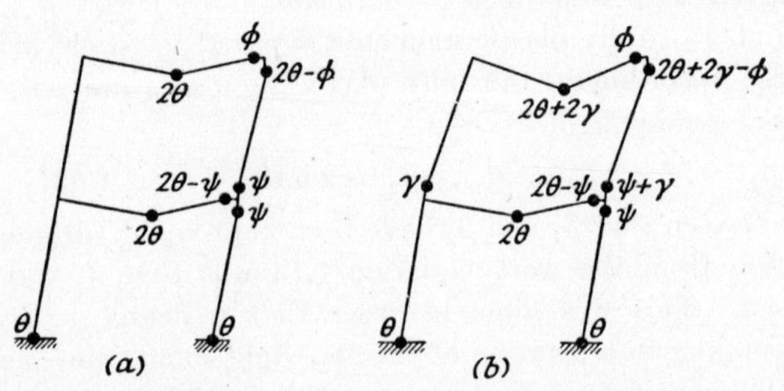

(a) Mechanism with three degrees of freedom. (b) Mechanism with four degrees of freedom.

Fig. 7.9. *Collapse mechanisms for two-storey single-bay rectangular frame.*

joint 3 must both be closing when viewed from within the frame, so that ϕ must lie between zero and 2θ; ψ must lie between the same limits so that the hinges at the joint 6,8,10 are in accordance with the requirements of rotational equilibrium for this joint. The work equation for this mechanism is

$$180\theta = \beta_1(2\theta + \phi) + \beta_2(2\theta - \phi + \psi) + \beta_3(4\theta - \psi) + \beta_4(2\theta + \psi) \quad . \quad . \quad . \quad 7.14$$

This equation can be checked by noting that θ, ϕ and ψ may be chosen independently within the limits which exist on ϕ and ψ. It follows that three equations can be derived from equation 7.14 by equating coefficients of θ, ϕ and ψ in turn. These three equations are

$$180 = 2\beta_1 + 2\beta_2 + 4\beta_3 + 2\beta_4$$
$$0 = \beta_1 - \beta_2$$
$$0 = \beta_2 - \beta_3 + \beta_4$$

These three equations are all consistent with the derived fully plastic moments.

Comparing the work equation 7.14 with the linearized weight equation 7.11 it is seen that for weight compatibility it must be possible to find values of the ratios of ϕ and ψ to θ such that

$$\frac{2\theta + \phi}{20} = \frac{2\theta - \phi + \psi}{30} = \frac{4\theta - \psi}{20} = \frac{2\theta + \psi}{30}.$$

Upon investigation it is found that these conditions cannot all be fulfilled, as might be expected from the consideration that there are three equations to be satisfied and only two ratios which can be varied. The closest agreement is obtained with $\phi = 0{\cdot}4\theta$ and $\psi = 1{\cdot}6\phi$; with these values the work equation becomes

$$180\theta = (2{\cdot}4\beta_1 + 3{\cdot}2\beta_2 + 2{\cdot}4\beta_3 + 3{\cdot}6\beta_4)\theta, \quad . \quad 7.15$$

in which the only coefficient in error is that of β_4, which is too high.

The fact that a work equation which is almost weight compatible can be formed for this design suggests that its weight must be very close to the minimum value, and in practice it would scarcely be necessary to improve the design any further. However, it is of interest to see how further improvements could be made if required.

Minimum weight design

At this stage it is best to discard the technique of making arbitrary adjustments to the ratios of the fully plastic moments, and instead attempt to guess a collapse mechanism with four degrees of freedom which is then tested for weight compatibility. Such a guess is best made by making use of the results obtained in the combining mechanisms analysis of the last design. Here it was found that the mechanism giving the highest value of M_p for this design was that of Fig. 7.9(*a*), this value of M_p being 10·59 tons ft. However, for the particular ratios of the fully plastic moments involved in this design, there are three other mechanisms for which the corresponding values of M_p are all 10 tons ft., which is very close to the value obtained for the actual collapse mechanism. These mechanisms are those involving plastic hinges at the following sections, as numbered in Fig. 7.8(*a*):

(i) 2, 3, 7, 8; combined beam and sidesway for upper storey only.

(ii) 9, 10, 11, 12; sidesway of lower storey.

(iii) 2, 3, 4, 6, 11, 12; sidesway of whole frame combined with upper beam.

There seems to be no *a priori* method for determining which of these three mechanisms will lead to an appropriate weight compatible mechanism when its hinge rotations and deflections are combined with those of the mechanism of Fig. 7.9(a) to furnish a fourth degree of freedom. In fact, mechanism (i) leads in this manner to the minimum weight design; this will be verified first, and then the effects of using the other mechanisms will be stated.

Fig. 7.9(b) shows the hinge rotations of mechanism (i), consisting of rotations of magnitude γ at sections 7 and 8 together with rotations of magnitude 2γ at sections 2 and 3, superposed on the hinge rotations of the mechanism of Fig. 7.9(a), thus furnishing a mechanism with four degrees of freedom θ, ϕ, ψ and γ. For this mechanism the work equation is

$$\begin{aligned}180\theta + 60\gamma &= \beta_1(2\theta + \phi + 2\gamma) + \beta_2(2\theta - \phi + 4\gamma + \psi) \\ &\quad + \beta_3(4\theta - \psi) + \beta_4(2\theta + \psi) \qquad . \quad 7.16\end{aligned}$$

Since this work equation involves four degrees of freedom, it is possible by equating in turn the coefficients of θ, ϕ, ψ and γ to determine the four unknown fully plastic moments. The equations thus obtained are

$$\begin{aligned}180 &= 2\beta_1 + 2\beta_2 + 4\beta_3 + 2\beta_4 \\ 0 &= \beta_1 - \beta_2 \\ 0 &= \beta_2 - \beta_3 + \beta_4 \\ 60 &= 2\beta_1 + 4\beta_2\end{aligned}$$

with the solution

$\beta_1 = \beta_2 = 10$ tons ft., $\beta_3 = 26{\cdot}67$ tons ft., $\beta_4 = 16{\cdot}67$ tons ft.

With these values of the fully plastic moments the value of the linearized weight function is given by

$$x = 20 \times 10 + 30 \times 10 + 20 \times 26{\cdot}67 + 30 \times 16{\cdot}67 = 1{,}533{\cdot}3,$$

a value which is only slightly less than the upper bound 1,535·3 obtained for the previous design.

Comparing the work equation 7.16 with the linearized weight equation 7.11, it is seen that for weight compatibility the following conditions must be fulfilled:

$$\frac{2\theta + \phi + 2\gamma}{20} = \frac{2\theta - \phi + 4\gamma + \psi}{30} = \frac{4\theta - \psi}{20} = \frac{2\theta + \psi}{30}$$

Solving these equations it is found that

$$\phi = \frac{4}{15}\theta \qquad \gamma = \frac{1}{15}\theta \qquad \psi = 1{\cdot}6\theta$$

As a check, if these values of ϕ, γ and ψ are substituted in equation 7.16, it is found that

$$184\theta = (2{\cdot}4\beta_1 + 3{\cdot}6\beta_2 + 2{\cdot}4\beta_3 + 3{\cdot}6\beta_4)\theta$$

This work equation is clearly weight compatible, and is also satisfied by the values of the fully plastic moments which were derived above. Moreover, each hinge rotation in the mechanism of Fig. 7.9(b) is seen to be of the correct sign; in this connection the most critical section is section 6 at the right-hand end of the lower beam, where the hogging rotation is $2\theta - \psi = 0{\cdot}4\theta$. It thus follows from Theorem 4 that the weight of this design is either less than or equal to the minimum weight; if a corresponding safe and statically admissible bending moment distribution can be found the design will in fact be of minimum weight.

The fully plastic moments in tons ft. corresponding to the plastic hinges shown in Fig. 7.9(b) are as follows:

$$M_2 = 10,\ M_3 = -10,\ M_5 = 26{\cdot}67,\ M_6 = -26{\cdot}67,\ M_7 = -10,$$
$$M_8 = 10,\ M_{10} = -16{\cdot}67,\ M_{11} = -16{\cdot}67,\ M_{12} = 16{\cdot}67.$$

The six equations of equilibrium are readily shown to be

$$\begin{aligned}
30 &= 2M_2 - M_1 - M_3\\
60 &= 2M_5 - M_4 - M_6\\
30 &= M_1 - M_7 + M_8 - M_3\\
60 &= M_9 - M_{11} + M_{12} - M_{10}\\
0 &= M_7 + M_4 - M_9\\
0 &= M_8 + M_6 - M_{10}
\end{aligned}$$

Substituting the above fully plastic moments in these equations it is found that the three remaining unknown bending moments have the values

$$M_1 = 0, \quad M_4 = 20, \quad M_9 = 10,$$

and that with these values the six equations of equilibrium are satisfied identically. Since none of these three bending moments exceeds the corresponding fully plastic moment, a safe and statically admissible distribution of bending moment has been found, and so it is concluded that the above design is the minimum weight design.

The effect of using mechanism (ii) to provide the fourth degree of freedom is that the hinge rotations required for weight compatibility produce a hinge rotation of negative sign at section 9, whereas the bending moment at this section is required to be equal to the positive fully plastic moment. A statical check reveals, however, that a safe and statically admissible distribution of bending moment can be found ; thus from Theorem 3 the weight of this design is greater than the minimum value. The fully plastic moments in this design are :

$$\beta_1 = \beta_2 = 11{\cdot}25 \text{ tons ft.}, \quad \beta_3 = 26{\cdot}25 \text{ tons ft.}, \quad \beta_4 = 15 \text{ tons ft.},$$

and the corresponding value of the linearized weight function is

$$x = 20 \times 11{\cdot}25 + 30 \times 11{\cdot}25 + 20 \times 26{\cdot}25 + 30 \times 15 = 1{,}537{\cdot}5,$$

a value which is very close to the minimum value of 1,533·3.

If mechanism (iii) is used, it is found that a weight compatible mechanism can be derived. The values of the fully plastic moments are, however,

$$\beta_1 = \beta_2 = 0, \quad \beta_3 = \beta_4 = 30 \text{ tons ft.}$$

and it is clearly impossible to derive a safe and statically admissible distribution of bending moment. The corresponding linearized weight function x is readily computed as 1,500 ; by Theorem 4 this is a lower bound on the minimum value of x.

It is of interest to note that Boulton [9] has suggested a method for determining the minimum weight design of multi-storey frames in which the top storey is first designed for minimum weight on the assumption that the remainder of the frame is completely rigid. Assuming the values of the fully plastic moments thus obtained for this storey, the storey immediately beneath it is then designed for minimum weight on the assumption that the remainder of the frame below this storey is completely rigid. Proceeding in this way, each storey is treated separately, and the final design is then checked for minimum weight in the usual way. This procedure was satisfactory in the cases considered by Boulton, and it also gives the correct solution in the example just discussed.

References

1. J. Foulkes. Minimum weight design and the theory of plastic collapse. *Quart. Appl. Math.*, **10**, 347 (1953).

2. J. FOULKES. The minimum weight design of structural frames. *Proc. Roy. Soc. A.*, **223,** 482 (1954).
3. M. R. HORNE. Determination of the shape of fixed-ended beams for maximum economy according to the plastic theory. *Prelim. Pubn. 4th Congr. Intern. Assn. Bridge and Struct. Engng.*, Cambridge, 111 (1952); see also *Final Rep.*, 119 (1952).
4. J. HEYMAN. Plastic analysis and design of steel-framed structures. *Prelim. Pubn. 4th Congr. Intern. Assn. Bridge and Struct. Engng.*, Cambridge, 95 (1952).
5. A. CHARNES, W. W. COOPER and A. HENDERSON. *An Introduction to Linear Programming.* John Wiley (N.Y.), Chapman & Hall (London), (1953).
6. W. PRAGER. Minimum weight design of a portal frame. Tech. Rep. C11–2, Brown Univ. (1955).
7. J. HEYMAN. Plastic design of beams and plane frames for minimum material consumption. *Quart. Appl. Math.*, **8,** 373 (1951).
8. J. HEYMAN. Plastic design of plane frames for minimum weight. *Struct. Engr.*, **31,** 125 (1953).
9. N. S. BOULTON. Discussion of " Plastic analysis and design of steel-framed structures ". *Final Rep. 4th Congr. Intern. Assn. Bridge and Struct. Engng.*, Cambridge, 113 (1952).

Examples

1. A beam ABC rests on three simple supports, A, B and C. $AB = BC = 12$ ft. Concentrated loads 6 tons and 5 tons are applied 4 ft. from A and 6 ft. from C, respectively. If the fully plastic moments of the spans AB and BC are β_1 and β_2, respectively, find the values of β_1 and β_2 in the minimum weight design.

2. A beam ABC rests on three simple supports A, B and C. $AB = l_1$, $BC = l_2$. Concentrated loads W_1 and W_2 are applied at the centres of the spans AB and BC, respectively. The beam is to be designed for minimum weight, the fully plastic moments of the spans AB and BC being β_1 and β_2, respectively. Construct a design chart from which the values of β_1 and β_2 in the minimum weight design can be read off for any values of the ratios $\dfrac{W_1 l_1}{W_2 l_2}$ and $\dfrac{l_1}{l_2}$.

3. A beam $ABCD$ rests on four simple supports A, B, C and D. $AB = 6$ ft., $BC = 8$ ft., $CD = 10$ ft. Concentrated loads W_1, W_2 and W_3 are applied at the centres of the spans AB, BC and CD, respectively, and the corresponding fully plastic moments of the beam in these spans are β_1, β_2 and β_3. Find the values of these fully plastic moments in the minimum weight designs for the following load combinations

(i) $W_1 = 4$ tons ; $W_2 = 3$ tons ; $W_3 = 3$ tons
(ii) $W_1 = 3$ tons ; $W_2 = 2$ tons ; $W_3 = 2$ tons

4. A fixed-base rectangular portal frame *ABCD* has stanchions *AB* and *DC* of equal height 10 ft. and a beam *BC* of length 20 ft. It is acted upon by a concentrated vertical load 4 tons at the centre of the beam, and a concentrated horizontal load 3 tons at *C* in the direction *BC*. If the fully plastic moments of the stanchions are each to be equal to β_1, and the fully plastic moment of the beam is β_2, find the values of β_1 and β_2 in the minimum weight design, confirming the calculations with reference to the chart in Fig. 7.7.

5. The fixed-base pitched roof portal frame whose dimensions and loading are as shown in Fig. 4.1(*a*) was analysed in Section 4.2 on the assumption that all the members of the frame possessed the same fully plastic moment. Show that if the two stanchions are required to have the same fully plastic moment β_1, while the two rafters are required to have the same fully plastic moment β_2, the minimum weight design is achieved when $\beta_1 = \beta_2$, as assumed in the analysis.

6. In the two-storey frame whose dimensions and loading are as shown in Fig. 4.7(*a*), the members *CA*, *AB* and *BD* are each to have the same fully plastic moment β_1, the members *CE* and *DF* are to have the same fully plastic moment β_2, and the fully plastic moment of *CD* is β_3. This frame was analysed in Section 4.3 assuming that $\beta_1 = M_p$, $\beta_2 = 2M_p$ and $\beta_3 = 3M_p$. Show that this analysis leads to the minimum weight design.

7. The two-bay rectangular frame whose dimensions are given in Fig. 4.4(*a*) is subjected to a horizontal load 2 tons as shown, but the vertical loads are halved, becoming 2 tons and 3 tons instead of 4 tons and 6 tons, respectively. If the two beams are each to have the same fully plastic moment β_1, and the three stanchions are each to have the same fully plastic moment β_2, find the values of β_1 and β_2 in the minimum weight design.

8. For the frame whose dimensions and loading were specified in example 4, the three members *AB*, *BC* and *CD* may all have different fully plastic moments β_1, β_2 and β_3, respectively. Find the values of these fully plastic moments in the minimum weight design.

CHAPTER

8

Variable Repeated Loading

8.1. Introduction

FOR MANY TYPES OF STRUCTURE, the loading will vary considerably during the lifetime of each structure. For example, the loads on a shed type of frame can often be classified under three headings, namely dead loading, snow loading and wind loading. The dead loading, consisting of the weight of the structure itself and its cladding, remains constant, but the snow and wind loads will be subject to continual variation. The magnitude of either of these loads at any particular instant cannot, of course, be foreseen. However, certain values for these loads are laid down in the appropriate specifications, for example in British Standard No. 449; these values, termed the *working loads*, are intended to represent the highest loads which may be expected to occur under normal working conditions. Thus the sequence of loading for a structure of this kind will normally consist of unpredictable variations in both the snow and wind loads, conditioned only by the fact that neither of these loads will be expected to exceed its working value. This type of loading, in which the limits of the various loads are specified but there is no other prior knowledge of the sequence of loading, is termed *variable repeated loading*.

It is possible, as first recognized by Grüning [1] and Kazinczy,[2] that under variable repeated loading a structure may fail due to the eventual development of excessive plastic flow in parts of the structure, even though none of the loadings applied is ever sufficiently severe to cause failure by plastic collapse. In fact, there are two possible types of failure which can occur. If the loads are essentially alternating in character, it is clearly possible that one or more of the members might be bent back and forth repeatedly so that yield occurred in the fibres alternately in tension and compression. Such behaviour is termed *alternating plasticity.* A condition of alternating plasticity in a structure would be expected to lead eventually to the fracture of a member. While a failure by alternating plasticity is due to the continuance of plastic flow in some part of the structure, the difference between this type of failure and a fatigue failure is one of degree rather than kind. Whereas fatigue failures are usually associated with a number of load reversals of the order of millions, alternating plasticity failures would occur with a number of load reversals of the order of hundreds or perhaps thousands, but the eventual fracture is probably of a similar nature in both cases. The only difference is that in the case of fatigue the computed peak stresses lie within the elastic range of the material, and plastic flow only occurs near points of high stress concentration, whereas in alternating plasticity appreciable portions of the material are repeatedly stressed beyond the elastic limit.

The other type of failure, which cannot be foreseen so easily, may occur if during the variable repeated loading a number of critical combinations of loads follow one another in fairly definite cycles. If the peak loads are all multiples of one of the loads W, it can be shown that if W exceeds a certain intensity W_s', increments of rotation at plastic hinges can take place at various cross-sections in the structure during each cycle of loading, these increments being *in the same sense* during each cycle of loading. If W, while exceeding W_s', is less than a higher critical value W_s, the increments which occur in the rotations at the plastic hinges during each cycle of loading become progressively smaller as the number of cycles of loading increases. Eventually a condition is reached in which no further changes in the plastic hinge rotations take place, and subsequent variations of the loads cause only elastic changes of bending moment in the structure. When this happens, the structure is said to have *shaken down*. If on the other hand W exceeds the critical value W_s, the structure never shakes down, and definite rotations take place at the plastic hinges during each cycle of loading. In fact, if the peak load intensities do not vary from cycle to cycle, a steady régime is established in which the increment in the rotation at any given hinge is the same in each cycle, so that in each cycle the deflections of the structure increase by a given amount. Thus if a sufficient number of cycles of loading takes place, unacceptably large deflections will be built up, rendering the structure useless.

The structure would then be said to have failed by *incremental collapse*. The critical load value W_s above which incremental collapse can occur is referred to as the incremental collapse load.

The phenomenon of incremental collapse was recognized by Grüning [1] in his theoretical studies related to the behaviour of redundant pin-jointed trusses. In this investigation it was assumed that each member of the truss possessed the ideal plastic type of load-extension relation of Fig. 1.4(*b*) for both tension and compression, so that the falling off of load in the compression members after buckling was not taken into account.

Section 8.2 of this chapter is devoted to a discussion of the behaviour of a simple rectangular portal frame when subjected to a number of repetitions of particular cycles of loading. It is shown how if certain simplifying assumptions regarding the bending moment-curvature relation are made, it is possible to trace the behaviour of the frame during the loading cycles by means of step-by-step calculations, similar to those described in Chapter 2. Two cycles of loading are considered in detail. These cycles of loading are such that if the peak loads exceed certain values, their repetition causes incremental collapse in one case and alternating plasticity in the other case. The calculations are of assistance in developing an insight into the nature of these two types of failure, particularly in the case of incremental collapse.

The various theorems concerning the values of the incremental collapse and alternating plasticity loads are then discussed in Sections 8.3 and 8.4. Two points emerge from this discussion. In the first place, it is shown that the calculation of alternating plasticity loads presents no theoretical difficulty, so that the subsequent discussion is concerned almost entirely with methods for calculating incremental collapse loads. In the second place, it appears that a close parallel exists between the theorems concerning the values of incremental collapse loads and the theorems which concern the values of plastic collapse loads, which were stated and discussed in Section 3.2. It follows that methods for the calculation of incremental collapse loads can be developed which are analogous with those given in Chapter 4 for the calculation of plastic collapse loads. Thus a trial and error procedure is applied to a simple example in Section 8.5, and a combining mechanisms method is described in Section 8.6.

In conclusion, a discussion is given in Section 8.7 of the significance of the phenomena of incremental collapse and alternating plasticity in relation to the method of plastic design. It is clearly imperative to design a structure so that the probability of the occurrence of a failure of any kind during its lifetime is acceptably small. In view of the fact that a structure subjected to variable repeated loading may fail by incremental collapse or alternating plasticity, apart from the possibility of a failure by plastic collapse due to a single overload, it would appear at first sight that it would be necessary to consider each of these three types of failure in any plastic design procedure. This, of course, would entail the calculation of the values of the loads which would cause these three types of failure. However, it can be shown that failures by incremental collapse or alternating plasticity are usually far less likely to occur in any given structure than a failure by plastic collapse. It is therefore only necessary to provide an adequate safeguard against the occurrence of plastic collapse in order that the safety of the structure in respect of failures by incremental collapse or alternating plasticity should be ensured.

8.2. Step-by-step calculations

The behaviour of a fixed-base rectangular portal frame when subjected to certain cycles of loading will now be discussed, so as to indicate how failures can occur due to both incremental collapse and alternating plasticity. The illustrative calculations which will be presented were carried out by the step-by-step method which was used in Chapter 2 to show how plastic hinges formed in succession when beams and simple frames were loaded to collapse. Thus it will be assumed that each member of the frame is of the same cross-section and material and possesses the ideal type of bending moment-curvature relation of Fig. 2.1. This implies that each member behaves elastically unless the fully plastic moment is attained at some cross-section, when a plastic hinge forms which can undergo rotation of any magnitude, the sense of the rotation being in conformity with the sense of the fully plastic moment. If the bending moment is reduced below its fully plastic value the plastic hinge rotation remains constant and elastic unloading occurs.

On this basis the behaviour of the frame during any cycle of loading can be traced very simply. Briefly, if increments of load are being applied to the frame which are causing rotation at a single plastic hinge at some cross-section, the *changes* of bending moment at this cross-section are zero so long as the rotation at this hinge is increasing, for the bending moment must then remain constant at its fully plastic value. Since the behaviour of the structure is elastic everywhere except at this hinge, the *increments* of bending moment throughout the structure due to given changes of load are identical with the bending moments which would be produced in this structure by the same load *changes* if a pin joint was placed at the cross-section where the plastic hinge is undergoing rotation and the whole structure behaved elastically. These bending moments are readily calculated by any of the orthodox methods of elastic structural analysis. The procedure can obviously be applied to cases where more than one plastic hinge is involved.

The changes of bending moment which take place during any prescribed cycle of loading may thus be calculated once the appropriate elastic bending moment distributions have been obtained for the effects of the various loads on the frame with pin joints placed at those sections where plastic hinges form. Each cycle of loading may be regarded as a series of steps. During each step the frame will either be behaving elastically or else rotations will be occurring at one or more plastic hinges. Transition from one step to the next takes place when a fresh plastic hinge forms or when rotation ceases at one or more hinges. The changes of load and bending moment in each step of the cycle are found from the appropriate elastic bending moment distributions.

Incremental collapse

The particular frame which is to be considered is shown in Fig. 8.1(*a*). All the members of this frame are uniform and have the same fully plastic moment M_p. The frame can be subjected to concentrated horizontal and vertical loads H and V at the positions shown. The first cycle of loading to be considered is of a kind which can produce incremental collapse. This cycle consists of the application and removal of horizontal and vertical loads each of magnitude W, followed by the application and removal of the horizontal load alone, also of magnitude W, and is illustrated in Fig. 8.1(*b*). The results of the step-by-step calculations show that if this cycle of loading is repeated a large number of times, the frame can behave in one of three ways, depending on the value of W.

In the first place, if W is less than $2{\cdot}737\frac{M_p}{l}$, the behaviour under cyclic loading is very simple, for after the first application of the horizontal and vertical loads together, no further plastic hinge rotation takes place at any section. Secondly, if W lies between the values $2{\cdot}737\frac{M_p}{l}$ and $2{\cdot}857\frac{M_p}{l}$, increments of plastic hinge rotation take place at various cross-sections during each

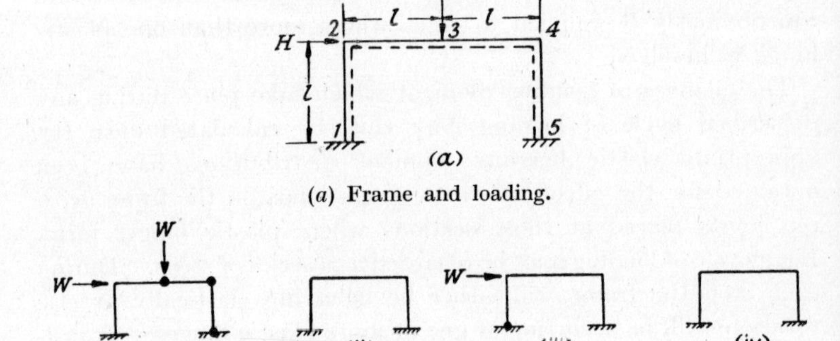

(*a*) Frame and loading.

(*b*) Cycle of loading and plastic hinge positions during incremental collapse.

Fig. 8.1. *Rectangular portal frame with cyclic loading.*

cycle of loading, each increment being in the same sense during each cycle. However, the magnitudes of the rotation increments decrease as the number of cycles increases in such a way that the total rotation at each plastic hinge, and thus the total deformation of the structure, is limited. Finally, if W exceeds $2{\cdot}857\frac{M_p}{l}$, increments of plastic hinge rotation take place at various cross-sections during each cycle of loading, and after a few cycles a steady state is reached in which the rotation increments which take place during each cycle are the same. Thus the increments of deflection during each cycle are the same, and so deflections of any magnitude can be built up after a large enough number of cycles, causing incremental collapse. It follows that $2{\cdot}857\frac{M_p}{l}$ is the incremental collapse value of W, which as stated in Section 8.1 is denoted by W_s, since this load is defined as the value of W above which incremental collapse can occur. Moreover, $2{\cdot}737\frac{M_p}{l}$ is the value W'_s of W above which cyclic loading effects first appear. The detailed calculations for three values of W, namely W'_s, W_s and a value greater than W_s will now be summarized.

Behaviour when $W = W'_s$

The frame now under discussion was analysed by the step-by-step method in Section 2.5 (see Fig. 2.8) for the case in which equal horizontal and vertical loads are increased steadily to the values which cause collapse, the results being summarized in Table 2.1. From this table it is seen that if W is equal to $2{\cdot}737\frac{M_p}{l}$, the first application of the horizontal and vertical loads in conjunction causes the formation and rotation of plastic hinges at the cross-sections 5 and 4 in turn. The distribution of bending moment following the application of these loads is given in the first row of Table 8.1. When these loads are removed, the behaviour of the entire frame during unloading is elastic, according to the assumed bending moment-curvature relation of Fig. 2.1. The elastic *changes* of bending moment which take place during the unloading are given in the second row of the table, and the third row gives the residual bending moments computed by summing the entries in the first two rows. These residual moments, existing in the structure when it is unloaded, are caused by the

TABLE 8.1

Single cycle of loading with $W = 2{\cdot}737\frac{M_p}{l}$

$\frac{\Delta Hl}{M_p}$	$\frac{\Delta Vl}{M_p}$	$\frac{Hl}{M_p}$	$\frac{Vl}{M_p}$	$\frac{M_1}{M_p}$	$\frac{M_2}{M_p}$	$\frac{M_3}{M_p}$	$\frac{M_4}{M_p}$	$\frac{M_5}{M_p}$
		2·737	2·737	−0·726	0·010	0·874	−1	1
−2·737	−2·737			0·582	0·034	−0·821	1·061	−1·129
		0	0	−0·144	0·044	0·053	0·061	−0·129
2·737	0			−0·856	0·513	0	−0·513	0·856
		2·737	0	−1	0·557	0·053	−0·452	0·727

plastic hinge rotations which occurred during the first loading, and would enable a further application of the horizontal and vertical loads together to be carried by wholly elastic changes of bending moment, since the unloading process was entirely elastic and therefore reversible.

The distribution of bending moment caused by the application of a horizontal load $H = 2{\cdot}737\frac{M_p}{l}$ by itself, on the assumption that the entire frame behaves elastically, is given in the fourth row of the table. It will be seen that the elastic bending moment at cross-section 1 caused by this load is $-0{\cdot}856M_p$. Since the residual bending moment at this cross-section is $-0{\cdot}144M_p$, the application of this load just brings the bending moment at this cross-section up to its fully plastic value without causing any plastic hinge rotation. The bending moments at the other cross-sections, given in the fifth row of the table, are all less than their fully plastic values. Thus the application of the horizontal load $H = 2{\cdot}737\frac{M_p}{l}$ does not cause any change in the plastic hinge rotations and therefore in the distribution of residual bending moment. It follows that no further increments in the plastic hinge rotations will take place if the cycle of loading is repeated, for a further application of the horizontal and vertical loads in conjunction would be resisted by wholly elastic changes of bending moment.

Behaviour when $W = W_s$

If W lies between the values $2{\cdot}737\frac{M_p}{l}$ and $2{\cdot}857\frac{M_p}{l}$ it is found that the residual bending moments established in the frame after the first application and removal of the horizontal and vertical loads in conjunction are such that the application of the horizontal load alone causes a plastic hinge to form and undergo rotation at cross-section 1. It is then found that when the horizontal load is removed, the residual bending moments have altered in such a way that if the horizontal and vertical loads are applied again further increments in the plastic hinge rotations take place at cross-sections 4 and 5. This effect is shown in Table 8.2, which gives the distribution of bending moment at each stage in the first two cycles of loading for $W = 2{\cdot}857\frac{M_p}{l}$. This particular value of W was chosen because it is the incremental collapse load W_s, as already stated.

TABLE 8.2

Two cycles of loading with $W = 2{\cdot}857\dfrac{M_p}{l}$

$\dfrac{Hl}{M_p}$	$\dfrac{Vl}{M_p}$	$\dfrac{M_1}{M_p}$	$\dfrac{M_2}{M_p}$	$\dfrac{M_3}{M_p}$	$\dfrac{M_4}{M_p}$	$\dfrac{M_5}{M_p}$
2·857	2·857	−0·828	0·029	0·943	−1	1
0	0	−0·221	0·065	0·086	0·107	−0·179
2·857	0	−1	0·610	0·074	−0·462	0·785
0	0	−0·107	0·074	0·074	0·074	−0·108
2·857	2·857	−0·794	0·063	0·960	−1	1
0	0	−0·187	0·099	0·103	0·107	−0·179
2·857	0	−1	0·642	0·095	−0·452	0·764
0	0	−0·107	0·106	0·095	0·084	−0·129

The elastic change of bending moment at cross-section 5 due to the application of the horizontal and vertical loads together is $1{\cdot}179M_p$. Since the first application of these loads causes the formation of a plastic hinge of positive sign at this cross-section, the residual moment after their removal is $(M_p - 1{\cdot}179M_p)$, or $-\ 0{\cdot}179M_p$, as given in the second row of the table. This residual bending moment would, of course, just enable a further application of the horizontal and vertical loads together to be supported elastically at this cross-section. However, after application of the horizontal load alone, this residual bending moment changes to $-\ 0{\cdot}108M_p$, as shown in the fourth row of the table. With this residual bending moment a further application of the horizontal and vertical loads could not be resisted by purely elastic changes of bending moment, for the bending moment at this cross-section would then become $(-\ 0{\cdot}108M_p + 1{\cdot}179M_p)$, or $1{\cdot}071M_p$.

This effect is due to the fact that the application of the horizontal load changes the residual moment at section 5 by a positive amount, thus reducing the change of bending moment required to bring this bending moment up to the positive fully plastic moment when the horizontal and vertical loads are applied together. It will be seen from the second and fourth rows of the table that a similar effect occurs at cross-section 4. In this case the application of the horizontal load changes the residual moment from

$0{\cdot}107M_p$ to $0{\cdot}074M_p$. This reduction favours the fresh formation and rotation of a plastic hinge of negative sign when the horizontal and vertical loads are applied together, for the elastic change of bending moment at this section due to the application of both the horizontal and vertical loads together is $-1{\cdot}107M_p$. Wholly elastic bending moment changes when the horizontal and vertical loads were reapplied would thus change the bending moment at this section to $(0{\cdot}074M_p - 1{\cdot}107M_p)$, or $-1{\cdot}033M_p$.

Since the bigger change of residual moment takes place at section 5, it is to be expected that when the horizontal and vertical loads are applied for the second time the fully plastic moment will be reached at this section before section 4. An elastic change of moment of $1{\cdot}108M_p$ is required to bring the moment at section 5 up to its fully plastic value, and an elastic change of $1{\cdot}179M_p$ is caused by the application of $H = V = 2{\cdot}857\frac{M_p}{l}$. Thus the value of W at which the bending moment at section 5 reaches the value M_p on the second application of the horizontal and vertical loads is $\frac{1{\cdot}108}{1{\cdot}179} \times 2{\cdot}857\frac{M_p}{l}$, or $2{\cdot}685\frac{M_p}{l}$. For this value of W the change in the bending moment at section 4 is $-\frac{2{\cdot}685}{2{\cdot}857} \times 1{\cdot}107M_p$, or $-1{\cdot}040M_p$, so that this bending moment becomes $(0{\cdot}074M_p - 1{\cdot}040M_p)$, or $-0{\cdot}966M_p$. On further investigation it is found that when H and V are increased above the value $2{\cdot}685\frac{M_p}{l}$ with the hinge at section 5 undergoing rotation, the bending moment at section 4 reaches the value $-M_p$ when $W = 2{\cdot}764\frac{M_p}{l}$. For the further increase of W to $2{\cdot}857\frac{M_p}{l}$ the plastic hinges at both sections 5 and 4 undergo rotation, and the final distribution of bending moment is shown in the fifth row of the table.

The bending moment at cross-section 1 caused by the application of the horizontal load alone of value $2{\cdot}857\frac{M_p}{l}$, assuming that the entire frame behaves elastically, is $-0{\cdot}893M_p$, so that after the first removal of the horizontal load alone the residual moment at this cross-section is $(-M_p + 0{\cdot}893M_p)$, or $-0{\cdot}107M_p$, as shown in the fourth row of the table. However, when the horizontal and vertical loads are then applied and removed a second time the further rotations at the plastic hinges at the cross-sections 4 and 5 cause the residual bending moment at this cross-section to change to $-0{\cdot}187M_p$. It follows that a second application of the horizontal load alone cannot be borne by wholly elastic changes of bending moment, for the bending moment which would be produced at the cross-section 1 would be $(-0{\cdot}187M_p - 0{\cdot}893M_p)$, or $-1{\cdot}080M_p$. The fully plastic moment would, in fact, be reached at this cross-section when the horizontal load had reached the value $\dfrac{0{\cdot}813}{0{\cdot}893} \times 2{\cdot}857\dfrac{M_p}{l}$, or $2{\cdot}603\dfrac{M_p}{l}$.

Thus in each cycle of loading the application of the horizontal and vertical loads together causes plastic hinge rotations at the cross-sections 4 and 5, and these rotations change the residual bending moment at cross-section 1 in such a way as to cause the further formation and rotation of a plastic hinge at this cross-section when the horizontal load is applied by itself. Moreover, this rotation of the plastic hinge at cross-section 1 changes the residual bending moments at cross-sections 4 and 5 in such a way as to cause the further formation and rotation of plastic hinges at these cross-sections when the horizontal and vertical loads are applied together.

When the step-by-step analysis is continued for further cycles of loading, it is found that the magnitudes of the increments in the plastic hinge rotations decrease according to a geometrical progression each time the loads are applied, so that after an infinite number of loading cycles the increments of rotation become indefinitely small. If this condition is reached no further changes take place in the residual bending moments, and any further applications of the loads, singly or together, are supported by purely elastic changes of bending moment. When this happens the structure has shaken down.

The bending moment distribution after each application of the horizontal and vertical loads together is given in Table 8.3, in which n denotes the number of applications of the horizontal and vertical loads in conjunction. It will be seen from this table that as the number of cycles of load increases, the value of M_3 approaches the value M_p asymptotically. After an infinite number of cycles, M_3 just reaches the value M_p when the horizontal and vertical loads are applied together, but no plastic hinge rotation occurs at this cross-section.

TABLE 8.3

Bending moment distribution when the horizontal and vertical loads are applied together

$$W = 2{\cdot}857\frac{M_p}{l}$$

n	$\frac{M_1}{M_p}$	$\frac{M_2}{M_p}$	$\frac{M_3}{M_p}$	$\frac{M_4}{M_p}$	$\frac{M_5}{M_p}$
1	−0·829	0·029	0·943	−1	1
2	−0·794	0·063	0·960	−1	1
3	−0·770	0·087	0·972	−1	1
4	−0·753	0·104	0·981	−1	1
5	−0·741	0·116	0·987	−1	1
6	−0·733	0·124	0·991	−1	1
7	−0·727	0·130	0·994	−1	1
8	−0·723	0·134	0·996	−1	1
9	−0·721	0·137	0·997	−1	1
10	−0·719	0·139	0·998	−1	1
∞	−0·714	0·143	1	−1	1

Behaviour when $W > W_s$

For any value of W between $2{\cdot}737\frac{M_p}{l}$ and $2{\cdot}857\frac{M_p}{l}$, a similar cyclic loading effect would be found, and after an infinite number of cycles the value of M_3 would be less than M_p when the horizontal and vertical loads were applied together. However, for any value of W exceeding $2{\cdot}857\frac{M_p}{l}$ the process of shake-down is prevented by the formation of a plastic hinge at cross-section 3 after some finite number of loading cycles. For instance, when $W = 2{\cdot}90\frac{M_p}{l}$, a plastic hinge forms at cross-section 3 after the third application of the horizontal and vertical loads together, as shown in Table 8.4.

It will be seen from this table that at the start of the fourth loading cycle the bending moment distribution is identical with that at the start of the third cycle, so that the behaviour in any subsequent cycle of loading would be exactly the same as the behaviour during the third cycle. In particular, the increments of rotation at the plastic hinges in each subsequent cycle would be the same, so that the same increments of deflection would take place in each cycle of loading. Thus if the condition of the structure at the end of any of the further cycles of loading is

TABLE 8.4

Bending moment distribution when the horizontal and vertical loads are applied together

$$W = 2{\cdot}90\frac{M_p}{l}$$

n	$\frac{M_1}{M_p}$	$\frac{M_2}{M_p}$	$\frac{M_3}{M_p}$	$\frac{M_4}{M_p}$	$\frac{M_5}{M_p}$
1	−0·865	0·035	0·968	−1	1
2	−0·818	0·082	0·991	−1	1
3	−0·800	0·100	1	−1	1
4	−0·800	0·100	1	−1	1

compared with its condition at the beginning of the same cycle, it will be found that the bending moment distribution will not have changed, but that increments in the rotations of the plastic hinges at the cross-sections 1, 3, 4 and 5 will have taken place. If hinges formed at each of these four sections simultaneously, the structure would be transformed into a mechanism. However, during incremental collapse these hinges do not all form simultaneously, but at different stages in the loading cycle, as indicated in Fig. 8.1(*b*). Their rotation increments are therefore constrained by the elastic action of other portions of the frame. Nevertheless, if the cycle of loading is repeated a sufficient number of times, deflections of any magnitude can be built up, so that failure occurs by incremental collapse.

Incremental collapse can in general only take place when during each cycle of loading increments of plastic hinge rotation occur at a sufficient number of cross-sections such that if hinges occurred simultaneously at all these cross-sections the structure would be transformed into a mechanism. This is because during incremental collapse the effect of a complete cycle must consist solely of causing increments in plastic hinge rotations, the bending moment distribution, and therefore the distribution of curvature along the members between the plastic hinge positions, being unaltered. Since the eventual deflections of the frame are indefinitely large, the requirements of compatibility can only be met if these increments of plastic hinge rotation are consistent with a mechanism motion, which of itself demands no curvature changes along the members. The mechanism which is associated with an incremental collapse failure in this way may be termed the incremental collapse mechanism. For this particular case the incremental collapse mechanism, with hinges at the cross-sections 1, 3, 4 and 5, is the same as the plastic collapse mechanism for the case in which $H = V = W$. As shown in Section 2.5, this is the combined mechanism illustrated in Fig. 2.9. It was also shown that the collapse load in this case is $W_c = 3\frac{M_p}{l}$, so that $\frac{W_s}{W_c} = \frac{2 \cdot 857}{3} = 0 \cdot 95$. Thus for this frame and loading incremental collapse cannot occur unless W is within 5% of the collapse load.

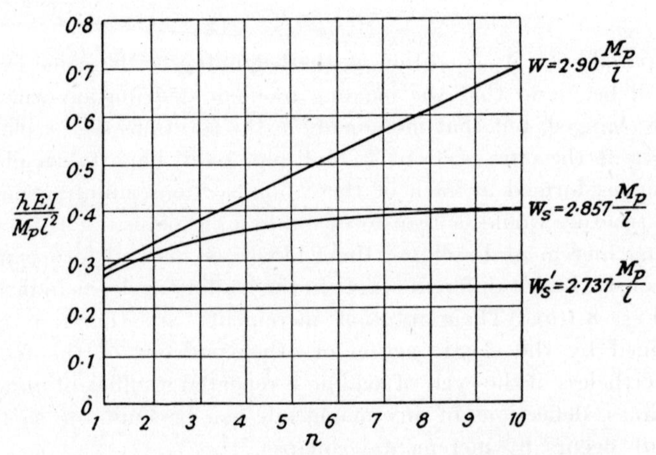

Fig. 8.2. Effect of cyclic loading on horizontal deflection.

The above calculations may best be summarized with reference to Fig. 8.2. This figure shows the horizontal deflection h at the top of either stanchion, expressed non-dimensionally as $\frac{hEI}{M_p l^2}$, plotted against n, the number of times that the horizontal and vertical loads have been applied in conjunction. Until W exceeds $W_s' = 2 \cdot 737\frac{M_p}{l}$, repetition of the loading cycle produces no increase in the deflection. For values of W between W_s' and $W_s = 2 \cdot 857\frac{M_p}{l}$, the deflection increases each time the loading cycle is repeated, but tends asymptotically to a definite limit as the number of loading cycles increases indefinitely. When W exceeds

W_s, a régime is reached after a certain number of loading cycles, depending on the value of W, in which the same increment in h occurs in each loading cycle, so that the deflection becomes indefinitely large if the loading cycle is repeated a large enough number of times.

Calculations of the kind summarized in Fig. 8.2 have been given by Horne [3] for the case of a fixed-ended beam of length l subjected to a load which is applied alternately at points distant $\frac{1}{3}l$ from either end of the beam.

The procedure explained above for investigating the behaviour of a frame subjected to cyclic loading would have little value as a method for determining incremental collapse loads. Its main purpose was to illustrate the behaviour of a typical frame under cyclic loading, and in particular to show how incremental collapse can occur. Moreover, it shows that the number of loading cycles required to produce considerable deflection increases during incremental collapse is quite small. For instance, when $W = 2{\cdot}90\dfrac{M_p}{l}$, the deflection occurring upon the first application of the loads is doubled after only seven cycles of loading.

General methods for the determination of incremental collapse loads which are valid for a more general type of bending moment-curvature relation than the idealized relation of Fig. 2.1 will be given in Sections 8.5 and 8.6. Moreover, in applying these general methods a knowledge of the actual programme of loading to which the structure will be subjected is not required, and it is merely necessary to know the extreme limits between which each load can vary.

Experimental evidence

Symonds [4] has tested miniature rectangular portal frames subjected to cycles of loading of the type shown in Fig. 8.1(b). These portal frames were of rigid welded construction, the members being $\frac{1}{4}$ in. square mild steel. The curves obtained, relating the growth of deflection to the number of cycles of loading at varying load intensities, agreed well with theoretical curves of the kind shown in Fig. 8.2, and in particular it was observed that the growth of deflection was limited until a load close to the calculated incremental collapse load was reached. In contrast, it must be recorded that Klöppel,[5] in testing a continuous beam on three supports, obtained results which appeared to show that no significant deflection increases took place at loads above the theoretical incremental collapse load. This beam was of length 3 metres, resting on simple supports at each end and at its centre. One span was subjected to a constant central concentrated load P, while the other span was subjected to a central concentrated load varying between the limits P and 200 kg. Assuming the shape factor of the beam to be 1·16, the predicted incremental collapse load was $P_s = 5{,}030$ kg., while the plastic collapse load was $P_c = 5{,}950$ kg. 700,000 pulsations of load were first applied at the calculated elastic limit load $P_e = 4{,}210$ kg., then 500,000 pulsations were applied at $P = 5{,}040$ kg., close to P_s, and finally 500,000 pulsations were applied at $P = 5{,}830$ kg., close to P_c. The eventual deflections recorded were little more than would have occurred if the beam had remained elastic. However, the value of the fully plastic moment used in calculating P_s and P_c was derived from tension test data, which as pointed out in Section 6.2 can lead to very erroneous results, both on account of the strain rate effect and because of the wide variations in yield stresses which are recorded from tensile specimens cut from different positions in the cross-sections of rolled steel joists. The absence of any comparative test results on simply supported beams under static loading renders the interpretation of the results of this single test very difficult. However, Massonet [6] has carried out a much more comprehensive series of tests on beams which were similarly supported and loaded, the investigation including comparative collapse tests, and the results were in close accordance with theoretical expectation.

Other repeated loading effects

A repeated loading effect of a somewhat different nature to incremental collapse or alternating plasticity has been investigated theoretically by Fritsche,[7] who considered a continuous beam resting on three supports, one at each end and one at its centre, which is subjected to equal loads centrally concentrated in each span. If these loads are increased so that yield occurs over the central support without the fully plastic moment being reached, and the loads are then removed, the beam will lift off the central support. If this support is raised so that it is in contact with the beam, and the loads are reapplied, further yield will take place over the support. Repetition of this process eventually causes the fully plastic moment to be attained over the support.

Another type of repeated loading effect was investigated experimentally by Patton and Gorbunow.[8] These authors tested simply supported beams subjected to a central concentrated load, and also continuous beams on three supports with one span subjected to a central concentrated load, the beams being of box and other sections built up by welding. In each test a given load was applied and removed a large number of times, of the order of a thousand or more, and the growth of deflection with number of applications of load was recorded. The procedure was repeated for successively larger values of the load. The object of the tests was to see whether the value of the collapse load predicted by the plastic theory was significantly affected by the adoption of this kind of loading. In general, it was observed that while there was some increase of deflection with number of load applications for loads below the predicted collapse loads, this increase was limited, whereas for loads above the predicted collapse loads much larger increases took place, apparently without limit.

Alternating plasticity

The behaviour of the frame of Fig. 8.1 when subjected to repetitions of another loading cycle which can cause alternating plasticity will now be considered. This loading cycle consists of the application and removal of the horizontal and vertical loads, each of magnitude W, followed by the application of the horizontal load by itself of magnitude W but with its direction reversed. It will be shown that if W exceeds a critical value W_a, alternating plasticity occurs.

For the particular value $2{\cdot}85\dfrac{M_p}{l}$ of W selected for illustrative purposes, the first application of the horizontal and vertical loads in conjunction causes the formation and rotation of plastic hinges at cross-sections 5 and 4 in succession, the distribution of bending moment after application of these loads being given in the first row of Table 8.5. The plastic hinge rotation at cross-section 5 during this loading can be shown to be $0{\cdot}103\dfrac{M_pl}{EI}$. When these loads are removed and the horizontal load is then applied alone in the opposite direction, it is found that a plastic hinge of negative sign is formed at cross-section 5, the *change* in the rotation at this hinge being $-0{\cdot}034\dfrac{M_pl}{EI}$. The corresponding bending moment distribution is given in the third row of Table 8.5. Removal of this load followed by a further application of the horizontal and vertical loads together is then found to cause the formation of a plastic hinge of positive sign at cross-section 5, and the *change* in the rotation at this hinge is $0{\cdot}034\dfrac{M_pl}{EI}$. At the same time the bending moment at cross-section 4 just attains the fully plastic value $-M_p$, but no increment of plastic hinge rotation occurs at this cross-section. The distribution of bending moment is shown in the fifth row of the table.

TABLE 8.5

Cycle of loading causing alternating plasticity

$$W = 2{\cdot}85\frac{M_p}{l}$$

$\frac{Hl}{M_p}$	$\frac{Vl}{M_p}$	$\frac{M_1}{M_p}$	$\frac{M_2}{M_p}$	$\frac{M_3}{M_p}$	$\frac{M_4}{M_p}$	$\frac{M_5}{M_p}$
2·85	2·85	−0·823	0·028	0·939	−1	1
0	0	−0·217	0·063	0·084	0·104	−0·176
−2·85	0	0·715	−0·491	0·077	0·645	−1
0	0	−0·176	0·044	0·077	0·110	−0·109
2·85	2·85	−0·823	0·028	0·939	−1	1

It is clear that a cycle of plastic deformation has now been established, for as will be seen from the table the bending moment distribution after the second application of the horizontal and vertical loads in conjunction is identical with that which occurred after their first application. In each subsequent cycle changes in the plastic hinge rotation of magnitude $0{\cdot}034\dfrac{M_pl}{EI}$ will occur at cross-section 5, the change being positive when the horizontal and vertical loads are applied simultaneously, and negative when the horizontal load is applied alone in the opposite direction. No other changes of plastic hinge rotation will occur anywhere in the structure.

In this particular example it is easy to see how to calculate the critical value W_a of W above which alternating plasticity would occur. If W was equal to W_a, then after the first application of the horizontal and vertical loads together, removal of the vertical load together with a reversal of the horizontal load would just be sufficient to change the bending moment at cross-section 5 from M_p to $-M_p$, on the assumption that the entire structure behaved elastically during this load change. This elastic change of bending moment is found to be $-0{\cdot}725Wl$, so that W_a is given by

$$-0{\cdot}725W_a l = -2M_p$$

$$W_a = 2{\cdot}759\frac{M_p}{l}$$

This value of W is termed the alternating plasticity load.

The general problem of the determination of alternating plasticity loads will be discussed in Sections 8.5 and 8.6. The analysis which will be given is valid for a more general type of bending moment curvature relation than the idealized type of relation of Fig. 2.1, and can also be made without a knowledge of the particular loading programme to which the structure will be subjected. A discussion of the experimental evidence concerning failure by alternating plasticity is included in Section 8.7.

8.3. Shake-down theorems

In Section 8.2 a detailed discussion was given of the behaviour of a particular frame when subjected to several repetitions of prescribed cycles of loading, on the assumption that the relation between bending moment and curvature for each member of the frame was of the idealized type of Fig. 2.1. It was shown that for a certain cycle of loading failure by incremental collapse would occur if W exceeded a critical value W_s, termed the incremental collapse load, and for another cycle of loading failure by alternating plasticity would occur if W exceeded another critical value W_a, termed the alternating plasticity load. These critical load values, W_s and W_a, were only evaluated for cases in which the structure was subjected to repetitions of definite cycles of loading, whereas as discussed in Section 8.1 the variable repeated loading to which many structures may be subjected is largely random in respect of the sequence of loading. Fortunately, it can be shown that the values of W_s and W_a are not dependent on the detailed nature of the cycles of loading to which a structure may be subjected. Provided that the extreme values of each of the loads on a structure are specified in terms of a single parameter W, it is possible to determine unique values of W_s and W_a which are independent of the particular sequence of loading which may occur. In this Section and in Section 8.4, theorems are presented which justify this statement, and which enable the values of W_s and W_a to be calculated.

General shake-down theorem

Failures by both incremental collapse and alternating plasticity may be said to be due to the indefinite continuance of plastic flow, for in both cases plastic flow in various parts of the structure never ceases, no matter how often and in what sequence the loads are applied. If the structure is not to fail in either of these ways it must eventually shake down by finding its way to a residual moment distribution which enables the structure to support all further variations of the loads between their prescribed limits by purely elastic changes of bending moment. The general shake-down theorem, which states the conditions under which a structure will eventually shake down, can be stated as follows:

General shake-down theorem. If any particular distribution of residual bending moment can be found which would enable all possible variations of the applied loads between their prescribed limits to be supported by purely elastic changes of bending moment, then the structure will eventually shake down, although the actual distribution of residual bending moment existing in the structure when it has shaken down will not necessarily be the particular distribution which has been found.

This theorem clearly states a necessary condition for shake-down to occur; its proof shows that this condition is also sufficient.

It is only proposed to give a brief outline of the proof of this theorem, the detailed proof being given in Appendix D. For this purpose the theorem must be formulated in mathematical terms; this formulation will lead to a better understanding of the theorem, and will enable a distinction to be drawn between the conditions governing the avoidance of incremental collapse and those governing the avoidance of alternating plasticity. Since the theorem is concerned with systems of residual bending moments and purely elastic changes of bending moment, it will be necessary to give formal definitions of these two terms.

The purely elastic changes of bending moment will be considered first. Consider any particular cross-section i in a member, and let $\mathscr{M}_i$ denote the bending moment at this cross-section due to any particular load combination, computed on the assumption that the structure carries this load combination by wholly elastic action. Under this assumption the principle of superposition may be applied, and it is therefore possible to calculate the maximum and minimum possible values of $\mathscr{M}_i$ when all possible variations of the external loads between their prescribed limits are considered. This calculation is performed by considering the effect of each load separately, and then superposing. For instance, suppose that a particular load W_j could vary between the limits W_j^{max} and W_j^{min}, and suppose that a unit load applied in the direction of the load W_j produced an elastic bending moment q_{ij} at the section i. Then the contribution to the maximum elastic bending moment at the section i due to W_j would be $q_{ij}W_j^{max}$ if q_{ij} were positive, and $q_{ij}W_j^{min}$ if q_{ij} were negative. The summation of similar terms for all the other loads would then give the maximum value of $\mathscr{M}_i$. The maximum and minimum values of $\mathscr{M}_i$ found in this way will be denoted by $\mathscr{M}_i^{max}$ and $\mathscr{M}_i^{min}$. When this superposition is carried out, any connection between two or more of the loads could be taken into account. Thus for example if it were known that two particular loads could never be applied simultaneously, but only separately, the smaller of the two terms in the summation corresponding to these loads would be omitted. If the maximum and minimum values of all the loads are specified as multiples of a single load W, the values of $\mathscr{M}_i^{max}$ and $\mathscr{M}_i^{min}$ will all be proportional to W, since the value of any one of the elastic bending moments will itself be proportional to W.

A formal definition can now be given for the term residual bending moment. At any arbitrary stage of the loading programme, let the actual bending moment at cross-section i be M_i, and let the elastic bending moment which would occur under the same loads be $\mathscr{M}_i$. The residual bending moment at this section, denoted by m_i, is then defined by the equation

$$m_i = M_i - \mathscr{M}_i \qquad . \quad . \quad . \qquad 8.1$$

The residual bending moments defined in this way might not be the actual residual moments that would arise in a frame if all the loads were removed, for it is possible that some yielding might take place upon their removal. In such a case the change of bending moment on unloading would not be $-\mathscr{M}_i$.

Since for any given set of loads both the actual set of bending moments M_i and the elastic set of bending moments $\mathscr{M}_i$ at all the cross-sections i of a frame must satisfy the equations of equilibrium for the given loads, it follows that the residual bending moments m_i must satisfy the equations of equilibrium for zero external loads. Any system of residual moments, whether real or hypothetical, which satisfies the appropriate equations of equilibrium for zero external load, is said to be *statically admissible.*

Shake-down theorem for ideal bending moment-curvature relation

The precise form taken by the shake-down theorem depends on the particular type of relation between bending moment and curvature which is assumed to hold for the members of the frame. In the first place, the theorem will be formulated for a frame whose members all behave according to the ideal type of relation of Fig. 2.1, in which the fully plastic moment is of magnitude M_p and the elastic range of bending moment is $2M_p$. The extension of the theorem to the more realistic case in which the elastic range of bending moment is less than $2M_p$ will be given later.

For a frame whose members all possess this idealized type of bending moment-curvature relation, the shake-down theorem takes the following form:

Simple shake-down theorem. If it is possible to find any distribution of residual bending moments $\bar{m}_i$ which satisfies at every cross-section i the conditions

$$\bar{m}_i + \mathscr{M}_i^{\max} \leqslant (M_p)_i \quad . \quad . \quad . \quad 8.2$$

$$\bar{m}_i + \mathscr{M}_i^{\min} \geqslant -(M_p)_i \quad . \quad . \quad . \quad 8.3$$

and which is statically admissible, the frame will eventually shake down, although the residual moments existing in the frame after it has shaken down will not necessarily be the distribution $\bar{m}_i$.

This theorem was first stated by Bleich,[9] who gave a proof for structures with not more than two redundancies. A general proof of the theorem for trusses with any number of redundancies was given by Melan,[10, 11] on the assumption that the load extension relation for each truss member was of the ideal plastic type shown in Fig. 1.4(*b*) for both tension and compression. This proof was naturally of considerable theoretical interest, but it could not be related to the behaviour of actual trusses under variable repeated loading because of the inadequacy of the ideal plastic assumption for compression members, as discussed in Appendix A. A simpler version of this proof for hypothetical pin-jointed trusses was given by Symonds and Prager,[12] who also discussed the behaviour of simple trusses of this type under variable repeated loading with the aid of a geometrical representation developed earlier by Prager.[13] Melan's proof has been adapted by Neal [14] to the case of a frame whose members possess the ideal type of bending moment-curvature relation of Fig. 2.1. Details of this proof are given in Appendix D, but its nature will be indicated briefly here.

To prove the theorem, it is supposed that a particular distribution of residual moments $\overline{m}_i$ has been found which satisfies the inequalities 8.2 and 8.3 and is statically admissible. A quantity E is defined by the equation

$$E = \int \frac{(m_i - \overline{m}_i)^2}{2(EI)_i} ds_i \quad . \quad . \quad . \qquad 8.4$$

in which m_i represents the actual residual moment at any cross-section i of the frame and at any stage of the loading programme, ds_i is an element of length of the member at section i, $(EI)_i$ is the flexural rigidity at this section, and the integration extends over all the members of the structure. Clearly E is a positive quantity; it can be thought of as a measure of the difference between the actual and hypothetical residual moment distributions, m_i and $\overline{m}_i$.

If a variation of the loads causes rotation at plastic hinges at one or more cross-sections, the actual residual moment distribution m_i will change, so that E will change. The proof of the theorem falls into two parts. In the first part of the proof, it is shown that E remains constant except when plastic hinge rotations are occurring, and that changes of E caused by plastic hinge rotations must always be negative. Since E itself cannot be negative, it follows that ultimately E must become zero or else settle down to some positive value and thereafter remain constant. In either case the frame would shake down.

Although this part of the proof demonstrates that plastic flow will ultimately cease provided that at least one distribution of residual moments $\bar{m}_i$ fulfilling the stated conditions can be found, it does not set any limit to the *amount* of plastic flow which may take place before the frame shakes down. The second part of the proof is therefore devoted to a demonstration that finite changes in the plastic hinge rotations must cause finite changes of E. Since the total change of E is limited, it follows that the total plastic hinge rotations which occur prior to shake-down are also limited, thus completing the proof.

In the proof of this theorem it is not necessary to impose any restriction on the loading programme apart from the condition that each of the loads can only vary within its prescribed limits. Apart from the assumption that each member behaves according to the ideal type of bending moment-curvature relation, the theorem therefore has the widest possible generality. It is also noteworthy that it is not necessary to assume that the frame is initially free from stress. Thus the presence of initial residual moments due to imperfect fit of members, sinking of supports, etc., has no influence on the conditions for shake-down.

If the extreme load limits are all specified in terms of a single load parameter W, it is clear that if W is imagined to be increased steadily from zero it will become progressively more difficult to satisfy the inequalities 8.2 and 8.3 at all cross-sections of the frame, for as already pointed out the values of the maximum and minimum elastic bending moments at any section are proportional to W. Thus a critical value of W must exist above which these inequalities can no longer be satisfied. If W exceeded this critical value the frame would not shake down, but would fail by either alternating plasticity or incremental collapse. It is important to know how to predict the type of failure that would occur if the shake-down conditions were not fulfilled, and such a prediction can be made very simply. It is merely necessary to note that a continued inequality can be written down for each cross-section i from the inequalities 8.2 and 8.3 as follows:

$$-(M_p)_i - \mathscr{M}_i^{\min} \leqslant \bar{m}_i \leqslant (M_p)_i - \mathscr{M}_i^{\max},$$

so that

$$\mathscr{M}_i^{\max} - \mathscr{M}_i^{\min} \leqslant 2(M_p)_i \cdot \qquad 8.5$$

While this inequality is contained in the inequalities 8.2 and

8.3, the advantage of extracting it in this form is that it can be seen there are two ways in which it could become impossible for the shake-down conditions to be fulfilled if W increased beyond a certain value. In the first place, the inequality 8.5 might not be fulfilled at a single cross-section. In this case it is clear that a failure by alternating plasticity would take place, since the condition 8.5 states that the elastic range of bending moment at any cross-section must be less than the available elastic range of behaviour of the member. Alternatively, it is possible that while the inequality 8.5 is not violated at any cross-section, two or more of the inequalities 8.2 and 8.3 could not be satisfied simultaneously. In this case plastic hinge rotations would occur at the corresponding cross-sections at different stages of the loading programme, and a failure by incremental collapse would result.

Shake-down theorem for more general bending moment-curvature relation

The ideal type of relation between bending moment and curvature which was illustrated in Fig. 2.1 is unrealistic, since beams of ductile material always show a range of bending moment between the first yielding and the development of full plasticity (see for example Fig. 1.1). The shake-down theorem has been extended by Neal [15] to cover frames whose members possess the more realistic type of bending moment-curvature relation of Fig. 8.3. In this relation it is assumed that the yield and

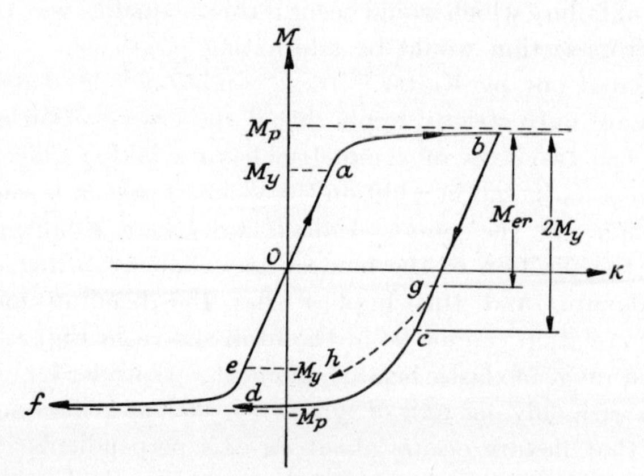

Fig. 8.3. *More general type of bending moment-curvature relation.*

fully plastic moments M_y and M_p are of the same magnitude for flexure in either sense. Thus an initially stress-free beam behaves elastically along *Oa* or *Oe*, and with further loading the slope of the bending moment-curvature relation is progressively reduced along *ab* or *ef* as the bending moment tends to M_p in magnitude. The elastic range of bending moment for the initially stress-free beam is $2M_y$, and it is assumed that the elastic range remains constant regardless of the history of loading. For example, if the bending moment is reduced after loading along *Oab*, the relation *bcd* is followed, where the elastic range of moment along *bc* is still $2M_y$.

For this type of bending moment-curvature relation the conditions for shake-down to occur are as follows:

$$\bar{m}_i + \mathscr{M}_i^{\max} \leqslant (M_p)_i \qquad 8.6$$

$$\bar{m}_i + \mathscr{M}_i^{\min} \geqslant -(M_p)_i \qquad 8.7$$

$$\mathscr{M}_i^{\max} - \mathscr{M}_i^{\min} \leqslant 2(M_y)_i \qquad 8.8$$

These conditions may be compared with the inequalities 8.2, 8.3 and 8.5, which are appropriate to frames whose members behave according to the ideal type of bending moment-curvature relation. It will be seen that the inequalities 8.6 and 8.7 are identical with the inequalities 8.2 and 8.3, and the type of failure which would occur if these inequalities were violated is incremental collapse. The inequality 8.8, which states that the elastic range of bending moment should not exceed the available elastic range of behaviour of the material, is similar to the inequality 8.5, and the type of failure which would occur if this inequality was violated at any cross-section would be alternating plasticity.

As pointed out by Koiter,[16] these conditions for shake-down to occur are only strictly applicable if the cross-section of each member has two axes of symmetry, flexure taking place about one of these axes, and in addition the yield stresses in tension and compression are the same. Unless both these conditions are fulfilled, the position of the neutral axis changes during elastic-plastic flexure, and this implies that the bending moment-curvature relation cannot be of the form shown in Fig. 8.3, with a constant range of elastic bending moment. Consider for example a section with only one axis of symmetry, such as a T-section, and suppose that flexure occurs about an axis perpendicular to this axis. If the behaviour is wholly elastic, the neutral axis passes through the centroid of the section, but if the section is fully plastic the neutral axis divides the section into equal areas, assuming the same yield stresses in tension and compression. The centroidal and equal area axes will not coincide, so that when the section is fully plastic the stresses immediately on either side of the centroidal axis will be of the same sign, and equal in magnitude to the yield stress. Thus if the direction of bending is reversed after the fully plastic moment is attained, the changes of bending moment and curvature during unloading cannot be related in the same way as during the elastic range, for the stress changes would then be required to vary linearly with distance from the centroidal axis. On one side of the centroidal axis the stress would be reduced below the yield value, but on the other side the stress would be increased above the yield value over that portion of the section between the centroidal and equal area axes, which by hypothesis is impossible. It follows that during unloading the neutral axis cannot be the centroidal axis, but will assume a position dictated by the requirement of zero axial force, subject to the condition that the stress changes must not imply an increase of stress above the yield value. The flexural rigidity during unloading will not be the same as in the elastic range, and the reduction of bending moment which can occur before yield again commences will not be the same as the initial range of elastic bending moment.

For such cases the conditions for shake-down to occur must be stated in terms of residual stress distributions across each critical section rather than simply in terms of distributions of residual bending moment in the structure, and are thus more complicated. However, the conditions expressed in the inequalities of 8.6, 8.7 and 8.8, covering as they do the case which often occurs of members with two axes of symmetry and equal yield stresses, are sufficient for many practical purposes.

The assumption of a constant elastic range of bending moment is not in accordance with the behaviour of actual beams. In the case of mild steel, for example, the elastic range in tensile-compressive tests is reduced considerably after yielding has occurred, as observed originally by Bauschinger,[17] and this causes a corresponding reduction in the elastic range of bending moment. A similar effect has been observed in tensile-compressive tests on light alloys, as for instance by Templin and Sturm.[18] Thus in

Fig. 8.3 the actual behaviour of a beam upon unloading from *b* would be according to a relation such as *bgh*, the elastic range of moment being M_{er}, less than $2M_y$. If experimental results indicated that no matter what the sequence of loading on a member might be, the available elastic range of bending moment would never fall below a value M_{er}^{min}, it might be conjectured that the conditions for shake-down to occur would be the conditions 8.6 and 8.7 with the condition 8.8 replaced by

$$\mathscr{M}_i^{max} - \mathscr{M}_i^{min} \leqslant M_{er}^{min} \quad . \quad . \quad . \quad 8.9$$

but this theorem has yet to be proved.

Melan [19] has extended the shake-down theorem to cover the case of continuous media, for which the stress-strain relations are appropriate generalizations of the ideal plastic type of stress-strain relation for a bar in tension or compression. Very few applications of this generalized theorem have been made. Symonds [20] has considered the case of a solid circular cylinder in combined tension and torsion, and Hodge [21] has discussed the case of a thick tube under internal pressure.

8.4. Incremental collapse load theorems

Much of the remainder of this chapter will be devoted to the development of methods for calculating incremental collapse loads. These methods depend on certain theorems relating to the values of incremental collapse loads which are closely analogous to those theorems concerning the values of plastic collapse loads which were stated in Section 3.2. Thus static, kinematic, and uniqueness theorems can be established for incremental collapse loads.

The static theorem is merely a restatement of that part of the general shake-down theorem which refers to the avoidance of incremental collapse, as expressed in the inequalities 8.6 and 8.7, and is as follows:

Static theorem of incremental collapse. If it is possible to find any distribution of residual bending moment throughout a frame to which the maximum and minimum elastic bending moments corresponding to a set of extreme load limits W can be added without exceeding the fully plastic moment at any cross-section, the value of W must be less than or equal to the incremental collapse load W_s.

The second theorem concerning incremental collapse arises from a consideration of the kinematics of possible incremental collapse mechanisms. Suppose that the actual incremental collapse mechanism is known, and let θ_k represent the hinge rotation which takes place at cross-section k during a small motion of this mechanism. If the loads on the frame are those corresponding to the incremental collapse load W_s, the frame will eventually shake down, and the residual moments m_i in the frame when it has shaken down will be such that when the maximum and minimum elastic moments are added to these residual moments, the fully plastic moment will just be attained at each cross-section k where a plastic hinge occurs in the incremental collapse mechanism. It follows that

$$m_k + \mathscr{M}_k^{\max} = (M_p)_k \text{ if } \theta_k > 0 \quad . \quad . \quad 8.10$$

$$m_k + \mathscr{M}_k^{\min} = -(M_p)_k \text{ if } \theta_k < 0 \quad . \quad . \quad 8.11$$

Since the m_k are in equilibrium with zero external loads, and the θ_k represent a set of hinge rotations for a mechanism, it follows from the Principle of Virtual Work that

$$\Sigma m_k \theta_k = 0 \quad . \quad . \quad . \quad . \quad 8.12$$

Since the m_k are known from equations 8.10 and 8.11, and the $\mathscr{M}_k^{\max}$ and $\mathscr{M}_k^{\min}$ are all known in terms of W, equation 8.12 will determine the value W_s of W above which incremental collapse would occur.

A computation similar to the above can be performed for any arbitrary choice of incremental collapse mechanism, and a corresponding value of W can be determined. The kinematic theorem, which is based on this fact, is as follows:

Kinematic theorem of incremental collapse. The value of W corresponding to any assumed mechanism of incremental collapse must either be greater than or equal to the incremental collapse load W_s.

A formal proof of this theorem is given in Appendix D, but it is also possible to deduce the theorem from the static theorem by means of a physical argument similar to that given in Section 3.2 for the corresponding plastic collapse theorem. Consider a frame and loading for which the incremental collapse load is W_s. If the fully plastic moment is imagined to be increased at one or more cross-sections of a frame by increasing the yield stress, thus leaving the elastic properties of the members unchanged, the incremental collapse load for the strengthened frame cannot thereby be reduced below the value W_s. This result is an obvious consequence of the static theorem, for the inequalities 8.6 and 8.7 can just be satisfied for the original frame when $W = W_s$ and *a fortiori* can be satisfied for the strengthened frame. If a mechanism is assumed which is not the actual incremental collapse mechanism, the frame may be imagined to be strengthened by increasing the yield stress indefinitely at all cross-sections except those where hinges occur in the assumed mechanism, where the fully plastic moments are left unchanged. The assumed incremental collapse mechanism would then undoubtedly be the actual incremental collapse mechanism for the frame strengthened in this way, and so from the result just stated the corresponding incremental collapse load could not be less than the actual incremental collapse load W_s for the original frame.

Finally, a uniqueness theorem can be stated. This theorem, like the corresponding theorem for plastic collapse loads, follows immediately as a deduction from the static and kinematic theorems and can be stated as follows:

Uniqueness theorem of incremental collapse. If for a given value of W a corresponding statically admissible distribution of residual bending moment can be found, such that when the maximum and minimum elastic bending moments corresponding to this value of W are added to the residual moment at every cross-section the fully plastic moment is never exceeded, but is attained at a sufficient number of cross-sections to transform the structure into a mechanism if hinges formed at all these cross-sections simultaneously, this value of W must be equal to the incremental collapse load W_s.

Methods for calculating incremental collapse loads which are closely analogous to those for calculating plastic collapse loads can be developed from the static, kinematic and uniqueness theorems. In Section 8.5 a description is given of a trial and error method based on the uniqueness theorem, and in Section 8.6 a combining mechanisms method based on the kinematic theorem is explained. While it would be possible to devise a method of calculation based on the static theorem which is analogous to the moment distribution method for calculating plastic collapse loads, such a method has not been developed as yet. The particular advantage of this method in the case of plastic collapse is the scope which it offers for design, but this advantage is lost for incremental collapse owing to the fact that elastic solutions must be obtained at the outset which depend on assigning sections to the members so as to establish their relative elastic stiffnesses.

8.5. Trial and error method

For structures with only a small number of redundancies, such as beams continuous over a few supports or single-bay portal frames, the distribution of residual moment as the incremental collapse load is just reached, and thence the value of the incremental collapse load itself, can usually be found by inspection or by one or two trials. H. Bleich [9] adopted this procedure in solving problems relating to continuous beams with not more than two redundancies, and to a pinned-base portal frame. F. Bleich [22, 23] also used the same method for similar problems, and Neal and Symonds [24] have given a trial and error solution for a fixed-base two-bay rectangular frame with six redundancies. The method will here be described in relation to a single-bay rectangular portal frame under variable horizontal and vertical loads.

The trial and error method is based on the uniqueness theorem, and is analogous to the trial and error method described in Section 4.2 for the calculation of plastic collapse loads. The first step in the analysis is to assume an incremental collapse mechanism, for which a corresponding value of the load W can be found at which incremental collapse would just occur if the correct choice of mechanism had been made. At this value of W, a distribution of residual moment would be attained such that when the maximum and minimum elastic moments for this value of W were superposed on the residual moments the fully plastic moment would just be attained at each of the cross-sections at which plastic hinges occur in the assumed incremental collapse mechanism. As already pointed out in Section 8.4, the residual moment may be calculated in terms of W at each of the plastic hinge positions in the assumed incremental collapse mechanism, using equations 8.10 and 8.11. By inserting the residual moments thus found in the equations of equilibrium for zero external loads, the value of W corresponding to the particular choice of mechanism can then be found. If the assumed incremental collapse mechanism is of the complete type, the equations of equilibrium for zero external loads will also suffice to determine the distribution of residual moment throughout the frame. The maximum and minimum elastic moments corresponding to the value of W which has been found are then added to these residual moments to determine the greatest and least possible bending moments which could occur for these values of the residual moments. If none of these extreme values of the bending moments exceeds the fully plastic moment, it follows from the uniqueness theorem that the correct incremental collapse mechanism has been chosen. If, however, one or more of these extreme values of the bending moments exceeds the fully plastic moment, then a fresh guess as to the incremental collapse mechanism must be made, and the procedure repeated. If the actual incremental collapse mechanism is partial, or indeed if it is desired at any stage of the analysis to investigate a partial type of mechanism, obvious difficulties arise which are similar to those which are encountered in applying the trial and error method to cases of partial plastic collapse; these difficulties were discussed in Section 4.2.

The method will now be described with reference to the fixed-base rectangular portal frame whose dimensions are shown in Fig. 8.4(a). The members of this frame are supposed to be of uniform section throughout, the fully plastic moment being M_p. The frame is subjected to horizontal and vertical loads, H and V, each of which can vary independently of one another between the limits shown in the figure. The problem is to determine the value W_s of W above which failure by incremental collapse will occur.

Since the bending moment must vary linearly between the five cross-sections numbered from 1 to 5 in Fig. 8.4(a), it follows that plastic hinges can only form at these cross-sections. It is therefore only necessary to consider conditions at these cross-sections, and not at intermediate sections in the spans. The equations of equilibrium for a frame of these dimensions have already been derived in Section 3.3 for the case in which horizontal and vertical loads $3W$ and $2W$, respectively, were applied. These equations were

$$3Wl = M_2 - M_1 + M_5 - M_4 \qquad . \quad . \quad . \quad 3.5$$

$$2Wl = 2M_3 - M_2 - M_4 \qquad . \quad . \quad . \quad 3.6$$

the sign convention being that a positive bending moment causes tension in the fibres of the member which are adjacent to the

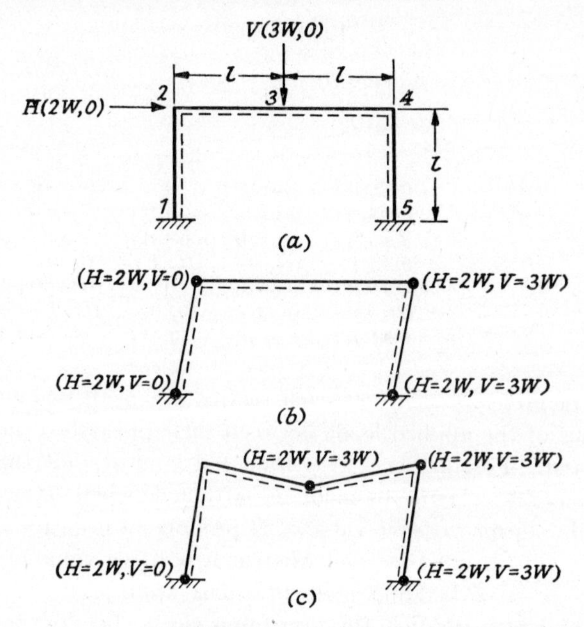

(*a*) Frame and loading.
(*b*) Assumed sidesway mechanism of incremental collapse. (*c*) Combined mechanism of incremental collapse.

Fig. 8.4. *Rectangular portal frame.*

dotted line in Fig. 8.4(*a*). The equations of equilibrium for zero external loads on the frame are thus :

$$m_2 - m_1 + m_5 - m_4 = 0 \quad . \quad . \quad . \quad 8.13$$

$$2m_3 - m_2 - m_4 = 0 \quad . \quad . \quad . \quad 8.14$$

These two equations could, of course, be derived directly by applying the Principle of Virtual Work to the sidesway and beam mechanisms for the unloaded frame.

The bending moments in the frame on the assumption of wholly elastic behaviour may be found by any of the conventional methods of elastic analysis. These bending moments are given in Table 8.6. In this table the first row shows the elastic bending moments caused by the application of the horizontal load $H = 2W$ alone, and the second row shows the elastic bending moments caused by the application of the vertical load $V = 3W$ alone. The third row of the table shows the maximum possible

TABLE 8.6

Elastic bending moments for rectangular portal frame

Cross-section	1	2	3	4	5
$H = 2W$	$-0{\cdot}625Wl$	$0{\cdot}375Wl$	0	$-0{\cdot}375Wl$	$0{\cdot}625Wl$
$V = 3W$	$0{\cdot}300Wl$	$-0{\cdot}600Wl$	$0{\cdot}900Wl$	$-0{\cdot}600Wl$	$0{\cdot}300Wl$
$\mathscr{M}^{max}$	$0{\cdot}300Wl$ $(0, V)$	$0{\cdot}375Wl$ $(H, 0)$	$0{\cdot}900Wl$ (H, V)	0 $(0, 0)$	$0{\cdot}925Wl$ (H, V)
$\mathscr{M}^{min}$	$-0{\cdot}625Wl$ $(H, 0)$	$-0{\cdot}600Wl$ $(0, V)$	0 $(0, 0)$	$-0{\cdot}975Wl$ (H, V)	0 $(0, 0)$
$\mathscr{M}^{max} - \mathscr{M}^{min}$	$0{\cdot}925Wl$	$0{\cdot}975Wl$	$0{\cdot}900Wl$	$0{\cdot}975Wl$	$0{\cdot}925Wl$

elastic bending moments at each section when all possible combinations of the applied loads between their prescribed limits are considered. For instance, at section 1 it is clear that the maximum possible elastic moment is attained when $H = 0$ and $V = 3W$, since a positive value of H produces a negative bending moment at this cross-section, whereas a positive value of V produces a positive bending moment. The appropriate load combinations which produce the maximum elastic bending moments are shown in brackets beneath the values of these bending moments. The fourth row of the table gives the minimum elastic bending moments, and the fifth row gives the range of elastic bending moment at each cross-section.

The analysis now proceeds by assuming an incremental collapse mechanism. Suppose that the first assumption is the sidesway mechanism shown in Fig. 8.4(*b*). In this mechanism, plastic hinges occur at the cross-sections 1, 2, 4 and 5, not simultaneously, but at different peak load combinations. The signs of the fully plastic moments at these hinges are as follows:

$$M_1 = -M_p\,,\ M_2 = M_p\,,\ M_4 = -M_p\,,\ M_5 = M_p$$

The corresponding residual moments may now be written down. For instance, at cross-section 1, where a plastic hinge of negative sign is formed, the minimum elastic moment must occur as the bending moment just reaches the value $-M_p$, so that $m_1 + \mathscr{M}_1^{min} = -M_p$. Thus from Table 8.6,

$$m_1 - 0{\cdot}625Wl = -M_p \quad . \quad . \quad . \quad 8.15$$

Similarly

$$m_2 + 0{\cdot}375Wl = M_p \quad . \quad . \quad . \quad 8.16$$

$$m_4 - 0{\cdot}975Wl = -M_p \quad . \quad . \quad . \quad 8.17$$

$$m_5 + 0{\cdot}925Wl = M_p \quad . \quad . \quad . \quad 8.18$$

The minimum elastic bending moment at cross-section 1 is seen from Table 8.6 to occur when the load combination $H = 2W$, $V = 0$ is applied. If the sidesway mechanism of Fig. 8.4(*b*) were the actual incremental collapse mechanism, it would follow that during the process of incremental collapse rotation at the plastic hinge at this cross-section would occur whenever this load combination was applied. The load combinations which would cause rotations at the other plastic hinges during incremental collapse may be determined in a similar manner, and these load combinations are indicated in Fig. 8.4(*b*) at the various plastic hinge positions.

When the values of the residual moments obtained from equations 8.15–8.18 are substituted in the equilibrium equation 8.13, it is found that

$$(M_p - 0{\cdot}375Wl) - (-M_p + 0{\cdot}625Wl) + (M_p - 0{\cdot}925Wl) - (-M_p + 0{\cdot}975Wl) = 0$$

$$2{\cdot}900Wl = 4M_p$$

$$W = 1{\cdot}379\frac{M_p}{l}$$

When this value of W is substituted in equations 8.15–8.18, values for the residual moments m_1, m_4, m_4 and m_5 are at once found. These values are

$$\begin{aligned} m_1 &= -0{\cdot}138M_p \\ m_2 &= 0{\cdot}483M_p \\ m_4 &= 0{\cdot}345M_p \\ m_5 &= -0{\cdot}276M_p \end{aligned}$$

Substituting for m_2 and m_4 in equation 8.14 the value of m_3 is then found to be:

$$m_3 = 0{\cdot}414M_p$$

Since all the residual moments are now known, the extreme possible values of the bending moments at each cross-section may be found, and the calculations are set out in Table 8.7. In this table, the second and third rows give the values of $\mathscr{M}^{\max}$ and $\mathscr{M}^{\min}$ when $W = 1{\cdot}379\frac{M_p}{l}$. When these values are added to the residual moments, which are given in the first row of the table, the corresponding values of the extreme bending moments, $M^{\max}$ and $M^{\min}$, are obtained. The values of $M^{\max}$ and $M^{\min}$ are

TABLE 8.7

Maximum and minimum bending moments in the sidesway mechanism. $W = 1{\cdot}379\dfrac{M_p}{l}$

Cross-section	1	2	3	4	5
m	$-0{\cdot}138M_p$	$0{\cdot}483M_p$	$0{\cdot}414M_p$	$0{\cdot}345M_p$	$-0{\cdot}276M_p$
$\mathscr{M}^{\max}$	$0{\cdot}414M_p$	$0{\cdot}517M_p$	$1{\cdot}241M_p$	0	$1{\cdot}276M_p$
$\mathscr{M}^{\min}$	$0{\cdot}862M_p$	$-0{\cdot}828M_p$	0	$-1{\cdot}345M_p$	0
$M^{\max}$	$0{\cdot}276M_p$	M_p	$1{\cdot}655M_p$	$0{\cdot}345M_p$	M_p
$M^{\min}$	$-M_p$	$-0{\cdot}345M_p$	$0{\cdot}414M_p$	$-M_p$	$-0{\cdot}276M_p$

given in the fourth and fifth rows of the table, and it will be seen that $M_3{}^{\max} = 1{\cdot}655M_p$. Thus the choice of incremental collapse mechanism was incorrect, since the residual moments and the value of W corresponding to this mechanism lead to an extreme value of a bending moment which exceeds the fully plastic moment.

A fresh guess as to the incremental collapse mechanism must now be made. Since the previous solution led to a condition in which the fully plastic moment was exceeded at the section 3, the fresh choice of mechanism should involve a plastic hinge at this section. Suppose that the combined mechanism of Fig. 8.4(c) is selected, with plastic hinges at the cross-sections 1, 3, 4 and 5. An analysis similar to that just described for the assumed sidesway mechanism shows that the corresponding value of W is $1{\cdot}132\dfrac{M_p}{l}$. The results of this analysis are summarized in Table 8.8, and it will be seen that none of the extreme values of the bending moments given in the fourth and fifth rows of the table exceeds the fully plastic moment. It follows from the uniqueness principle that this solution is correct. Thus the incremental collapse load W_s is $1{\cdot}132\dfrac{M_p}{l}$, and the mechanism of incremental collapse is the combined mechanism shown in Fig. 8.4(c).

The load combinations which would cause rotations at the various plastic hinges during incremental collapse, deduced in the same manner as for the assumed sidesway mechanism, are indicated in Fig. 8.4(c) at the four plastic hinge positions.

The calculation of the incremental collapse load W_s can usually be shortened by first calculating the plastic collapse load W_c for

TABLE 8.8

Maximum and minimum bending moments in the combined mechanism. $W = 1{\cdot}132\dfrac{M_p}{l}$

Cross-section	1	2	3	4	5
m	$-0{\cdot}292M_p$	$-0{\cdot}141M_p$	$-0{\cdot}019M_p$	$0{\cdot}104M_p$	$-0{\cdot}047M_p$
$\mathscr{M}^{max}$	$0{\cdot}340M_p$	$0{\cdot}425M_p$	$1{\cdot}019M_p$	0	$1{\cdot}047M_p$
$\mathscr{M}^{min}$	$-0{\cdot}708M_p$	$-0{\cdot}679M_p$	0	$-1{\cdot}104M_p$	0
M^{max}	$0{\cdot}048M_p$	$0{\cdot}284M_p$	M_p	$0{\cdot}104M_p$	M_p
M^{min}	$-M_p$	$-0{\cdot}820M_p$	$-0{\cdot}019M_p$	$-M_p$	$-0{\cdot}047M_p$

the worst possible combination of loads, for the value of W_s cannot exceed W_c. This is because if $W = W_s$, there must exist a distribution of residual bending moment which enables *all* the possible load combinations to be borne by purely elastic bending moment changes such that the fully plastic moment is not exceeded anywhere in the frame. It follows that the *worst* possible load combination can be carried in this way, giving rise to a resulting distribution of bending moment in which the fully plastic moment is nowhere exceeded. From the static theorem of plastic collapse, the load W_s cannot therefore exceed the plastic collapse load W_c, so that

$$W_s \leqslant W_c$$

In this particular case the plastic collapse load W_c when the maximum horizontal and vertical loads, $2W$ and $3W$, are applied simultaneously is found to be $1{\cdot}2\dfrac{M_p}{l}$. For the sidesway mechanism of incremental collapse which was first analysed, the corresponding value of W was found to be $1{\cdot}379\dfrac{M_p}{l}$, which is greater than the value of W_c. This mechanism could therefore have been rejected without calculating the values of M^{max} and M^{min} at every cross-section, as in Table 8.7, thus eliminating a considerable amount of labour.

Since the incremental collapse load W_s was found to be $1{\cdot}132\dfrac{M_p}{l}$, it follows that for this case W_c exceeeds W_s by about 6%.

It is interesting to note that an estimate can be made of the limiting deflections developed if an infinite number of cycles of load is applied with the value of W equal to W_s, the method being similar to that described in Section 5.5 for the estimation of the deflections in a structure at the point of collapse. Consider for the moment the rectangular frame shown in Fig. 8.1(*a*), whose behaviour under the repetition of the cycle of loading illustrated in Fig. 8.1(*b*) was discussed in Section 8.2. Table 8.3 shows that for this frame and cyclic loading, with $W = W_s$, the bending moment at section 3 which occurs when the horizontal and vertical loads are applied simultaneously tends asymptotically to M_p as the number of loading cycles tends to infinity, but there is never any plastic hinge rotation at this section. Thus when the frame has shaken down, there are undetermined hinge rotations at sections 1, 4 and 5, but zero rotation at section 3. In the example which has just been considered there will similarly be zero hinge rotation and therefore continuity at one of the sections 1, 3, 4 or 5 at which hinges form in the incremental collapse mechanism. This is the counterpart of the continuity which occurs at the section where the last hinge forms at plastic collapse.

Since the distribution of residual bending moment after shakedown has occurred with W equal to W_s is known, it is possible by assuming continuity at any one of the sections 1, 3, 4 or 5 to calculate the hinge rotations at the other three sections and the deflections when the frame is free from load by using the slope-deflection equations. If these hinge rotations are not all of the correct sign an appropriate mechanism motion can be superposed to bring the rotation at one of the hinges to zero and give all the other hinge rotations the correct sign, and the deflections amended accordingly.

Finally, it may be remarked that the alternating plasticity load W_a can be found very simply by noting in Table 8.6 that the greatest elastic range of bending moment is $0{\cdot}975Wl$. If the available range of elastic bending moment is assumed to be $2M_y = \frac{2M_p}{\alpha}$, where α is the shape factor, the alternating plasticity load is given by

$$0{\cdot}975W_a l = \frac{2M_p}{\alpha}$$

$$W_a = 2{\cdot}051\frac{M_p}{\alpha l}$$

8.6. Method of combining mechanisms

The limitations of the trial and error method for cases in which incremental collapse mechanisms of the partial type need to be investigated have already been referred to. The method of combining mechanisms [25] for determining incremental collapse loads presents no additional difficulties when mechanisms of the partial type are encountered, and therefore fulfils the same role in the calculation of incremental collapse loads as the corresponding method fulfilled in the calculation of plastic collapse loads.

For any assumed incremental collapse mechanism, a corresponding value of W can be found by an application of the Principle of Virtual Work, and the kinematic theorem states that the actual incremental collapse load W_s is the smallest value of W thus obtained when all possible incremental collapse mechanisms are examined. The method of combining mechanisms consists simply of deriving values of W corresponding to the independent mechanisms and their most likely combinations, until it is thought that the lowest possible value of W has been found. The result is then subjected to a statical check by determining the corresponding residual moment distribution and verifying that the maximum and minimum elastic bending moments can be added to the residual moment at every cross-section without exceeding the fully plastic moment. The method is thus seen to differ from the method of combining mechanisms for calculating plastic collapse loads only in the details of its application.

The method will be illustrated with reference to the particular frame whose dimensions and loading are illustrated in Fig. 8.5(*a*). In this frame all the joints are assumed to be rigid, and the stanchions are assumed to be rigidly built in at their feet. A horizontal load H is applied at beam level, which can vary between limits of zero and W, and there are uniformly distributed vertical loads P and Q acting on the beams which can each vary between limits of zero and $4W$. These three loads, H, P and Q, can each vary independently of one another between their prescribed limits. The fully plastic moment of each stanchion is M_p, and that of each beam is $2M_p$. Since for members of geometrically similar cross-section the moment of inertia varies as the fully plastic moment raised to the power of $\frac{4}{3}$, the ratio of the moment of inertia of a beam to that of a stanchion is assumed to be 2·5, which is approximately $(2)^{4/3}$.

In this frame there are ten cross-sections, numbered from 1 to 10 in Fig. 8.5(a), at which plastic hinges can occur. The positions of the plastic hinges which may occur in the beams beneath the uniformly distributed loads are not known *a priori*. However, in the first instance, when considering the independent mechanisms and their combinations, it will be assumed that if plastic

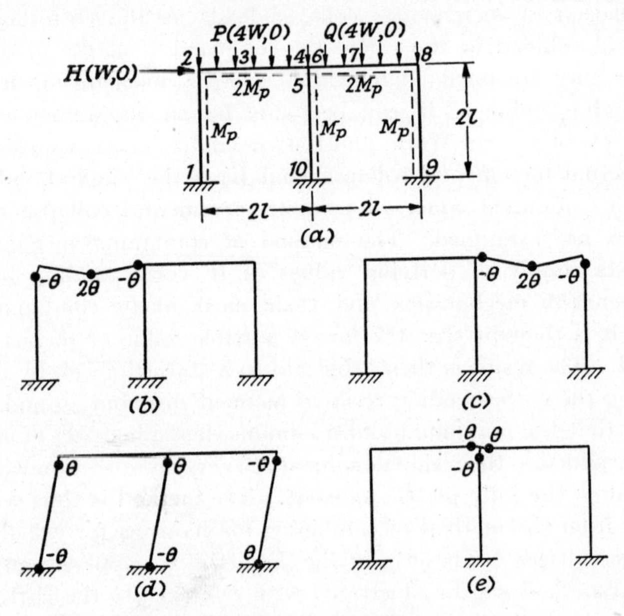

(a) Frame and loading.
(b) Left-hand beam mechanism. (c) Right-hand beam mechanism.
(d) Sidesway mechanism. (e) Joint rotation mechanism.

Fig. 8.5. *Two-bay single-storey rectangular portal frame.*

hinges occur in the incremental collapse mechanisms within the spans of the beams, these hinges will occur in mid-span.

The elastic solutions for this frame for the three cases in which each load is applied singly may be obtained by any of the orthodox elastic methods. These solutions are given in Table 8.9, which also gives the maximum and minimum elastic bending moment at each cross-section, together with the elastic range of bending moment. The usual sign convention is adopted for the bending moments, namely, that positive bending moments cause tension in the fibres of the members adjacent to the dotted lines in Fig. 8.5(*a*).

TABLE 8.9

Elastic bending moments for two-bay single-storey rectangular portal frame

Cross-section	1	2	3	4	5
$H = W$	$-0{\cdot}339Wl$	$0{\cdot}287Wl$	$0{\cdot}051Wl$	$-0{\cdot}184Wl$	$0{\cdot}368Wl$
$P = 4W$	$0{\cdot}091Wl$	$-0{\cdot}224Wl$	$0{\cdot}608Wl$	$-0{\cdot}560Wl$	$0{\cdot}216Wl$
$Q = 4W$	$0{\cdot}004Wl$	$0{\cdot}033Wl$	$-0{\cdot}156Wl$	$-0{\cdot}344Wl$	$-0{\cdot}216Wl$
$\mathscr{M}^{max}$	$0{\cdot}095Wl$	$0{\cdot}320Wl$	$0{\cdot}659Wl$	0	$0{\cdot}584Wl$
	$(0, P, Q)$	$(H, 0, Q)$	$(H, P, 0)$	$(0, 0, 0)$	$(H, P, 0)$
$\mathscr{M}^{min}$	$-0{\cdot}339Wl$	$-0{\cdot}224Wl$	$-0{\cdot}156Wl$	$-1{\cdot}088Wl$	$-0{\cdot}216Wl$
	$(H, 0, 0)$	$(0, P, 0)$	$(0, 0, Q)$	(H, P, Q)	$(0, 0, Q)$
$\mathscr{M}^{max} - \mathscr{M}^{min}$	$0{\cdot}434Wl$	$0{\cdot}544Wl$	$0{\cdot}815Wl$	$1{\cdot}088Wl$	$0{\cdot}800Wl$

Cross-section	6	7	8	9	10
$H = W$	$0{\cdot}184Wl$	$-0{\cdot}051Wl$	$-0{\cdot}287Wl$	$0{\cdot}339Wl$	$-0{\cdot}380Wl$
$P = 4W$	$-0{\cdot}344Wl$	$-0{\cdot}156Wl$	$0{\cdot}033Wl$	$0{\cdot}004Wl$	$-0{\cdot}128Wl$
$Q = 4W$	$-0{\cdot}560Wl$	$0{\cdot}608Wl$	$-0{\cdot}224Wl$	$0{\cdot}091Wl$	$0{\cdot}128Wl$
$\mathscr{M}^{max}$	$0{\cdot}184Wl$	$0{\cdot}608Wl$	$0{\cdot}033Wl$	$0{\cdot}434Wl$	$0{\cdot}128Wl$
	$(H, 0, 0)$	$(0, 0, Q)$	$(0, P, 0)$	(H, P, Q)	$(0, 0, Q)$
$\mathscr{M}^{min}$	$-0{\cdot}904Wl$	$-0{\cdot}207Wl$	$-0{\cdot}511Wl$	0	$-0{\cdot}508Wl$
	$(0, P, Q)$	$(H, P, 0)$	$(H, 0, Q)$	$(0, 0, 0)$	$(H, P, 0)$
$\mathscr{M}^{max} - \mathscr{M}^{min}$	$1{\cdot}088Wl$	$0{\cdot}815Wl$	$0{\cdot}544Wl$	$0{\cdot}434Wl$	$0{\cdot}636Wl$

The frame has six redundancies, and since ten bending moments must be known in order to specify the bending moment distribution throughout the frame there must be four independent mechanisms. Two of these mechanisms are the simple beam mechanisms shown in Figs. 8.5(*b*) and (*c*), and there is one sidesway mechanism, as shown in Fig. 8.5(*d*). The fourth independent mechanism is the joint rotation shown in Fig. 8.5(*e*). The hinge rotations involved in each of these mechanisms are shown in both magnitude and sign in each of these figures, each hinge rotation being given in terms of a single parameter θ. The reason for specifying the signs of the plastic hinge rotations will appear later; the sign convention adopted is similar to that adopted for the bending moments, and is that positive hinge rotations cause extensions of the fibres of the members adjacent to the dotted lines in Fig. 8.5(*a*).

Consider first the beam mechanism illustrated in Fig. 8.5(*b*). Suppose that this was the correct incremental collapse mechanism, and that the value of the load W was just equal to its incremental collapse value. Then at the cross-section 2, where a negative plastic hinge is shown, the fully plastic moment $-M_p$ would just be attained at the particular load combination which causes the minimum elastic bending moment at this cross-section. The value of this minimum elastic bending moment is seen from Table 8.9 to be $-0{\cdot}224Wl$, and it follows that

$$m_2 - 0{\cdot}224Wl = -M_p$$

Similarly
$$m_3 + 0{\cdot}659Wl = 2M_p$$
$$m_4 - 1{\cdot}088Wl = -2M_p$$

The corresponding value of W can now be found by applying the Principle of Virtual Work. For any assumed mechanism in which the hinge rotation at the cross-section k is θ_k, it follows from the Principle of Virtual Work, equation 8.12, that

$$\Sigma m_k \theta_k = 0,$$

where m_k represents the residual moment at cross-section k obtained from any distribution of residual moment which satisfies the conditions of equilibrium with zero external load. The summation in this equation covers all those positions where plastic hinges occur in the assumed mechanism. If the hinge rotations of the mechanism of Fig. 8.5(*b*) are substituted in this equation, together with the residual moments just obtained, it is found that

$$-\theta(-M_p + 0{\cdot}224Wl) + 2\theta(2M_p - 0{\cdot}659Wl)$$
$$-\theta(-2M_p + 1{\cdot}088Wl) = 0$$

$$2{\cdot}630Wl\theta = 7M_p\theta \qquad . \quad 8.19$$

$$W = 2{\cdot}662\frac{M_p}{l}$$

The calculation of the value of W which corresponds to any assumed mechanism can be made in a more systematic manner. The plastic hinge rotation at any typical hinge position k in an assumed incremental collapse mechanism will be a multiple of a parameter θ, say $\alpha_k\theta$, where α_k is a numerical coefficient. If the fully plastic moment at this hinge position is $(M_p)_k$, the corresponding residual moment at this section will be given by one of the two equations 8.10 or 8.11,

$$m_k = (M_p)_k - \mathscr{M}_k^{\max} \quad \text{if } \alpha_k > 0$$
$$m_k = -(M_p)_k - \mathscr{M}_k^{\min} \text{ if } \alpha_k < 0$$

The virtual work equation for the assumed incremental collapse mechanism is thus :

$$\Sigma\left\{\begin{matrix} (M_p)_k - \mathscr{M}_k^{\max} \\ -(M_p)_k - \mathscr{M}_k^{\min} \end{matrix}\right\}\alpha_k\theta = 0,$$

where the summation includes all the cross-sections k where plastic hinges occur in the assumed mechanism. The terms within the brackets are alternatives, each positive α_k being multiplied by $[(M_p)_k - \mathscr{M}_k^{\max}]$ and each negative α_k being multiplied by $[-(M_p)_k - \mathscr{M}_k^{\min}]$. Thus each term $\alpha_k\left\{\begin{matrix} (M_p)_k \\ -(M_p)_k \end{matrix}\right\}$ is positive, regardless of the sign of α_k. The equation can therefore be rearranged as follows :

$$\Sigma\alpha_k\left\{\begin{matrix} \mathscr{M}_k^{\max} \\ \mathscr{M}_k^{\min} \end{matrix}\right\}\theta = \Sigma\,|\alpha_k|(M_p)_k\theta \qquad . \quad . \quad 8.20$$

In the alternatives on the left-hand side of equation 8.20, each positive α_k will be multiplied by $\mathscr{M}_k^{\max}$, and each negative α_k will be multiplied by $\mathscr{M}_k^{\min}$. The right-hand side of this equation is seen to be identical with the work which would be absorbed at the plastic hinges if the assumed mechanism were regarded as a plastic collapse mechanism rather than an incremental collapse mechanism.

Equation 8.20 enables the virtual work equation for any assumed incremental collapse mechanism to be written down directly. Consider for instance the beam mechanism of Fig. 8.5(*c*). For this mechanism, $\alpha_6 = -1$, $\alpha_7 = 2$ and $\alpha_8 = -1$, and the fully plastic moments at the cross-sections 6, 7 and 8 are of magnitude $2M_p$, $2M_p$ and M_p, respectively. Applying equation 8.20, it is found that

$$(-\mathscr{M}_6^{\min} + 2\mathscr{M}_7^{\max} - \mathscr{M}_8^{\min})\theta = 7M_p\theta$$

Inserting in this equation the values of $\mathscr{M}_6^{\min}$, $\mathscr{M}_7^{\max}$ and $\mathscr{M}_8^{\min}$ from Table 8.9, it is found that

$$(0{\cdot}904Wl + 2 \times 0{\cdot}608Wl + 0{\cdot}511Wl)\theta = 7M_p\theta$$
$$2{\cdot}631Wl\theta = 7M_p\theta$$
$$W = 2{\cdot}661\frac{M_p}{l} \qquad 8.21$$

Application of equation 8.20 to the sidesway mechanism of Fig. 8.5(d) leads to the result

$$2{\cdot}696Wl\theta = 6M_p\theta$$

$$W = 2{\cdot}226\frac{M_p}{l} \quad . \quad . \quad . \quad . \quad 8.22$$

For the joint rotation mechanism of Fig. 8.5(e), equation 8.20 gives

$$1{\cdot}488Wl\theta = 5M_p\theta$$

$$W = 3{\cdot}360\frac{M_p}{l} \quad . \quad . \quad . \quad . \quad 8.23$$

At first sight it may be surprising that a value of W corresponding to the joint rotation mechanism can be obtained. It will be recalled that in the case of plastic collapse a joint rotation mechanism rarely has any physical significance in itself, for unless an external couple is applied at the joint the work equation for a joint rotation mechanism does not include any term for the work done by the external loads, and therefore corresponds to zero values for the required fully plastic moments. However, this is not the case where incremental collapse is concerned, the reason being that the hinge rotations are not all required to take place simultaneously, but only at different stages during the loading programme. The joint rotation mechanism would, in fact, be the actual incremental collapse mechanism for this structure if the fully plastic moments of the members were increased sufficiently at every cross-section except 4, 5 and 6 by increasing the yield stress, so as to leave the elastic properties of the members unchanged.

Of the four independent mechanisms, the sidesway mechanism of Fig. 8.5(d) gave the lowest value of W, namely $2{\cdot}226\frac{M_p}{l}$. It now remains to investigate possible combinations of the independent mechanisms to see whether a still lower value of W can be found. Consider first the combination of the beam mechanism of Fig. 8.5(b) with the sidesway mechanism. If these two mechanisms are combined, the hinge at the cross-section 2 is cancelled. The virtual work equation for the resulting mechanism, which is as shown in Fig. 8.6(a), could be obtained directly, but it is quicker to derive this equation from the virtual work equations

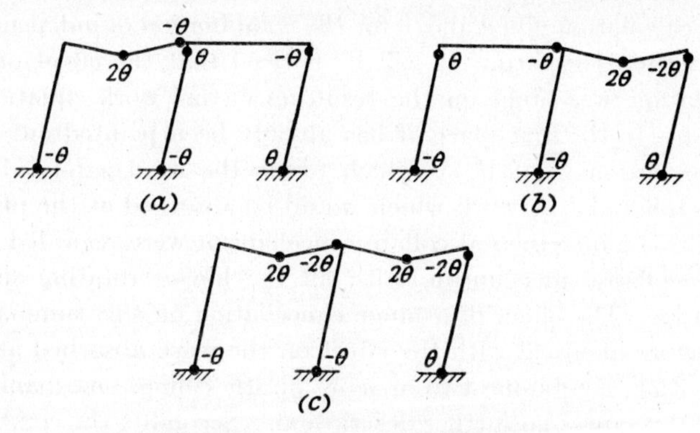

Fig. 8.6. Combined mechanisms.

8.19 and 8.22, which refer to the two independent mechanisms which are being combined. These equations were:

$$2{\cdot}630Wl\theta = 7M_p\theta \qquad . \quad . \quad . \quad 8.19$$

$$2{\cdot}696Wl\theta = 6M_p\theta \qquad . \quad . \quad . \quad 8.22$$

The right-hand side of each of these equations included a term $M_p\theta$ for the work absorbed at the plastic hinge at cross-section 2. When these mechanisms are combined, there will thus be a reduction in the work absorbed of $2M_p\theta$. Moreover, the left-hand sides of equations 8.19 and 8.22 include terms $-\mathscr{M}_2^{\min}\theta$ and $\mathscr{M}_2^{\max}\theta$, respectively, so that when the corresponding mechanisms are combined a term $(\mathscr{M}_2^{\max} - \mathscr{M}_2^{\min})\theta$ is cancelled in the resulting virtual work equation. This term is seen at once from Table 8.9 to be $0{\cdot}544Wl\theta$. The resulting virtual work equation is thus obtained as follows:

$$2{\cdot}630Wl\theta = 7M_p\theta \qquad . \quad . \quad . \quad 8.19$$

$$2{\cdot}696Wl\theta = 6M_p\theta \qquad . \quad . \quad . \quad 8.22$$

$$5{\cdot}326Wl\theta - 0{\cdot}544Wl\theta = 13M_p\theta - 2M_p\theta$$

$$4{\cdot}782Wl\theta = 11M_p\theta$$

$$W = 2{\cdot}300\frac{M_p}{l} \qquad . \quad . \quad . \quad 8.24$$

This value of W is no improvement over the value $2{\cdot}226\dfrac{M_p}{l}$ obtained for the sidesway mechanism.

The calculation given above for the combination of independent mechanisms is typical. It will be noticed that the effect of the cancellation of a hinge on the resulting virtual work equation is twofold. In the first place, it has already been pointed out that the summation $\Sigma|\alpha_k|(M_p)_k\theta$ which forms the right-hand side of equation 8.20 is the work which would be absorbed at the plastic hinges if the incremental collapse mechanism were regarded as a plastic collapse mechanism, with all the hinges rotating simultaneously. The effect of a hinge cancellation on this summation is therefore identical with the effect on the work absorbed at the hinges when combining two or more plastic collapse mechanisms, and thus requires no further description. Secondly, the combination of mechanisms just carried out illustrates the point that the summation $\Sigma\alpha_k\begin{Bmatrix}\mathscr{M}_k^{\max}\\ \mathscr{M}_k^{\min}\end{Bmatrix}\theta$ is affected by a hinge cancellation. This is in contrast with the case of plastic collapse, where the terms for the work done by the applied loads are always additive. When incremental collapse mechanisms are combined so as to eliminate a hinge at some cross-section j, one of the combining mechanisms will involve a positive hinge rotation $|\alpha_j|\theta$, with which the maximum elastic moment $\mathscr{M}_j^{\max}$ is associated, and another of the mechanisms will involve a negative hinge rotation $-|\alpha_j|\theta$, with which the minimum elastic moment $\mathscr{M}_j^{\min}$ is associated. It follows that in respect of the cancelled hinge, a term

$$|\alpha_j|(\mathscr{M}_j^{\max} - \mathscr{M}_j^{\min})\theta$$

must be subtracted from the sum of the left-hand sides of the virtual work equations for the combining mechanisms in forming the virtual work equation for the combined mechanism.

Another possible combination of the independent mechanisms is illustrated in Fig. 8.6(b). This mechanism is obtained by adding the displacements and hinge rotations of the beam mechanism of Fig. 8.5(c) and the sidesway mechanism of Fig. 8.5(d), and then adding the joint rotation mechanism of Fig. 8.5(e) so as to cancel the hinge rotations at the cross-sections 5 and 6. The cancellation of the hinge rotation of magnitude θ at cross-section 5, where the fully plastic moment is M_p, reduces the work absorbed at the plastic hinges by $2M_p\theta$. Moreover, the cancellation of the hinge rotation of magnitude θ at cross-section 6, where the fully plastic moment is $2M_p$, reduces the work absorbed by $4M_p\theta$. Also, $|\alpha_5| = |\alpha_6| = 1$, and from Table 8.9 $(\mathscr{M}_5^{\max} - \mathscr{M}_5^{\min})$

$= 0{\cdot}800Wl$ and $(\mathscr{M}_6^{\max} - \mathscr{M}_6^{\min}) = 1{\cdot}088Wl$. Hence the resulting virtual work equation for the combined mechanism of Fig. 8.6(*b*) is obtained as follows:

$$2{\cdot}631Wl\theta = 7M_p\theta \quad . \quad . \quad . \quad 8.21$$
$$2{\cdot}696Wl\theta = 6M_p\theta \quad . \quad . \quad . \quad 8.22$$
$$1{\cdot}488Wl\theta = 5M_p\theta \quad . \quad . \quad . \quad 8.23$$

$$6{\cdot}815Wl\theta - 0{\cdot}800Wl\theta - 1{\cdot}088Wl\theta = 18M_p\theta - 2M_p\theta - 4M_p\theta$$
$$4{\cdot}927Wl\theta = 12M_p\theta$$
$$W = 2{\cdot}436\frac{M_p}{l} \quad . \quad . \quad 8.25$$

Again this value of W is no improvement over the value $2{\cdot}226\frac{M_p}{l}$ obtained for the sidesway mechanism.

The only remaining combination of the independent mechanisms is shown in Fig. 8.6(*c*). This mechanism is obtained by adding the displacements and hinge rotations of the combined mechanism of Fig. 8.6(*b*) to those of the beam mechanism of Fig. 8.5(*b*), thus cancelling the hinge rotation of magnitude θ at cross-section 2, where the fully plastic moment is M_p. From Table 8.9,

$$(\mathscr{M}_2^{\max} - \mathscr{M}_2^{\min}) = 0{\cdot}544Wl.$$

The virtual work equation for the combined mechanism of Fig. 8.6(*c*) is therefore obtained as follows:

$$4{\cdot}927Wl\theta = 12M_p\theta \quad . \quad . \quad . \quad 8.25$$
$$2{\cdot}630Wl\theta = 7M_p\theta \quad . \quad . \quad . \quad 8.19$$

$$7{\cdot}557Wl\theta - 0{\cdot}544Wl\theta = 19M_p\theta - 2M_p\theta$$
$$7{\cdot}013Wl\theta = 17M_p\theta$$
$$W = 2{\cdot}424\frac{M_p}{l} \quad . \quad . \quad 8.26$$

The value of W obtained is again no improvement on the value $2{\cdot}226\frac{M_p}{l}$ obtained for the sidesway mechanism. Since there are no other possible combinations to be tested, it is concluded that the actual incremental collapse mechanism is the sidesway mechanism, subject only to the proviso that so far it has been assumed that any plastic hinges which may form in the beams beneath the uniformly distributed loads P and Q will occur at mid-span. Thus the value of W found to correspond to the mechanism of Fig. 8.6(a) was $2{\cdot}300\frac{M_p}{l}$, assuming that the hinge in the left-hand beam occurred at mid-span. In fact, the correct location of this hinge would be the position which minimized the corresponding value of W, which might thereby be reduced below the value $2{\cdot}226\frac{M_p}{l}$ for the sidesway mechanism. However, it will be assumed that this is not the case; this assumption is in fact confirmed by a statical check for the sidesway mechanism.

Check by statics

The statical check takes the form of constructing the maximum and minimum elastic bending moment diagrams for the whole frame, with $W = 2{\cdot}226\frac{M_p}{l}$, and then verifying that it is possible to construct a residual bending moment diagram upon which the maximum and minimum elastic bending moments can be superposed without exceeding the fully plastic moment at any cross-section. Consider first the construction of the maximum and minimum elastic bending moment diagrams for the left-hand beam. Fig. 8.7(a) shows the elastic bending moment diagrams for this beam due to the separate application of the three loads $H = W$, $P = 4W$ and $Q = 4W$, positive bending moments being plotted as ordinates below the datum line. Since each of the loads can vary independently between zero and its prescribed maximum value, the maximum elastic bending moment at any cross-section is the sum of those bending moments, due to the peak values of the individual loads, which are positive. Similarly, the minimum elastic bending moment at any cross-section is the sum of those bending moments, due to the peak values of the individual loads, which are negative. Curves showing the maximum and minimum elastic bending moments are given in Fig. 8.7(b).

The maximum and minimum elastic bending moment diagrams for the frame, obtained in this way, are shown as the full lines in Fig. 8.8. In this diagram the ordinates of maximum and minimum elastic bending moment are set off at right angles to each member. The sign convention for these bending moments is that a positive ordinate is plotted on the same side of a member

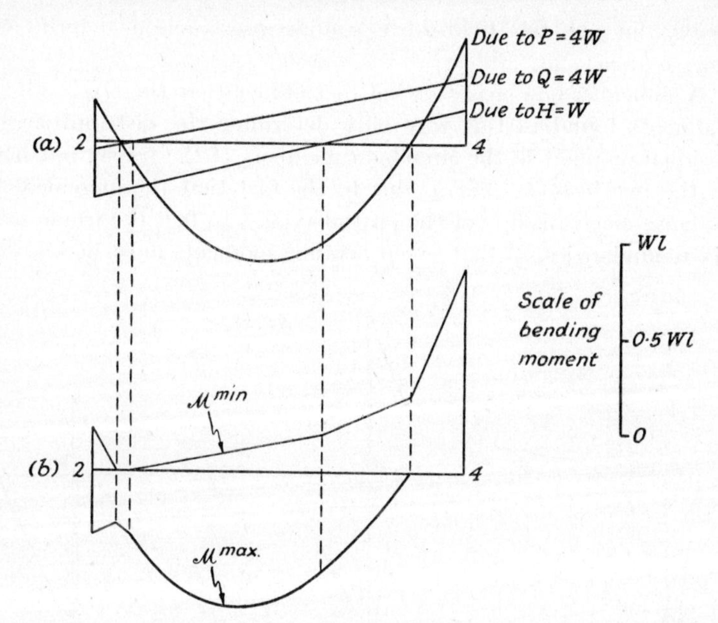

Fig. 8.7. Maximum and minimum elastic bending moments for left-hand beam.

as the dotted line in Fig. 8.5(*a*). A residual moment diagram is also indicated by the dotted lines in Fig. 8.8. For convenience, the residual moments are plotted with their sign changed. Since the actual bending moment at any cross-section is obtained by adding the residual moment to the elastic moment, it follows that it will also be obtained by subtracting from the elastic moment the residual moment with its sign changed. Thus the peak bending moments in Fig. 8.8 are found by regarding the residual moment diagram as a base from which the peak bending moments are measured.

In the sidesway mechanism of incremental collapse plastic hinges occur at the cross-sections 1, 2, 5, 8, 9 and 10. This enables the residual moment at each of these cross-sections to be determined. For instance, a negative hinge occurs in the incremental collapse mechanism at the cross-section 1, where the fully plastic moment is of magnitude M_p. Thus $m_1 + \mathscr{M}_1^{\min} = -M_p$. The value of m_1 is found graphically by setting off M_p from the minimum elastic bending moment at the cross-section 1, as indicated by the heavy line in the figure. A similar construction determines m_2, m_5, m_8, m_9 and m_{10}.

A difficulty now arises owing to the fact that the six residual moments found in this way only determine the distribution of residual moment in the stanchion members of the frame, but not in the two beams. This is due to the fact that the incremental collapse mechanism is of the partial type. In fact the frame has six redundancies, so that seven bending moments must be known

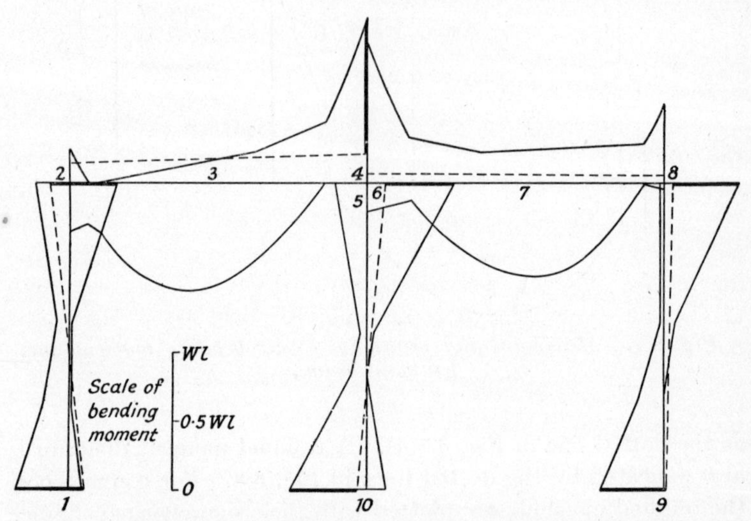

Fig. 8.8. ***Maximum and minimum elastic bending moment distribution, and residual bending moment distribution in two-bay single-storey rectangular frame at incremental collapse load.***

in order to determine any bending moment distribution completely. It is thus possible to make an arbitrary choice of the residual moment at any one cross-section in a beam, after which the distribution of residual moment throughout the remainder of the frame will be determined from the equations of equilibrium under zero external load.

Guidance as to this arbitrary choice is best afforded by examining the mechanism which gave the next lowest value of W. This mechanism was the mechanism shown in Fig. 8.6(a), with hinges in the left-hand beam at the sections 3 and 4, for which the corresponding value of W was only $2{\cdot}300\frac{M_p}{l}$. This suggests that when W is $2{\cdot}226\frac{M_p}{l}$, as in the sidesway collapse mechanism, it will only just be possible to find a residual moment line for this beam which would not give rise to actual bending moments exceeding the fully plastic moment $2M_p$ at the cross-sections 3 and 4. The arbitrary choice of residual moment is therefore made at the cross-section 4, where the residual moment is selected so that the fully plastic moment $-2M_p$ would just be attained at this cross-section when the elastic bending moment had its minimum value. With this value of m_4, it is found that the greatest actual bending moment in the left-hand beam does not exceed the fully plastic moment $2M_p$ anywhere within the span. Since m_5 is already known, m_6 is at once found by considering the equilibrium of the joint 456, and the value of m_6 thus obtained is such that the fully plastic moment is not exceeded anywhere within the span of the right-hand beam. Thus it is possible to find at least one residual bending moment distribution corresponding to the sidesway mechanism such that when the maximum and minimum elastic bending moments are added the fully plastic moment is nowhere exceeded. It follows from the uniqueness theorem that the incremental collapse mechanism is the sidesway mechanism, and that the incremental collapse load W_s is $2{\cdot}226\frac{M_p}{l}$.

The plastic collapse load W_c for this frame under the worst combination of loads, when H, P and Q have their maximum values, can be shown to be $2{\cdot}684\frac{M_p}{l}$, the mechanism of collapse being that of Fig. 8.6(a). Thus in this case W_c exceeds W_s by about 21%.

In conclusion, the load combinations which would cause rotations at the various plastic hinges in the incremental collapse mechanism may be determined in the manner already described. These combinations are as follows:

Cross-section	*Load combination*
1	$H = W,\ P = 0,\quad Q = 0$
2	$H = W,\ P = 0,\quad Q = 4W$
5	$H = W,\ P = 4W,\ Q = 0$
8	$H = W,\ P = 0,\quad Q = 4W$
9	$H = W,\ P = 4W,\ Q = 4W$
10	$H = W,\ P = 4W,\ Q = 0$

The alternating plasticity load W_a is found by noting from Table 8.9 that at those sections where the fully plastic moment is

M_p the greatest range of elastic moment is $0{\cdot}800Wl$ at section 5, whereas the greatest range of elastic moment at the remaining sections where the fully plastic moment is $2M_p$ is $1{\cdot}088Wl$ at sections 4 and 6. Evidently alternating plasticity would first take place at section 5, and W_a is given by

$$0{\cdot}800W_a l = 2\frac{M_p}{\alpha}$$

$$W_a = 2{\cdot}5\frac{M_p}{\alpha l}$$

As already mentioned, no method for the calculation of incremental collapse loads analogous to the plastic moment distribution method for determining plastic collapse loads has been developed. The only method of calculation based on the static theorem is that due to Symonds and Neal,[26] who showed that incremental collapse and alternating plasticity loads could be obtained by solving the inequalities 8.6, 8.7 and 8.8, taking into account the conditions of equilibrium existing between the residual moments at the various possible plastic hinge positions. However, this method is cumbersome in use, and the methods described in this and the preceding Section are much swifter in application. Heyman [27] has shown that this method can be adapted to solving the problem of designing frames for minimum weight if the criterion of design is the provision of a load factor against incremental collapse or alternating plasticity.

8.7. Relation to design

The normal procedure in plastic design is to proportion the members of the frame so that if the worst combination of the working loads were multiplied by a specified load factor, failure by plastic collapse would just occur. The value of the load factor must be chosen so that the probability of a failure by plastic collapse actually occurring during the lifetime of the structure is sufficiently remote to be acceptable. However, since many buildings and other structures are subjected to variable repeated loading, it is necessary to ensure in addition that the probabilities of failure by incremental collapse or alternating plasticity are also acceptably small. This would apparently require the calculation of the incremental collapse and alternating plasticity loads, and since neither of these loads can be calculated without first deriving the elastic bending moment distributions in the frame for the various loads acting independently, the design procedure would become very lengthy. Fortunately, however, this is not necessary, for it appears that if a frame is designed by the plastic methods to a certain load factor against plastic collapse, the probability of a failure by plastic collapse actually occurring, while very small, is yet far greater than the probability of a failure by either incremental collapse or alternating plasticity.

The reason for this is best understood by considering a specific example. Thus for the rectangular portal frame of Fig. 8.1(a), it was shown that the worst combination of loads as far as plastic collapse is concerned consists of the application of both the horizontal and vertical loads, each of these loads having its maximum value W. Plastic collapse would then occur if W had the value $W_c = 3\frac{M_p}{l}$. If this frame were designed to a load factor of 2, the working loads would be $H = V = 1{\cdot}5\frac{M_p}{l}$. Considering first the case of incremental collapse, it was shown in Section 8.2 that if both H and V could vary between the same limits of zero and W, incremental collapse could not occur unless W exceeded the value $W_s = 2{\cdot}857\frac{M_p}{l}$, so that the load factor against incremental collapse would be $\frac{2{\cdot}857}{1{\cdot}5} = 1{\cdot}9$. The question is therefore whether this load factor of 1·9 for incremental collapse provides an even greater margin of safety against the occurrence of this type of failure than the safety margin furnished by the load factor of 2 against plastic collapse.

The crucial point here is that a failure by incremental collapse could not occur unless at least one *cycle* of loading took place in which the peak intensities of the loads *exceeded* the incremental collapse load. A cycle of loading in this case consists of the application of the load combination $H = W, V = W$, its removal, and the application of the load combination $H = W$, $V = O$, so that two peak load combinations are involved in each cycle. The probability of the occurrence of one of these peak load combinations, assumed equal to the same value p in either case, will be very small if W is greater than W_s, for there is a load factor of 1·9 against incremental collapse, and the working loads themselves represent the highest loads which are expected to occur in normal usage. The probability of the occurrence of two peak load combinations, constituting one cycle of loading, is the product of the probabilities of the occurrence of the two separate combinations, and is thus p^2 if W exceeds W_s in each case. Since the load factor against plastic collapse is 2, as against 1·9 for incremental collapse, the probability p' of the occurrence of the single peak load combination $H = V = W_c$ which would cause plastic collapse must be rather smaller than p, but both these probabilities are so small that p' will be far greater than p^2. Thus the probability of the occurrence of a failure by plastic collapse, while acceptably small, is far greater than the probability of the occurrence of a failure by incremental collapse. This rather crude argument is strengthened when it is realized that unless W is appreciably greater than W_s, several cycles of loading are required to produce unacceptably large deflections, as can be seen from Fig. 8.2.

The argument is not convincing when related to a single specific example, but it can be seen that the important factor is the ratio of the incremental collapse load to the plastic collapse load. In the above case these loads were $2{\cdot}857\frac{M_p}{l}$ and $3\frac{M_p}{l}$, respectively, so that W_c exceeded W_s by only 5%. For the examples considered in Sections 8.5 and 8.6 (see Figs. 8.4 and 8.5), W_c exceeded W_s by 6% and 21% respectively. These values are typical, and it appears that W_c rarely exceeds W_s by more than about 25% unless the structure is of unusual proportions or subjected to unlikely load combinations. Moreover, the difference between W_c and W_s naturally becomes less as the ratio of dead to live load is increased, and in the above cases there was no dead load. The above argument has been given generality by Horne,[28] who by making certain assumptions has shown that for any structure and loading the probability of failure by plastic collapse is always greater than the probability of failure by incremental collapse. This runs counter to the view of Dutheil,[29, 30] who proposed that frames should be designed on the basis of the incremental collapse load W_s, on the grounds that this critical load is always less than the plastic collapse load W_c. However, Dutheil coupled with this proposal a suggestion that W_s should be calculated empirically as $\frac{W_c}{\alpha}$, where α is the shape factor. In essence, therefore, Dutheil's proposed design method is actually based on the plastic collapse load.

A further point is that if a structure is on the verge of failure by plastic collapse, little warning of the imminence of failure is available, since the deflections will in general still be of reasonable magnitude. However, a failure by incremental collapse is gradual, the deflections building up cumulatively during the lifetime of the structure, so that ample warning of the progress of this type of failure is available. This implies that for incremental collapse a lower load factor than that provided against plastic collapse would be acceptable.

Turning now to the question of failure by alternating plasticity, it seems probable that failures of this type are unlikely to occur unless the number of cycles of loading is of the order of magnitude of 100–1,000, or even greater, depending of course on the intensity of the peak loads. Relatively little experimental work has been carried out to determine the number of cycles of reversed bending required to cause failure when the range of bending stress exceeds the elastic range. Tests have been carried out at the University of Sheffield [31, 32, 33] on simply supported steel beams of rectangular and I-section subjected to centrally concentrated alternating loads. These experiments showed that as many as 50 cycles of load reversals such that the central bending moment varied between the limits $\pm M_p$ could be withstood without any signs of fracture. Lazard [34, 35] has reported similar tests on *IPN*200 joists in which it appeared that plain joists and joists with drilled holes showed no tendency to fracture under severe alternating bending. However, joists with punched holes were found to fracture after about six load reversals.

While it cannot be stated that there are no cases in which the design should be based on the provision of a load factor against alternating plasticity, there are many types of structure for which no consideration need be given to this phenomenon. At one end of the scale lies the structure such as the portal type of shed frame for which only a very small number of peak loadings would be expected during its lifetime. At the other extreme the expected number of important load fluctuations would be very large, so that the dominant design consideration is that of fatigue. Intermediate cases would appear to be rare.

The conclusion can therefore be drawn that unless a design is governed by fatigue, the recommended plastic design procedure, in which a specified load factor is provided against plastic collapse, will generally result in a structure which is far less likely to fail by incremental collapse or alternating plasticity than by plastic collapse under a single overload. It is interesting to note, however, that Parkes [36] has shown that the stresses set up in aircraft wings due to repeated temperature changes can cause failures by incremental collapse or alternating plasticity, and the possibility of the occurrence of these types of failure has become an important criterion in the design of high speed aircraft.

References

1. M. Grüning. *Die Tragfähigkeit statisch unbestimmten Tragwerke aus Stahl bei beliebig haufig wiederholter Belastung.* Julius Springer, Berlin, 1926.
2. G. v. Kazinczy. Die Weiterentwicklung der Elastizitätstheorie. *Technika,* Budapest (1931).
3. M. R. Horne. The effect of variable repeated loads in the plastic theory of structures. *Research* (*Engng. Struct. Suppl.*), *Colston Papers*, **2,** 141 (1949).
4. P. S. Symonds. Cyclic loading tests on small frames. *Final Report, 4th Congr. Intern. Assn. Bridge and Struct. Engng.*, 109, Cambridge (1953).
5. K. Klöppel. Beitrag zur Frage der Ausnutzbarkeit der Plastizität bei dauerbeanspruchten Durchlaufträgern. *Final Report, 2nd Congr. Intern. Assn. Bridge and Struct. Engng.*, **77,** Berlin (1939).
6. C. Massonet. Essais d'adaptation et de stabilisation plastiques sur les poutrelles laminées *Proc. Intern. Assn. Bridge and Struct. Engng.*, **13,** 239 (1953). See also *Ossat. métall.*, **19,** 318 (1954).
7. J. Fritsche. Die Tragfähigkeit von Balken aus Baustahl bei beliebig oft wiederholter Belastung. *Bauingenieur,* **12,** 827 (1931).
8. E. O. Patton and B. N. Gorbunow. *The carrying capacity of welded beams, which are deformed plastically by repeated loads* (in Russian). Vidavnitstvo Vseukrainskoy Akademii Nauk, Kiev (1935).
9. H. Bleich. Über die Bemessung statisch unbestimmter Stahltragwerke unter Berücksichtigung des elastisch-plastischen Verhaltens des Baustoffes. *Bauingenieur,* **13,** 261 (1932).
10. E. Melan. Theorie statisch unbestimmter Systeme. *Prelim.*

Pubn. 2nd Congr. Intern. Assn. Bridge and Struct. Engng., 43, Berlin (1936).
11. E. MELAN. Die Theorie statisch unbestimmter Systeme aus ideal plastischen Baustoff. *S.B. Akad. Wiss. Wien.* (*Abt.* IIa), **145**, 195 (1936).
12. P. S. SYMONDS and W. PRAGER. Elastic-plastic analysis of structures subjected to loads varying arbitrarily between prescribed limits. *J. Appl. Mech.*, **17**, 315 (1950).
13. W. PRAGER. Problem types in the theory of perfectly plastic materials. *J. Aero Sci.*, **15**, 337 (1948).
14. B. G. NEAL. The behaviour of framed structures under repeated loading. *Quart. J. Mech. Appl. Math.*, **4**, 78 (1951).
15. B. G. NEAL. Plastic collapse and shakedown theorems for structures of strain-hardening material. *J. Aero. Sci.*, **17**, 297 (1950).
16. W. T. KOITER. Some remarks on plastic shakedown theorems. *8th Intern. Congr. Theor. and Appl. Mech.*, Istanbul (1952).
17. J. BAUSCHINGER. Die Veränderungen der Elastizi tätsgrenze. *Mitt. mech.-tech. Lab. tech. Hochschule*, München (1886).
18. R. L. TEMPLIN and R. G. STURM. Some stress-strain studies of metals. *J. Aero. Sci.*, **7**, 189 (1940).
19. E. MELAN. Der Spannungszustand eines " Mises-Henckyscher " Kontinuums bei verändlicher Belastung. *S.B. Akad. Wiss. Wien.* (*Abt.* IIa), **147**, 73 (1938).
20. P. S. SYMONDS. Shakedown in continuous media. *J. Appl. Mech.*, **18**, 85 (1951).
21. P. G. HODGE, Jr. Shake-down of elastic-plastic structures. *Residual Stresses in Metals and Metal Construction.* (Ed. W. R. Osgood), Reinhold, N.Y., 163 (1954).
22. F. BLEICH. La ductilité de l'acier. Son application au dimensionnenment des systemes hyperstatiques. *Ossat. métall.*, **3**, 93 (1934).
23. F. BLEICH. Bemessung statisch unbestimmter Systeme nach der Plastizitätstheorie (Traglastverfahren). *Prelim. Pubn. 2nd Congr. Intern. Assn. Bridge and Struct. Engng*, 131, Berlin (1936).
24. B. G. NEAL and P. S. SYMONDS. A method for calculating the failure load for a framed structure subjected to fluctuating loads. *J. Instn. Civ. Engrs.*, **35**, 186 (1950).
25. P. S. SYMONDS and B. G. NEAL. Recent progress in the plastic methods of structural analysis. *J. Franklin Inst.*, **252**, 383, 469 (1951).
26. P. S. SYMONDS and B. G. NEAL. The calculation of failure loads on plane frames under arbitrary loading programmes. *J. Instn. Civ. Engrs.*, **35**, 41 (1950).
27. J. HEYMAN. Plastic design of beams and plane frames for minimum material consumption. *Quart. Appl. Math.*, **8**, 373 (1951).

28. M. R. HORNE. The effect of variable repeated loads in building structures designed by the plastic theory. *Proc. Intern. Assn. Bridge and Struct. Engng.*, **14,** 53 (1954).
29. J. DUTHEIL. L'exploitation des phénomène d'adaptation dans les ossatures en acier doux. *Ann. Inst. Tech. Bât. Trav. Publ.*, No. 2, Jan., 1948.
30. J. DUTHEIL. La conception des ossatures métalliques basée sur la déformation plastique. *Ossat. métall.*, **14,** 143 (1949).
31. H. CORKER. The effect of variable repeated loads on a simply supported rectangular beam in the elasto-plastic range. *Report, Dept. of Civ. Engng., Univ. of Sheffield*, 1953.
32. G. A. HOYLAND. The behaviour of simply supported I section mild steel beams under reversed bending in the plastic range. *Report, Dept. of Civ. Engng., Univ. of Sheffield*, 1953.
33. C. SANDELL. The effects of variable repeated loads on simply supported rectangular beams loaded in the elasto-plastic range. *Report, Dept. of Civ. Engng., Univ. of Sheffield*, 1954.
34. A. LAZARD. Plastification of bending plate-web girders in mild steel. *Prelim. Pubn., 4th Congr. Intern. Assn. Bridge and Struct. Engng.*, 123, Cambridge (1952).
35. A. LAZARD. The effect of plastic yield in bending on mild steel plate girders. *Struct. Engr.*, **32,** 49 (1954).
36. E. W. PARKES. Wings under repeated thermal stress. *Aircraft Engineering*, **26,** 402 (1954).

Examples

1. In Klöppel's test, which was referred to in Section 8.2, a uniform beam ABC was continuous over three simple supports A, B and C. $AB = BC = 150$ cm. A constant load P was applied at the centre of the span AB, while at the centre of the span BC a load was continuously varied between the limits P and 200 kg. The fully plastic moment of the beam was $1{\cdot}488 \times 10^5$ kg. cm. Find the values of the incremental collapse, alternating plasticity and plastic collapse loads. For the alternating plasticity load assume a shape factor 1·16.

It may be assumed that a concentrated load W kg. applied at the centre of one of the spans produces a hogging bending moment $\frac{900}{64}W$ kg. cm. at the central support B if the entire beam behaves elastically.

2. A uniform beam $ABCD$ whose fully plastic moment is M_p is continuous over four simple supports A, B, C and D. $AB = BC = CD = l$. Concentrated loads are applied at the mid-points of each of the three spans. The load on the span AB can vary between the limits zero and $2W$, while the loads on the spans BC and CD can each vary between the limits zero and W. Each of the three loads can vary independently of the other two loads. Find the incremental collapse and plastic collapse loads, and also the alternating plasticity load, assuming an elastic range $2M_p$. Check the calculation for the incremental collapse load by drawing the elastic bending moment diagram for the effect of each load acting separately, thence constructing a diagram showing the variation of the maximum and minimum elastic bending moments along the beam, and drawing a residual moment diagram with the signs of the residual moments reversed.

It may be assumed that if the entire beam behaves elastically a concentrated load W applied at the centre of BC produces hogging bending moments $0{\cdot}075Wl$ over the supports B and C, and that a concentrated load W applied at the centre of CD produces a hogging bending moment $0{\cdot}1Wl$ at C and a sagging bending moment $0{\cdot}025Wl$ at B.

3. A uniform fixed-ended beam AB whose fully plastic moment is M_p carries a concentrated load W which rolls back and forth along the beam. Plot the elastic bending moment diagram for various positions of the load, and hence construct a diagram showing the variation of the maximum and minimum possible elastic bending moments at every section for all possible positions of the load. Determine the incremental collapse load, and compare its value with the plastic collapse load. Find also the incremental collapse and plastic collapse loads if the beam carries in addition a uniformly distributed dead load W. Show that the alternating plasticity load is unaffected by the dead load, and calculate its value.

It may be assumed that if a concentrated load W acts at a distance λl from A the hogging bending moments at A and B are $Wl\lambda(1-\lambda)^2$ and $Wl\lambda^2(1-\lambda)$, respectively, provided that the beam is everywhere elastic.

4. The uniform fixed-base rectangular portal frame whose dimensions are shown in Fig. 8.4(a) is composed of members which all have a fully plastic moment M_p. The horizontal load H can vary between the limits zero and W, while the vertical load V can vary independently between the limits zero and $2W$. Find the incremental collapse and plastic collapse loads, and show that the incremental collapse load is unaltered if V is held constant at the value $2W$ while H varies between zero and W. Determine the alternating plasticity load in each case if the elastic range of bending moment is $2M_p$.

The appropriate elastic solutions can be derived from Table 8.6.

5. For the frame of example 4, estimate the horizontal deflection at the point of application of H and the vertical deflection at the point of application of V if shake-down has occurred after a large number of cycles of loading with $W = W_s$ and the loads have then been removed from the frame. The effects of strain-hardening and the spread of plastic zones along the members may be neglected.

6. A uniform beam is of Tee-section, width a and depth $1{\cdot}24a$. The section of the beam is composed of two similar rectangles whose sides are a and $0{\cdot}24a$, and the stress-strain relation for the material is the ideal plastic relation of Fig. 1.4(*b*). The beam is subjected to a bending moment about an axis parallel to the flange of the Tee, causing flexure in the plane of the web. Find the position of the neutral axis if the bending moment is such that the beam behaves elastically, and show that if the bending moment is increased to its fully plastic value the neutral axis moves parallel to itself through a distance $0{\cdot}19a$ to the equal area axis. If the bending moment is reduced from the fully plastic moment, show that the neutral axis immediately assumes a new position in the web at a distance $0{\cdot}2a$ from the equal area axis, so that the stress remains constant at the yield stress over a length of the web equal to this value. Hence show that upon unloading from the fully plastic moment the elastic range of bending moment is 0·36% greater than the elastic range for the initially unstressed beam, whereas the flexural rigidity is 0·88% less.

APPENDIX A

Plastic Theory and Trusses

For trusses the basic property of each member is the relation between axial force S and axial deformation e. The basic hypotheses which would be required by a simple plastic theory for trusses would therefore be concerned with the axial force-deformation relations for the truss members, which would be required to be of the same general form as the bending moment-curvature relation of Fig. 1.2. In particular, the force in a typical truss member would be bounded by limiting values S_p in tension and $-S_p$ in compression, and increments in the axial force and axial deformation would be required to be always of the same sign. The basic hypotheses could be summarized by two conditions similar to the conditions 1.1 and 1.2, namely,

$$-S_p \leqslant S \leqslant S_p \quad . \quad . \quad . \quad . \qquad \text{A.1}$$

$$\frac{dS}{de} \geqslant 0 \quad . \quad . \quad . \quad . \qquad \text{A.2}$$

The counterpart of the plastic hinge would be the condition of a member which had reached its limiting force in one sense or the other, and could then undergo an unlimited amount of axial deformation while the force in the member remained constant. This represents the behaviour of a mild steel member very well in tension, provided that the strains produced are not too great, but the behaviour in compression is much more complex because of the buckling phenomena which almost invariably occur. The tests which were carried out by von Karman[1] on mild steel struts of rectangular cross-section, for which some of the results are reproduced in Fig. A.1, illustrate the dependence of the load-deformation relation of a pin-ended compression member on its slenderness ratio $\frac{l}{r}$. It will be seen from the figure that for high slenderness ratios, when the buckling is elastic and large deformations begin to occur at roughly the Euler critical load, the curves resemble that of the basic hypothesis, although elastic unloading does not occur when the direction of loading is reversed. However, for lower slenderness ratios when yield occurs simultaneously with or even before the buckling, as in the upper curves in the figure, the load reaches a maximum and then drops off more or less abruptly to much smaller values. As pointed out by von Karman, this reduction of load with increase of deformation in cases where yield occurs during or before the buckling process is capable of a simple explanation. When there is an appreciable lateral deflection at the centre of a pin-ended strut, the central section

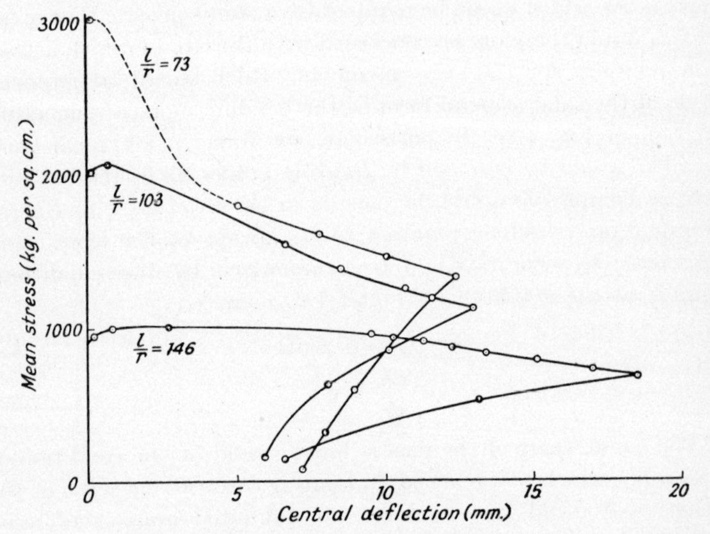

Fig. A.1. *Actual axial force-lateral deformation relation for pin-ended struts* (after von Karman).

must sustain not only the axial thrust but a bending moment whose value is the product of the axial thrust and the lateral deflection. If this central section has more or less fully yielded, an increase in the bending moment which it is called upon to carry is of necessity accompanied by a decrease of the thrust which can be borne. Thus as the lateral deflection at the centre of the strut increases, so increasing the lever arm of the thrust, a reduction in this thrust must occur, for if the thrust remained constant the bending moment would increase.

While the above discussion was related to the central lateral deflection rather than the axial deflection, it will be appreciated that these two deflections are intimately related, and an increase in the former deflection will always cause an increase in the latter. A further point is that in actual trusses the rigidity of the joints would modify these curves somewhat, but it is unlikely that their general character would be altered.

The reason for the exclusion of trusses from treatment by the simple plastic theory is now apparent, for it is seen that the basic hypotheses concerning the load-deformation relation for truss members are not obeyed by compression members. The fact that the magnitude of the limiting axial force in compression is not constant but depends on the slenderness ratio is not of great importance, for the plastic theory could easily be modified to take account of this effect. However, the reduction of compressive load with increase of deformation which takes place invalidates the vital postulate that a member in which the axial compressive force was equal to its limiting value could deform to an indefinite extent under constant load.

Although the simple plastic theory cannot be applied to practical trusses, hypothetical truss systems for which the individual bars behave in accordance with the conditions A.1 and A.2 have been studied theoretically, and several useful and important results have been obtained in this way.[2, 3]

References

1. T. v. Karman. Untersuchungen über Knickfestigkeit. *Mitt. über Forschungsarbeiten*, No. 81 (1910).
2. W. Prager. Problem types in the theory of perfectly plastic materials. *J. Aero. Sci.*, **15**, 337 (1948).
3. P. S. Symonds and W. Prager. Elastic-plastic analysis of structures subjected to loads varying arbitrarily between prescribed limits. *J. Appl. Mech.*, **17**, 315 (1950).

APPENDIX B

Plastic Moduli of British Standard Beams

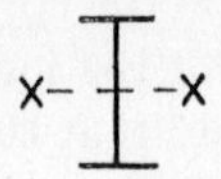

MAJOR AXIS

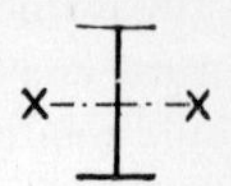

Reference Mark	Dimensions A×B	Weight per foot (lb.)	Area (in.^{2})	Plastic modulus about Major Axis (in.^{3})	Reduced Values of Plastic Modulus under Axial Load (in.^{3})		
					Lower values of n	Change formula at $n=$	Higher values of n
BSB 101	$3\times1\frac{1}{2}$	4	1·18	1·309	$1{\cdot}309-2{\cdot}16n^2$	0·334	$0{\cdot}2411(1-n)\,(6{\cdot}32+n)$
BSB 102	3×3	8·5	2·52	2·979	$2{\cdot}979-7{\cdot}95n^2$	0·179	$0{\cdot}5662(1-n)\,(5{\cdot}68+n)$
BSB 103	$4\times1\frac{3}{4}$	5	1·47	2·153	$2{\cdot}153-3{\cdot}17n^2$	0·401	$0{\cdot}3266(1-n)\,(8{\cdot}00+n)$
BSB 104	4×3	10	2·94	4·539	$4{\cdot}539-9{\cdot}00n^2$	0·264	$0{\cdot}7634(1-n)\,(6{\cdot}70+n)$
BSB 105	$4\frac{3}{4}\times1\frac{3}{4}$	6·5	1·91	3·335	$3{\cdot}335-5{\cdot}06n^2$	0·382	$0{\cdot}538(1-n)\,(7{\cdot}43+n)$
BSB 106	5×3	11	3·26	6·318	$6{\cdot}318-12{\cdot}03n^2$	0·283	$0{\cdot}929(1-n)\,(7{\cdot}76+n)$
BSB 107	$5\times4\frac{1}{2}$	20	5·88	11·653	$11{\cdot}653-29{\cdot}81n^2$	0·192	$2{\cdot}027(1-n)\,(6{\cdot}25+n)$
BSB 108	6×3	12	3·53	8·084	$8{\cdot}084-13{\cdot}56n^2$	0·337	$1{\cdot}094(1-n)\,(8{\cdot}69+n)$
BSB 109	$6\times4\frac{1}{2}$	20	5·89	13·46	$13{\cdot}46-23{\cdot}4n^2$	0·325	$2{\cdot}072(1-n)\,(7{\cdot}53+n)$
BSB 110	6×5	25	7·37	17·00	$17{\cdot}00-33{\cdot}1n^2$	0·271	$2{\cdot}878(1-n)\,(6{\cdot}67+n)$
BSB 111	7×4	16	4·75	12·90	$12{\cdot}90-22{\cdot}5n^2$	0·323	$1{\cdot}519(1-n)\,(9{\cdot}93+n)$
BSB 112	8×4	18	5·30	16·02	$16{\cdot}02-25{\cdot}0n^2$	0·376	$1{\cdot}882(1-n)\,(10{\cdot}26+n)$
BSB 113	8×5	28	8·28	25·80	$25{\cdot}80-48{\cdot}9n^2$	0·286	$3{\cdot}604(1-n)\,(8{\cdot}19+n)$
BSB 114	8×6	35	10·30	32·94	$32{\cdot}94-75{\cdot}7n^2$	0·224	$4{\cdot}670(1-n)\,(7{\cdot}82+n)$
BSB 115	9×4	21	6·18	20·85	$20{\cdot}85-31{\cdot}8n^2$	0·389	$2{\cdot}517(1-n)\,(10{\cdot}04+n)$
BSB 116	9×7	50	14·71	53·18	$53{\cdot}18-135{\cdot}2n^2$	0·197	$8{\cdot}092(1-n)\,(7{\cdot}18+n)$
BSB 117	$10\times4\frac{1}{2}$	25	7·35	28·07	$28{\cdot}07-45{\cdot}1n^2$	0·363	$3{\cdot}174(1-n)\,(10{\cdot}58+n)$
BSB 118	10×5	30	8·85	33·68	$33{\cdot}68-54{\cdot}4n^2$	0·358	$4{\cdot}138(1-n)\,(9{\cdot}70+n)$
BSB 119	10×6	40	11·77	46·74	$46{\cdot}74-96{\cdot}2n^2$	0·259	$6{\cdot}049(1-n)\,(8{\cdot}73+n)$
BSB 120	10×8	55	16·18	65·65	$65{\cdot}65-163{\cdot}5n^2$	0·205	$8{\cdot}725(1-n)\,(8{\cdot}27+n)$
BSB 121	12×5	32	9·45	42·50	$42{\cdot}50-63{\cdot}7n^2$	0·400	$4{\cdot}71(1-n)\,(11{\cdot}02+n)$
BSB 122	12×6	44	13·00	60·46	$60{\cdot}46-105{\cdot}7n^2$	0·322	$7{\cdot}37(1-n)\,(9{\cdot}58+n)$
BSB 123	12×6	54	15·89	72·72	$72{\cdot}72-126{\cdot}2n^2$	0·319	$10{\cdot}83(1-n)\,(7{\cdot}80+n)$
BSB 124	12×8	65	19·12	92·46	$92{\cdot}46-212{\cdot}5n^2$	0·226	$12{\cdot}00(1-n)\,(8{\cdot}56+n)$
BSB 125	13×5	35	10·30	50·28	$50{\cdot}28-75{\cdot}7n^2$	0·398	$5{\cdot}55(1-n)\,(11{\cdot}05+n)$
BSB 126	14×6	46	13·59	72·53	$72{\cdot}53-115{\cdot}4n^2$	0·368	$8{\cdot}07(1-n)\,(10{\cdot}79+n)$
BSB 127	14×6	57	16·78	88·43	$88{\cdot}43-140{\cdot}7n^2$	0·363	$12{\cdot}08(1-n)\,(8{\cdot}72+n)$
BSB 128	14×8	70	20·59	114·59	$114{\cdot}59-230{\cdot}3n^2$	0·269	$13{\cdot}88(1-n)\,(9{\cdot}38+n)$
BSB 129	15×5	42	12·36	66·96	$66{\cdot}96-90{\cdot}9n^2$	0·463	$7{\cdot}94(1-n)\,(10{\cdot}67+n)$
BSB 130	15×6	45	13·24	75·20	$75{\cdot}20-115{\cdot}2n^2$	0·390	$7{\cdot}70(1-n)\,(11{\cdot}88+n)$
BSB 131	16×6	50	14·71	88·79	$88{\cdot}79-135{\cdot}1n^2$	0·393	$9{\cdot}41(1-n)\,(11{\cdot}49+n)$
BSB 132	16×6	62	18·21	106·23	$106{\cdot}23-150{\cdot}6n^2$	0·429	$14{\cdot}25(1-n)\,(9{\cdot}22+n)$
BSB 133	16×8	75	22·06	138·6	$138{\cdot}6-253{\cdot}5n^2$	0·304	$15{\cdot}91(1-n)\,(10{\cdot}09+n)$
BSB 134	18×6	55	16·18	108·1	$108{\cdot}1-155{\cdot}8n^2$	0·425	$11{\cdot}36(1-n)\,(11{\cdot}82+n)$
BSB 135	18×7	75	22·09	148·3	$148{\cdot}3-221{\cdot}8n^2$	0·399	$18{\cdot}05(1-n)\,(10{\cdot}01+n)$
BSB 136	18×8	80	23·53	163·8	$163{\cdot}8-276{\cdot}7n^2$	0·339	$18{\cdot}07(1-n)\,(10{\cdot}72+n)$
BSB 137	$20\times6\frac{1}{2}$	65	19·12	141·8	$141{\cdot}8-203{\cdot}0n^2$	0·430	$14{\cdot}63(1-n)\,(12{\cdot}07+n)$
BSB 138	$20\times7\frac{1}{2}$	89	26·19	194·3	$194{\cdot}3-285{\cdot}7n^2$	0·409	$23{\cdot}63(1-n)\,(10{\cdot}08+n)$
BSB 139	22×7	75	22·06	177·2	$177{\cdot}2-243{\cdot}4n^2$	0·458	$18{\cdot}16(1-n)\,(12{\cdot}36+n)$
BSB 140	$24\times7\frac{1}{2}$	95	27·94	245·6	$245{\cdot}6-342{\cdot}3n^2$	0·446	$26{\cdot}91(1-n)\,(11{\cdot}46+n)$

Note : $n = \dfrac{N}{N_p}$, where N = magnitude of axial force.
N_p = fully plastic thrust.

TABLE I

Plastic Moduli of British Standard Beams

Reference Mark	Dimensions A×B	Weight per foot (lb.)	Area (in.^{2})	Plastic modulus about Minor Axis (in.^{3})	Reduced Values of Plastic Modulus under Axial Load (in.^{3})		
					Lower values of n	Change formula at $n=$	Higher values of n
BSB 101	$3\times1\frac{1}{2}$	4	1·18	0·280	0·280–0·115n^2	0·408	0·585(1−n) (0·345+n)
BSB 102	3×3	8·5	2·52	1·372	1·372–0·530n^2	0·238	1·980(1−n) (0·652+n)
BSB 103	$4\times1\frac{3}{4}$	5	1·47	0·365	0·365–0·135n^2	0·463	0·932(1−n) (0·208+n)
BSB 104	4×3	10	2·94	1·471	1·471–0·540n^2	0·327	2·594(1−n) (0·483+n)
BSB 105	$4\frac{3}{4}\times1\frac{3}{4}$	6·5	1·91	0·505	0·505–0·192n^2	0·448	1·219(1−n) (0·245+n)
BSB 106	5×3	11	3·26	1·602	1·602–0·529n^2	0·338	2·977(1−n) (0·444+n)
BSB 107	$5\times4\frac{1}{2}$	20	5·88	4·795	4·795–1·729n^2	0·247	7·130(1−n) (0·626+n)
BSB 108	6×3	12	3·53	1·626	1·626–0·520n^2	0·391	3·501(1−n) (0·335+n)
BSB 109	$6\times4\frac{1}{2}$	20	5·89	4·087	4·087–1·446n^2	0·377	8·25(1−n) (0·378+n)
BSB 110	6×5	25	7·37	6·089	6·089–2·259n^2	0·334	10·90(1−n) (0·470+n)
BSB 111	7×4	16	4·75	2·850	2·850–0·804n^2	0·369	5·94(1−n) (0·362+n)
BSB 112	8×4	18	5·30	2·991	2·991–0·876n^2	0·423	7·23(1−n) (0·256+n)
BSB 113	8×5	28	8·28	6·75	6·75–2·14n^2	0·338	12·67(1−n) (0·438+n)
BSB 114	8×6	35	10·30	10·71	10·71–3·31n^2	0·272	17·27(1−n) (0·560+n)
BSB 115	9×4	21	6·18	3·51	3·51–1·06n^2	0·437	8·80(1−n) (0·231+n)
BSB 116	9×7	50	14·71	18·67	18·67–6·01n^2	0·245	28·20(1−n) (0·615+n)
BSB 117	$10\times4\frac{1}{2}$	25	7·35	4·84	4·84–1·35n^2	0·408	11·29(1−n) (0·282+n)
BSB 118	10×5	30	8·85	6·55	6·55–1·96n^2	0·407	14·99(1−n) (0·293+n)
BSB 119	10×6	40	11·77	11·89	11·89–3·46n^2	0·306	20·95(1−n) (0·489+n)
BSB 120	10×8	55	16·18	22·59	22·59–6·54n^2	0·247	34·87(1−n) (0·598+n)
BSB 121	12×5	32	9·45	6·56	6·56–1·86n^2	0·445	17·11(1−n) (0·207+n)
BSB 122	12×6	44	13·00	12·20	12·20–3·52n^2	0·369	25·34(1−n) (0·364+n)
BSB 123	12×6	54	15·89	15·47	15·47–5·26n^2	0·378	31·64(1−n) (0·370+n)
BSB 124	12×8	65	19·12	26·63	26·63–7·62n^2	0·270	43·29(1−n) (0·555+n)
BSB 125	13×5	35	10·30	7·26	7·26–2·04n^2	0·442	18·81(1−n) (0·212+n)
BSB 126	14×6	46	13·59	11·94	11·94–3·30n^2	0·412	28·31(1−n) (0·272+n)
BSB 127	14×6	57	16·78	15·42	15·42–5·02n^2	0·417	35·64(1−n) (0·283+n)
BSB 128	14×8	70	20·59	27·35	27·35–7·57n^2	0·313	49·46(1−n) (0·470+n)
BSB 129	15×5	42	12·36	8·08	8·08–2·54n^2	0·510	25·58(1−n) (0·082+n)
BSB 130	15×6	45	13·24	11·15	11·15–2·92n^2	0·431	28·31(1−n) (0·227+n)
BSB 131	16×6	50	14·71	11·53	11·53–3·38n^2	0·435	32·06(1−n) (0·221+n)
BSB 132	16×6	62	18·21	15·30	15·30–5·18n^2	0·483	43·11(1−n) (0·149+n)
BSB 133	16×8	75	22·06	28·12	28·12–7·60n^2	0·348	55·89(1−n) (0·398+n)
BSB 134	18×6	55	16·18	13·24	13·24–3·64n^2	0·467	37·48(1−n) (0·156+n)
BSB 135	18×7	75	22·09	22·25	22·25–6·78n^2	0·448	57·59(1−n) (0·209+n)
BSB 136	18×8	80	23·53	28·72	28·72–7·69n^2	0·383	62·87(1−n) (0·328+n)
BSB 137	$20\times6\frac{1}{2}$	65	19·12	16·84	16·84–4·47n^2	0·471	48·38(1−n) (0·147+n)
BSB 138	$20\times7\frac{1}{2}$	89	26·19	27·94	27·94–8·57n^2	0·458	74·59(1−n) (0·189+n)
BSB 139	22×7	75	22·06	19·96	19·96–5·53n^2	0·499	62·91(1−n) (0·090+n)
BSB 140	$24\times7\frac{1}{2}$	95	27·94	28·11	28·11–8·13n^2	0·490	84·83(1−n) (0·115+n)

Note : $n = \dfrac{N}{N_p}$, where N = magnitude of axial force.
N_p = fully plastic thrust.

TABLE 2

APPENDIX C

Proofs of Plastic Collapse Theorems

Principle of Virtual Work

In establishing the various plastic collapse theorems use is made of the Principle of Virtual Work. This Principle is concerned with any distribution of bending moments in a frame which satisfies all the requirements of *equilibrium* with prescribed loads, and also with any distribution of curvatures and hinge rotations in the members of a frame which satisfies all the geometrical requirements of *compatibility* with prescribed external deflections, but it must be emphasized that the distribution of bending moments need not be related to the distribution of curvatures and hinge rotations as cause and effect.

Suppose that a given frame is imagined to be acted upon by a given set of loads, and let the load at a given position j have horizontal and vertical components H_j^* and V_j^*, respectively. There will be a certain number of equations of equilibrium which must be obeyed by any bending moment distribution in the frame if statical equilibrium is to be preserved. Let M_i^* denote the bending moment at cross-section i in *any* distribution of bending moment in the frame which satisfies all the conditions of equilibrium between the bending moments and the given loads. This distribution of bending moment need not be the actual distribution which would arise if the given loads were applied to the frame.

Suppose also that the frame is imagined to deflect in such a way that the deflection at any position j where a load was imagined to be applied has horizontal and vertical components h_j^{**} and v_j^{**}, respectively. Let κ_i^{**} denote the curvature at any cross-section i in the distorted frame. The *only* requirement placed upon the curvature distribution κ_i^{**} is that it should satisfy the geometrical requirements of compatibility with the deflections h_j^{**} and v_j^{**}.

The Principle of Virtual Work then states that

$$\int M_i^* \kappa_i^{**} \, ds_i = \Sigma(H_j^* h_j^{**} + V_j^* v_j^{**})$$

where ds_i is an element of length of the member at section i. The integration in this equation is understood to cover all the members of the frame, and the summation covers all those positions where external loads are imagined to be applied.

In cases where the hypothetical deflected form of the frame includes hinge rotations at a number of sections k, the product $\kappa_k^{**}\, ds_k$ becomes finite and equal to the hinge rotation θ_k^{**} at each such section. It is convenient to separate these sections from those where the hypothetical curvatures are finite, leading to the equation

$$\int M_i^* \kappa_i^{**} ds_i + \Sigma M_k^* \theta_k^{**} = \Sigma (H_j^* h_j^{**} + V_j^* v_j^{**}) \quad . \quad \text{C.1}$$

Constancy of curvatures during plastic collapse

A state of plastic collapse is defined as one in which the deflections of the frame can continue to increase while the external loads remain constant. From this definition it can be shown that during collapse the distribution of bending moment in the frame remains unaltered as the deflections increase. To prove this, let the increment which occurs in the curvature at any cross-section i during a definite small interval of time in which plastic collapse is occurring be $\delta\kappa_i$, and let the increment which occurs in the plastic hinge rotation at any cross-section k during the same interval be $\delta\theta_k$. During this same interval suppose that the bending moments at the cross-sections i and k change from M_i and M_k to $M_i + \delta M_i$ and $M_k + \delta M_k$, respectively. Since the loads are all constant during collapse the *changes* of bending moment δM_i and δM_k must satisfy the conditions of equilibrium with zero external load. Using these changes of bending moment in the virtual work equation C.1, together with the compatible curvature changes $\delta\kappa_i$ and hinge rotations $\delta\theta_k$, it is found that

$$\int \delta M_i\, \delta\kappa_i\, ds_i + \Sigma \delta M_k\, \delta\theta_k = 0 \quad . \quad . \quad \text{C.2}$$

At any plastic hinge, rotation can only occur if the bending moment remains constant at its fully plastic value, so that $\delta M_k = 0$, whereas elsewhere $\delta\theta_k = 0$. Thus in equation C.2, every term $\delta M_k\, \delta\theta_k$ must be zero, so that

$$\int \delta M_i\, \delta\kappa_i\, ds_i = 0 \quad . \quad . \quad . \quad . \quad \text{C.3}$$

It was postulated in Section 1.2 that increments of curvature and bending moment must always be of the same sign, unless at a particular section the bending moment is equal to the fully plastic moment and a plastic hinge is undergoing rotation under constant bending moment. This fundamental hypothesis was indicated in Fig. 1.2 and summarized in the inequalities 1.1 and 1.2. It follows that at all those sections i where hinges are not undergoing rotation, the product $\delta M_i \, \delta\kappa_i$ must fulfil the condition

$$\delta M_i \, \delta\kappa_i \geqslant 0$$

It follows at once from equation C.3 that δM_i and thus $\delta\kappa_i$ must be zero at every cross-section where a plastic hinge is not undergoing rotation. Thus during plastic collapse the bending moment distribution remains unaltered, and the increases of deflection are due solely to the rotations which occur at the plastic hinges.

Proof of the Static Theorem

The Static Theorem was stated in Section 3.2 as follows:

For a given frame and loading, if there exists any distribution of bending moment throughout the frame which is both safe and statically admissible with a set of loads W, the value of W must be less than or equal to the collapse load W_c.

In this statement of the theorem, it is assumed that the value of each load is specified as a multiple of one of the loads W. The theorem is established by means of a *reductio ad absurdum* argument. Suppose that, contrary to the theorem, a safe distribution of bending moments throughout the frame could be found which was statically admissible with the set of loads γW_c, where γ is any factor greater than unity by which each load is imagined to be multiplied. Let the bending moment in this hypothetical distribution at any section k where a hinge occurs in the actual collapse mechanism be denoted by M'_k. Since the equations of equilibrium are linear in the loads and bending moments, it follows that the distribution of bending moments $\dfrac{M'_k}{\gamma}$ would satisfy the conditions of equilibrium with the set of loads W_c. If the actual bending moment at any section k where a hinge occurs in the actual collapse mechanism is denoted by M_k, the distribution of bending moments M_k must also satisfy the conditions of equilibrium with the set of loads W_c, so that the distribution of bending moments $\left(M_k - \frac{M'_k}{\gamma}\right)$ would satisfy the conditions of equilibrium with zero external loads. This latter distribution of bending moments will be used in the equation of virtual work, equation C.1. In this equation the changes of plastic hinge rotation $\delta\theta_k$ which actually occur during a small motion of the collapse mechanism will also be used, these changes being compatible with zero change of curvature at every other cross-section. It follows that

$$\sum\left(M_k - \frac{M'_k}{\gamma}\right)\delta\theta_k = 0 \quad . \quad . \quad . \quad \text{C.4}$$

the summation extending over all those sections k where plastic hinges occur in the collapse mechanism.

At any plastic hinge where $\delta\theta_k$ is positive, the actual bending moment M_k must be equal to the fully plastic moment M_p. Since the hypothetical bending moment M'_k was safe as well as being statically admissible with the loads γW_c, M'_k cannot exceed M_p, so that $\frac{M'_k}{\gamma}$ must be less than M_p. It follows that

$$\left(M_k - \frac{M'_k}{\gamma}\right) > 0,$$

and since $\delta\theta_k$ was assumed to be positive,

$$\left(M_k - \frac{M'_k}{\gamma}\right)\delta\theta_k > 0$$

Similarly, at any plastic hinge where $\delta\theta_k$ is negative, $M_k = -M_p$ and $M'_k \geqslant -M_p$, so that $\left(M_k - \frac{M'_k}{\gamma}\right) < 0$. It follows that in this case also

$$\left(M_k - \frac{M'_k}{\gamma}\right)\delta\theta_k > 0$$

Thus every term of the summation in equation C.4 must be positive, which is impossible, thus establishing the theorem.

Proof of the Kinematic Theorem

The Kinematic Theorem was stated in Section 3.2 as follows:

For a given frame subjected to a set of loads W, the value of W which is found to correspond to any assumed mechanism must be either greater than or equal to the collapse load W_c.

This theorem is also established by means of a *reductio ad absurdum* argument. Suppose that, contrary to the theorem, a mechanism existed for which the corresponding value of W was βW_c, where β is a positive factor which is less than unity. In this mechanism let there be plastic hinges at a number of cross-sections k, the rotation of the plastic hinge at section k during a small motion of the hypothetical collapse mechanism being $\delta\theta_k''$. The bending moment at section k when collapse actually occurs at the load W_c is denoted by M_k, the bending moment distribution M_k being statically admissible with the set of loads W_c. The bending moment distribution βM_k would therefore satisfy all the conditions of equilibrium with the set of loads βW_c. If M_k'' denotes the fully plastic moment at section k corresponding to the hinge rotation in the hypothetical mechanism, the distribution of bending moments M_k'' would also satisfy all the conditions of equilibrium with the set of loads βW_c. It follows that the distribution of bending moments $(M_k'' - \beta M_k)$ would satisfy the conditions of equilibrium with zero external loads, and this distribution will be used in the equation of virtual work, equation C.1. In this equation the changes of plastic hinge rotation $\delta\theta_k''$ which would occur during a small motion of the hypothetical mechanism will also be used, these changes being compatible with zero changes of curvature at every other cross-section. It follows that

$$\Sigma(M_k'' - \beta M_k)\delta\theta_k'' = 0 \quad . \quad . \quad . \quad \text{C.5}$$

If $\delta\theta_k''$ is positive, $M_k'' = M_p$. Since $M_k \leqslant M_p$, and $\beta < 1$, it follows that $\beta M_k < M_p$. Thus

$$(M_k'' - \beta M_k)\delta\theta_k'' > 0$$

This same result is obtained if $\delta\theta_k''$ is negative. Thus every term of the summation in equation C.5 is positive, which is impossible, so that the theorem is established.

APPENDIX D

Proof of Shake-down Theorem

The shake-down theorem for frames whose members behave according to the ideal type of bending moment-curvature relation of Fig. 2.1 was stated in Section 8.3 as follows:

If it is possible to find any distribution of residual bending moments $\bar{m}_i$ which satisfies at every cross-section i the conditions

$$\bar{m}_i + \mathscr{M}_i^{\max} \leqslant (M_p)_i \quad . \quad . \quad . \quad 8.2$$

$$\bar{m}_i + \mathscr{M}_i^{\min} \geqslant -(M_p)_i . \quad . \quad . \quad 8.3$$

and which is statically admissible, the frame will eventually shake down, although the residual moments existing in the frame after it has shaken down will not necessarily be the distribution $\bar{m}_i$.

As stated in Section 8.3 the theorem is proved by supposing that a particular distribution of residual moments $\bar{m}_i$ has been found which satisfies the inequalities 8.2 and 8.3 and is statically admissible with zero external loads. A quantity E is defined by the equation

$$E = \int \frac{(m_i - \bar{m}_i)^2}{2(EI)_i} ds_i \quad . \quad . \quad . \quad 8.4$$

here m_i represents the actual residual moment at cross-section i ring any stage of the loading, and $(EI)_i$ and ds_i are the flexural idity and element of length respectively at this section. The egration in equation 8.4 covers all the members of the frame. om the form of equation 8.4 it is evident that E is essentially positive quantity; it can be thought of as a measure of the difference between the actual and hypothetical residual moment distributions, m_i and $\bar{m}_i$.

As pointed out in Section 8.3, m_i is defined by the equation

$$m_i = M_i - \mathscr{M}_i . \quad . \quad . \quad . \quad 8.1$$

where M_i is the actual bending moment at section i at the particular stage of the loading programme in question, and $\mathscr{M}_i$ is the elastic bending moment which would be produced at this section by the same loads. Since the distributions of bending moment M_i and $\mathscr{M}_i$ must both satisfy the conditions of equilibrium with the same set of external loads, the distribution of residual moments m_i must satisfy the conditions of equilibrium with zero external loads.

Suppose now that during a definite small interval of time small changes occur in the applied loads, causing the actual bending moment at section i to change by an amount δM_i from M_i to $M_i + \delta M_i$, the corresponding changes in the elastic and residual bending moments at this section being $\delta\mathscr{M}_i$ and δm_i, respectively. During this interval suppose that rotations are occurring at various plastic hinges, the change of rotation at a typical section k being denoted by $\delta\theta_k$. The change in E during this interval is obtained by differentiating equation 8.4, and is given by

$$\delta E = \int (m_i - \bar{m}_i)\frac{\delta m_i}{(EI)_i} ds_i \, . \qquad \text{D.1}$$

This value of δE is now transformed by using the equation of virtual work, stated in Appendix C as equation C.1. In this equation the distribution of bending moments $(m_i - \bar{m}_i)$ will be used. This distribution must satisfy the conditions of equilibrium with zero external loads, since both the distributions m_i and $\bar{m}_i$ satisfy these conditions. For a compatible distribution of curvatures and hinge rotations it is noted that the actual changes of curvature $\dfrac{\delta M_i}{(EI)_i}$ must be compatible with the actual changes of rotation $\delta\theta_k$ at the plastic hinges. Moreover, the changes of curvature $\dfrac{\delta\mathscr{M}_i}{(EI)_i}$ which would have taken place if t whole frame had remained elastic would be compatible with ze hinge rotations. It follows that the changes of curvat $\left(\dfrac{\delta M_i - \delta\mathscr{M}_i}{(EI)_i}\right)$, which from equation 8.1 are equal to $\dfrac{\delta m_i}{(EI)_i}$, mu be compatible with the actual changes $\delta\theta_k$ in the rotations at the plastic hinges. Using these compatible changes of curvature and hinge rotation in equation C.1, together with the distribution of bending moments $(m_i - \bar{m}_i)$ which satisfies the conditions of equilibrium with zero external loads, it is found that

$$\int (m_i - \bar{m}_i)\frac{\delta m_i}{(EI)_i} ds_i + \Sigma(m_k - \bar{m}_k)\delta\theta_k = 0 \, , \qquad \text{D.2}$$

the summation in this equation covering all those sections k where rotations are occurring at plastic hinges during the interval considered.

From equations D.1 and D.2 it follows that

$$\delta E = -\Sigma(m_k - \bar{m}_k)\delta\theta_k \quad . \quad . \quad . \quad \text{D.3}$$

Suppose now that at a particular section k,

$$(m_k - \bar{m}_k) < 0 \quad . \quad . \quad . \quad \text{D.4}$$

Using the inequality 8.2, the following continued inequality results

$$m_k < \bar{m}_k \leqslant (M_p)_k - \mathscr{M}_k^{\max}$$

$$m_k + \mathscr{M}_k^{\max} < (M_p)_k$$

From equation 8.1, it is seen that

$$m_k + \mathscr{M}_k^{\max} = M_k^{\max}$$

where $M_k^{\max}$ is the maximum possible actual bending moment which could arise at section k if the residual moment was equal to m_k. It follows that

$$M_k^{\max} < (M_p)_k$$

This result shows that the plastic hinge which is undergoing rotation at this section must be of negative sign, so that $\delta\theta_k < 0$, for $\delta\theta_k$ could only be positive if M_k was equal to $(M_p)_k$. Using the inequality D.4, it follows that

$$(m_k - \bar{m}_k)\delta\theta_k > 0 \quad . \quad . \quad . \quad . \quad \text{D.5}$$

By a similar argument it can be shown that if $(m_k - \bar{m}_k) > 0$, $\delta\theta_k$ must be positive, so that the inequality D.5 also holds true in this case. It can therefore be concluded that

$$(m_k - \bar{m}_k)\delta\theta_k \geqslant 0, \quad . \quad . \quad . \quad \text{D.6}$$

the equality sign covering those sections at which $m_k = \bar{m}_k$. Comparing this condition with equation D.3, it is seen that

$$\delta E \leqslant 0 \quad . \quad . \quad . \quad . \quad \text{D.7}$$

It is evident from equation D.3 that δE will be zero if no plastic hinges undergo rotation during the interval considered, for the $\delta\theta_k$ will then all be zero. It has therefore been shown that E decreases whenever any plastic hinges undergo rotation, but remains constant when the behaviour is wholly elastic. Since E is essentially a positive quantity, it follows that E must either eventually become zero, in which case the distributions m_i and $\bar{m}_i$

would be identical, or else settle down at some positive value and thereafter remain unchanged. In either case the frame would then have shaken down, thus establishing the theorem.

Finite nature of deflections if shake-down occurs

It is unfortunately impossible to extend this theorem to specify a finite upper bound on the magnitudes of the deflections and hinge rotations which can be developed during shake-down if the conditions 8.2 and 8.3 are fulfilled. The reason for this is that these conditions include the static theorem of plastic collapse as a special case. When these conditions can only just be fulfilled at the incremental collapse load W_s, one of the two conditions 8.2 and 8.3 will become an equality at certain cross-sections j, these sections being such that if hinges occurred at all these sections simultaneously the frame would be transformed into a mechanism. Usually, the load combinations producing the corresponding maximum or minimum elastic bending moments at these sections, $\mathcal{M}_j^{\max}$ or $\mathcal{M}_j^{\min}$, will not all be the same. In such cases if W exceeded W_s plastic hinges would form and undergo rotation at these sections during the variable repeated loading, but not all simultaneously, so that failure by incremental collapse would result. However, in certain special cases the load combinations producing the maximum or minimum elastic bending moments at the cross-sections j will all be the same. Plastic hinges would then form simultaneously at all these sections when W was equal to W_s, and plastic collapse would then occur, so that the plastic collapse load W_c would be equal to W_s. These special cases are easily recognized, and if they are excluded from consideration it is possible to show that if the shake-down conditions are fulfilled the total plastic hinge rotations and deflections which can be developed during the shake-down process must be finite.

If a given change $\delta\theta_k$ occurs in the plastic hinge rotation at a particular cross-section k in a structure, the value of the residual moment at any other cross-section i will be changed by an amount δm_i which is linearly proportional to $\delta\theta_k$, so that

$$\delta m_i = \lambda_{ik}\,\delta\theta_k\,,$$

where λ_{ik} is an influence coefficient whose value can be determined by any of the orthodox methods of elastic structural analysis. The change δm_i in the residual moment at the cross-section i due to changes in the plastic hinge rotations at a number of cross-sections k is thus given by

$$\delta m_i = \Sigma \lambda_{ik}\, \delta\theta_k, \qquad . \quad . \quad . \quad . \quad \text{D.8}$$

where the summation includes all those cross-sections at which plastic hinges are undergoing rotation in the interval considered.

The influence coefficients λ_{ik} are in general of finite magnitude, so that from equation D.8 it might be concluded that the changes in the residual moments are of the same order of magnitude as the changes in the plastic hinge rotations, a conclusion which is true provided that the summation in equation D.8 is not zero for every section i of the frame. It will now be shown that this cannot occur unless the $\delta\theta_k$ correspond to a mechanism motion, a condition which has been excluded because it is assumed that W_s is less than W_c.

It has already been shown that the changes of plastic hinge rotation $\delta\theta_k$ are compatible with the curvature changes $\dfrac{\delta m_i}{(EI)_i}$. Using these compatible hinge rotations and curvature changes in the equation of virtual work, equation C.1, together with the actual changes of residual moment δm_i which satisfy the conditions of equilibrium with zero external loads, it is found that

$$\int \frac{(\delta m_i)^2}{(EI)_i} ds_i + \Sigma \delta m_k\, \delta\theta_k = 0 \qquad . \quad . \quad \text{D.9}$$

If the $\delta\theta_k$ corresponded to a small motion of a mechanism, they would be compatible with zero curvature changes everywhere in the frame, and so from the virtual work equation it would follow that

$$\Sigma \delta m_k\, \delta\theta_k = 0$$

Then from equation D.9,

$$\int \frac{(\delta m_i)^2}{(EI)_i} ds_i = 0$$

and this equation could only be fulfilled if δm_i was everywhere zero.

Thus if rotations are occurring simultaneously at various plastic hinges, the residual bending moments must undergo changes of the same order of magnitude in accordance with equation D.8 unless the hinge rotations correspond to a mechanism motion, a condition excluded from consideration by the proviso that W_s is less than W_c. From equation D.1 it is seen that the change in E during a given interval is of the same order of magnitude as the changes δm_i in the residual moments, and thus of the same order as the changes $\delta\theta_k$ in the plastic hinge rotations. Since the total possible change of E is finite during shake-down, it follows that the total changes in the plastic hinge rotations $\delta\theta_k$ must also be finite, and this evidently implies that the total deflections which can develop during shake-down are also finite.

ANSWERS TO EXAMPLES

Chapter 1

2. $0 \cdot 6h$, $0 \cdot 32 f_L$. **3.** 1·80. **4.** $0 \cdot 1b$ from centre, $0 \cdot 3b^3 f_L$. **5.** 0·741.

Chapter 2

1. $8\dfrac{M_p}{l}$, $4 \cdot 5\dfrac{M_p}{l}$. **2.** $9\dfrac{M_p}{l}$. **4.** $H = -0 \cdot 365\dfrac{M_p}{l}$, $V = -0 \cdot 033\dfrac{M_p}{l}$.

Chapter 3

1. 5 tons, 1·33 tons, 2·33 tons. **2.** $(6 + 4\sqrt{2})\dfrac{M_p}{l}$. **3.** 4·29 tons ft., 7·78 tons ft., 5·15 tons ft. **4.** $\dfrac{576}{49}\dfrac{M_p}{l}$, $9\dfrac{M_p}{l}$, $6\dfrac{M_p}{l}$. **5.** $0 \cdot 6l$. **7.** $4\dfrac{M_p}{l}$, $4\dfrac{M_p}{l}$, $3\dfrac{M_p}{l}$, $2\dfrac{M_p}{l}$, $\dfrac{4}{3}\dfrac{M_p}{l}$. **8.** $1 \cdot 5\dfrac{M_p}{l}$. **9.** $2 \cdot 13M_p$, $1 \cdot 88\dfrac{M_p}{l}$. **10.** $2 \cdot 97\dfrac{M_p}{l}$, $1 \cdot 011M_p$, $2 \cdot 94\dfrac{M_p}{l}$. **11.** $12\dfrac{M_p}{l}$, $(4 + 2\sqrt{3})\dfrac{M_p}{l}$. **12.** $8\frac{7}{16}$ tons ft., $7\frac{6}{7}$ tons ft. **13.** $Wl(3 - 2\sqrt{2})$.

Chapter 4

1. 2·59 ft., 5·74 tons ft. **2.** 6·32 tons ft. (hinge 3·1 ft. from apex), 7·60 tons ft. (hinge 1·9 ft. from apex). **3.** 5.56 tons ft. (hinge 2·4 ft. from apex). **4.** 13·12 tons ft. (hinge 2·8 ft. from apex). Same value when wind load acts. **5.** 6·03 tons ft. (hinge 11·5 ft. from apex). **6.** 5 tons ft., 4·75 tons ft. **7.** $\left(\dfrac{8}{8 - \sqrt{3}}\right)\dfrac{M_p}{l}$. **8.** 28·5 tons ft., 4·7 ft. to 36 ft. **9.** $2\frac{8}{11}$ tons ft., $4\frac{2}{7}$ tons ft., 2·4% increase. **10.** 6·6 tons ft., 4·6 tons ft. **11.** 6·25 tons ft., 5·41 tons. **12.** $5\frac{1}{7}$ tons ft. **13.** 17·29 tons ft. (hinge 9·4 ft. from apex), 22·34 tons ft. (hinge 6·0 ft. from apex), 23·48 tons ft. (hinges 7·5 ft. and 6·1 ft. from apex).

Chapter 5

2. $\dfrac{M_p l^2}{18EI}$. **3.** $\dfrac{13M_p l^2}{12EI}$, $\dfrac{4M_p l^2}{27EI}$. **4.** $\dfrac{5M_p l^2}{6EI}$, $\dfrac{5M_p l^2}{18EI}$.

Chapter 6

1. 448·4 tons in., 378·4 tons in., 406·1 tons in., 342·7 tons in. **2.** 44·9 tons in., 43·2 tons in., 38·0 tons in., 39·5 tons in. **3.** 2·06, 1·93. **4.** 2·05, 16·1 tons, 8·2 tons, 0·8%, 1·9%.

Chapter 7

1. $\beta_1 = 12\frac{2}{3}$ tons ft., $\beta_2 = 10$ tons ft.

2. If $W_1l_1 > W_2l_2$:— $l_1 < 2l_2$; $\beta_1 = \left(\frac{W_1l_1}{4} - \frac{W_2l_2}{12}\right)$, $\beta_2 = \frac{W_2l_2}{6}$

$$l_1 > 2l_2;\ \beta_1 = \beta_2 = \frac{W_1l_1}{6}$$

If $W_2l_2 > W_1l_1$, interchange suffices.

3. (i) 4, 4, 5·5 tons ft., (ii) 3, 3, 3·5 tons ft.

4. $\beta_1 = \beta_2 = 11\frac{2}{3}$ tons ft.

7. $\beta_1 - 9\frac{1}{21}$ tons ft., $\beta_2 = 7\frac{1}{7}$ tons ft.

8. $\beta_1 = 1\frac{1}{4}$ tons ft., $\beta_2 = \beta_3 = 13\frac{3}{4}$ tons ft.

Chapter 8

1. $P_s = 5{,}040$ kg., $P_a = 8{,}620$ kg., $P_c = 5{,}950$ kg.

2. $W_s = 2{\cdot}727\frac{M_p}{l}$, $W_c = 3\frac{M_p}{l}$, $W_a = 4{\cdot}444\frac{M_p}{l}$.

3. $W_s = 7{\cdot}322\frac{M_p}{l}$, $W_c = 8\frac{M_p}{l}$, $W_s = 5{\cdot}023\frac{M_p}{l}$, $W_c = 5{\cdot}333\frac{M_p}{l}$,

$W_a = 13{\cdot}5\frac{M_p}{l}$.

4. $W_s = 1{\cdot}829\frac{M_p}{l}$, $W_c = 2\frac{M_p}{l}$ (unchanged if V constant),

$W_a = 3{\cdot}333\frac{M_p}{l}$, $W_a = 6{\cdot}4\frac{M_p}{l}$.

5. $h = 0{\cdot}086\frac{M_pl^2}{EI}$, $v = 0{\cdot}302\frac{M_pl^2}{EI}$.

AUTHOR INDEX

SUBJECT INDEX